Approximation Theory in the Central Limit Theorem

Mathematics and Its Applications (*Soviet Series*)

Volume 32

Approximation Theory in the Central Limit Theorem

Exact Results in Banach Spaces

by

V. Paulauskas

and

A. Račkauskas

Department of Mathematical Analysis,
Vilnius University, Vilnius, U.S.S.R.

Kluwer Academic Publishers

Dordrecht / Boston / London

Library of Congress Cataloging in Publication Data **CIP**

Paulauskas, V. Ĭ. (Vigantas Ĭonovich)
Approximation theory in central limit theorems--
exact results in Banach spaces.

(Mathematics and its applications. Soviet series)
Translation of: Tochnost′ approksimat͡sii v t͡sentral′noĭ
peredel′noĭ teoreme v banakhovykh prostranstvakh.
Bibliography: p.
Includes index.
1. Asymptotic distribution (Probability theory)
2. Banach spaces. 3. Central limit theorem. 4. Convergence. I. Rachkauskas, A. I͡U. (Alfredas I͡Urgevich)
II. Title. III. Series: Mathematics and its applications
(Kluwer Academic Publishers). Soviet series.
QA273.6.P3813 1989 519.2 88-23101

ISBN: 90-277-2825-9

Published by Kluwer Academic Publishers,
P.O. Box 17, 3300 AA Dordrecht, The Netherlands.

Kluwer Academic Publishers incorporates
the publishing programmes of
D. Reidel, Martinus Nijhoff, Dr W. Junk and MTP Press.

Sold and distributed in the U.S.A. and Canada
by Kluwer Academic Publishers,
101 Philip Drive, Norwell, MA 02061, U.S.A.

In all other countries, sold and distributed
by Kluwer Academic Publishers Group,
P.O. Box 322, 3300 AH Dordrecht, The Netherlands.

Translated by B. Svecevičius and V. Paulauskas

Original publication:
Точность аппроксимации в центральной предельной
теореме в банаховых пространствах
published by Mokslas, Vilnius, © 1987

Printed in The Netherlands

To our reviving Homeland
LITHUANIA

Contents

Editor's Preface ix

Preface xi

Introduction xiii

List of Notations xvii

Chapter 1. General Questions of the Distribution Theory in Banach Spaces 1

1.1. Random Elements in Banach Spaces 1

1.2. Main Facts on Gaussian Measures in Banach Spaces 5

1.3. Weak Convergence of Probability Measures and Some Inequalities 7

Chapter 2. Some Questions of Non-Linear Analysis 11

2.1. Differentiation in Normed Spaces 11

2.2. Smooth Approximation in a Banach Space 22

2.3. Supplements 41

Chapter 3. The Central Limit Theorem in Banach Spaces 44

3.1. Operators of Type p 44

3.2. The Central Limit Theorem 52

3.3. Supplements 59

Chapter 4. Gaussian Measure of ε-Strip of Some Sets 62

4.1. General Remarks 62

4.2. Distribution Density of the Norm of a Gaussian Element in $\mathbf{L}_p$ 63

4.3. Distribution Density of the Norm of a Gaussian Element in $\mathbf{c}_0$ 67

4.4. Supplements 75

Chapter 5. Estimates of Rate of Convergence in the Central Limit Theorem 80

5.1. Estimates of the Rate of Convergence on Sets 80

5.2. Estimates of the Rate of Convergence in the Central Limit Theorem in Some Banach Spaces 110

5.3. Estimates of the Rate of Convergence in the Prokhorov Metric and the Bounded Lipschitz Metric 122

5.4. Supplements 135

Bibliographical Notes 139

References 144

Index 154

SERIES EDITOR'S PREFACE

'Et moi, ..., si j'avait su comment en revenir, je n'y serais point allé.'

Jules Verne

The series is divergent; therefore we may be able to do something with it.

O. Heaviside

One service mathematics has rendered the human race. It has put common sense back where it belongs, on the topmost shelf next to the dusty canister labelled 'discarded non-sense'.

Eric T. Bell

Mathematics is a tool for thought. A highly necessary tool in a world where both feedback and non-linearities abound. Similarly, all kinds of parts of mathematics serve as tools for other parts and for other sciences.

Applying a simple rewriting rule to the quote on the right above one finds such statements as: 'One service topology has rendered mathematical physics ...'; 'One service logic has rendered computer science ...'; 'One service category theory has rendered mathematics ...'. All arguably true. And all statements obtainable this way form part of the raison d'être of this series.

This series, *Mathematics and Its Applications*, started in 1977. Now that over one hundred volumes have appeared it seems opportune to reexamine its scope. At the time I wrote

> "Growing specialization and diversification have brought a host of monographs and textbooks on increasingly specialized topics. However, the 'tree' of knowledge of mathematics and related fields does not grow only by putting forth new branches. It also happens, quite often in fact, that branches which were thought to be completely disparate are suddenly seen to be related. Further, the kind and level of sophistication of mathematics applied in various sciences has changed drastically in recent years: measure theory is used (non-trivially) in regional and theoretical economics; algebraic geometry interacts with physics; the Minkowsky lemma, coding theory and the structure of water meet one another in packing and covering theory; quantum fields, crystal defects and mathematical programming profit from homotopy theory; Lie algebras are relevant to filtering; and prediction and electrical engineering can use Stein spaces. And in addition to this there are such new emerging subdisciplines as 'experimental mathematics', 'CFD', 'completely integrable systems', 'chaos, synergetics and large-scale order', which are almost impossible to fit into the existing classification schemes. They draw upon widely different sections of mathematics."

By and large, all this still applies today. It is still true that at first sight mathematics seems rather fragmented and that to find, see, and exploit the deeper underlying interrelations more effort is needed and so are books that can help mathematicians and scientists do so. Accordingly MIA will continue to try to make such books available.

If anything, the description I gave in 1977 is now an understatement. To the examples of interaction areas one should add string theory where Riemann surfaces, algebraic geometry, modular functions, knots, quantum field theory, Kac-Moody algebras, monstrous moonshine (and more) all come together. And to the examples of things which can be usefully applied let me add the topic 'finite geometry'; a combination of words which sounds like it might not even exist, let alone be applicable. And yet it is being applied: to statistics via designs, to radar/sonar detection arrays (via finite projective planes), and to bus connections of VLSI chips (via difference sets). There seems to be no part of (so-called pure) mathematics that is not in immediate danger of being applied. And, accordingly, the applied mathematician needs to be aware of much more. Besides analysis and numerics, the traditional workhorses, he may need all kinds of combinatorics, algebra, probability, and so on.

In addition, the applied scientist needs to cope increasingly with the nonlinear world and the extra mathematical sophistication that this requires. For that is where the rewards are. Linear models are honest and a bit sad and depressing: proportional efforts and results. It is in the non-linear world that infinitesimal inputs may result in macroscopic outputs (or vice versa). To appreciate what I am hinting at: if electronics were linear we would have no fun with transistors and computers; we would have no TV; in fact you would not be reading these lines.

There is also no safety in ignoring such outlandish things as nonstandard analysis, superspace and anticommuting integration, p-adic and ultrametric space. All three have applications in both electrical engineering and physics. Once, complex numbers were equally outlandish, but they frequently proved the shortest path between 'real' results. Similarly, the first two topics named have already provided a number of 'wormhole' paths. There is no telling where all this is leading - fortunately.

Thus the original scope of the series, which for various (sound) reasons now comprises five sub-series: white (Japan), yellow (China), red (USSR), blue (Eastern Europe), and green (everything else), still applies. It has been enlarged a bit to include books treating of the tools from one subdiscipline which are used in others. Thus the series still aims at books dealing with:

- a central concept which plays an important role in several different mathematical and/or scientific specialization areas;
- new applications of the results and ideas from one area of scientific endeavour into another;
- influences which the results, problems and concepts of one field of enquiry have, and have had, on the development of another.

The laws of large numbers, or, better, central limit theorems for sums of identically distributed random variables, are basic to (both applications and foundations of) probability and statistics. Given the frequent and natural role of Hilbert and Banach spaces as the spaces of values for such random variables, the question of how well these sums are approximated by Gaussian distributions becomes of fundamental importance. The present volume presents an up-to-date unified treatment of this topic for the case of Banach spaces.

The complementary topic of rates of convergence to the Gaussian limit law and the topic of large deviations in a Hilbert space is the subject of a forthcoming monograph by Prokhorov, Sazonov and Jurinsky (1990, translation in this series 1991). Together the two books should give the reader a very good idea of what is known at present concerning these matters.

The shortest path between two truths in the real domain passes through the complex domain.

J. Hadamard

La physique ne nous donne pas seulement l'occasion de résoudre des problèmes ... elle nous fait pressentir la solution.

H. Poincaré

Never lend books, for no one ever returns them; the only books I have in my library are books that other folk have lent me.

Anatole France

The function of an expert is not to be more right than other people, but to be wrong for more sophisticated reasons.

David Butler

Bussum, June 1989 Michiel Hazewinkel

Preface

The intention of this book is to analyse the accuracy in approximating the distribution of sums of independent identically distributed random elements in a Banach space by Gaussian distributions. During the past two decades quite a number of results, resembling the final ones, have been obtained in this field. At the same time a close relation has been found between the estimates of the rate of convergence in the central limit theorem and some other branches of mathematics (in particular, non-linear analysis). We have tried to systematize the known results by presenting them in a unified form. An 'operator language' is used for this purpose.

Due to limited space, it was not possible to touch upon certain trends in the asymptotic analysis of distributions of random element sums such as the rate of convergence to stable laws, or asymptotic expansions in the central limit theorem, approaching infinitely divisible laws in Banach spaces. In the future, the above questions, no doubt, may comprise a separate book.

Even concerning rates of convergence to the Gaussian law, not all questions have been discussed to the same extent. It is the hope to the authors, however, that the present book and the forthcoming monograph of J.V. Prokhorov, V.V. Sazonov and V.V. Jurinskii on the estimates of convergence rate and large deviations in a Hilbert space will provide the reader with a present-day review of research in this field.

The book consists of five chapters. Each chapter is divided into sections. At the end of Chapters 2–5 'Supplements' are provided which present short formulations of the results not included in the main text, but closely related to it.

The sign □ denotes the end of a proof. The following abbreviations are used in the text: a.s. = almost surely; i.i.d. = independent identically distributed; r.v. = random variable; CLT = the central limit theorem, **B**−r.e. = random element in the space **B**. In each chapter theorems, lemmas, propositions and formulas are marked by two numbers: (4.3), for instance, denotes formula number 3 in Section 4 of the given chapter. If a different chapter is referred to, three numbers are used, e.g. (5.4.3), the first number denoting the number of the chapter.

The authors' express their sincere thanks to their teachers, Prof. J. Kubilius and Prof. V. Statulevicius, for constant attention to the present work.

We also thank Dr. V.Yu. Bentkus who read all the proofs with the utmost care and gave a number of remarks that contributed much to the improvement of the book. Last, but not least, we express our gratitude to allow colleagues who helped us prepare the book for publication.

Introduction

The problem of estimating the rate of convergence in the central limit theorem in Banach spaces has arisen following Yu.V. Prokhorov's fundamental work [167], which has proved to be of great significance to many subjects concerning limit theorems in infinite-dimensional spaces. It should be noted that, at that time, investigations of estimates of convergence rates in the central limit theorem in finite-dimensional Euclidean spaces were but in their infancy. It is, therefore, not surprising that the early estimates in Hilbert space obtained by N.N. Vakhanya and N.P. Kandelaki [214], [97], by use of the finite-dimensional approximation method followed by a transition to the limit in dimension, had a logarithmic order of decrease with respect to the number of summands.

A considerable increase in interest in estimates of the convergence rate in general Banach spaces was particularly stimulated by the appearance of the works of Yu.V. Prokhorov and V.V. Sazonov [168], [187], which showed that the problem of approximation accuracy in statistical distributions may be formulated as a problem on the estimate of the convergence rate in the central limit theorem in some Banach or Hilbert space for summands of a special structure. In particular, for such summands in Hilbert space, a power estimate of the convergence rate was obtained by the use of a finite-dimensional approximation [187], [186]. A great impetus in this field was provided by the work by J. Kuelbs and T. Kurtz [103]. The method suggested by these authors is no longer associated with finite-dimensional approximation, and the estimate of the rate of convergence in the central limit theorem in Hilbert space under natural moment conditions has a power order.

During the past decade, considerable progress has been achieved in estimates of the convergence rate. Some trends have been identified and the results obtained for some problems are similar to the final ones. Methods of research based on functional analysis have been devised, and regularities have been revealed which relate only to infinite-dimensional spaces and are conditioned by their geometry.

At present, to our belief, the following three trends of research in asymptotic analysis of sums of independent random elements in Banach space may be distinguished.

1. Estimate of convergence rate on sets and construction of asymptotic expansions, in view of sufficient smoothness of a set or a functional under consideration.
2. Estimate of convergence rate on sets with a non-smooth boundary in general Banach spaces.
3. Estimate of convergence rate in different metrics.

The present book is intended mainly to address the second and third group of problems, as it is to these problems that the studies of Vilnius mathematicians have given the greatest attention.

The idea on which the method of compositions is based is simple. It can be explained without any complex notations. Let us suppose that we must estimate the difference of two measures F (distribution of a sum) and G (limit distribution) on a certain subset A of a Banach space $\mathbf{B}$, i.e. we must estimate the quantity

$$|F(A) - G(A)| = \left|\int_{\mathbf{B}} \chi_A(x)(F - G)(dx)\right|,$$

where χ_A is the indicator function of the set A. The first problem is to substitute the discontinuous function χ_A by the sufficiently smooth function $g_{a,\varepsilon}$ which coincides with χ_A everywhere, with the exception of the set

$$(\partial A)_\varepsilon = \left\{x \in B : \inf_{y \in \partial A} \|x - y\| < \varepsilon\right\}$$

of the ε-neighbourhood, of the boundary ∂A of the set A. Here $\varepsilon > 0$ is a certain parameter, defining the accuracy of approximation that is chosen in the last step of estimating the quantity $|F(A) - G(A)|$. At this substitution two terms are obtained: the integral of the smooth function

$$\left|\int_B g_{A,\varepsilon}(x)(F - G)(dx)\right|,$$

which is usually estimated by expanding the function $g_{A,\varepsilon}$ into a Taylor series, and the quantity $G((\partial A)_\varepsilon)$ whose estimate is the second problem.

A part of this book is devoted to the study of the above problems, on whose solution depends the rate of convergence in the central limit theorem.

The book contains an introduction and five chapters followed by bibliographical remarks and references.

The first chapter is auxiliary: definitions, concepts and statements of the theory of the distributions in Banach spaces necessary for further presentation are collected here.

The problem of constructing smooth functions 'well approximating' the indicator function of the given set in a Banach space is considered in the second chapter.

In addition to standard concepts of differentiation in the sense of Fréchet and Gateaux, differentiation with respect to an operator has been considered. Though this conception is formally equivalent to the conception of differentiation in directions from a certain subspace, 'operator language' has been chosen in the present work as it is found to be preferable to the 'language of embeddings'.

The central limit theorem in Banach spaces is considered in the third chapter. Ever since the term 'central limit theorem' appeared, this notion was understood to be the convergence of distributions of sums of independent identically distributed summands to the Gaussian law. At present, by this notion we mean convergence of the distributions of sums $\Sigma_{i=1}^{k_n} X_{ni}$, where $\{X_{ni}, i = 1, ..., k_n\}$, $n \geq 1$, is a sequence of series of independent random variables or elements which are uniformly asymptotically negligible. The studies under such a general design in separable Banach spaces are considered in a monograph by A. Araujo and E. Gine [3], where the central limit theorem in Banach spaces has been discussed in detail. In the present work, by the central limit theorem we mean a simple case of convergence of distributions of sums of i.i.d. random elements to the Gaussian law. This is done for two reasons: firstly, the presentation of the general scheme would take up twice as much space in the book; and, secondly, many facts connecting the geometric properties of Banach space with limit theorems are easily seen in this simple situation. The main results in Chapter 3, the central limit theorem in spaces $\mathbf{C}(S)$ and $\mathbf{c}_0$ in particular, are obtained by the use of type-2 operators. In the fourth chapter, the Gaussian measure of an ε-strip of sets in a Banach space is considered. It is well known that in finite-dimensional space the Gaussian measure of an ε-strip of a convex set admits a uniform estimate of the order ε. But even in an infinite-dimensional Hilbert space, however, the class of balls has no such property. This chapter presents mainly results confirming the essential difference in the problem under consideration in finite-dimensional and infinite-dimensional spaces. In particular, constructed is an example of a Gaussian r.e. in the space $\mathbf{c}_0$, the distribution of the norm of which has unbounded density.

Chapter 5, having taken up almost half of the book, is the main one. The estimates of the rate of convergence in the central limit theorem in Banach spaces are presented in this chapter.

The main result in Section 5.1 shows that for the summands with third finite moment, the estimate of convergence rate of the order $O(n^{-1/6})$ on the sets with non-smooth boundary takes place under two conditions: (a) the condition for the behaviour of Gaussian measure of an ε-strip of sets, and (b) the existence of smooth approximation of a set's indicator function of a set. It is shown by examples that both conditions are necessary and are independent of each other. Besides, in the general case, the order $O(n^{-1/6})$ is not improvable. This makes an essential difference in comparison to the case of the finite-dimensional Banach space.

It is in this section that a special method is proposed for estimating the convergence rate on sets for random elements that are mixtures of Gaussian measures. The section ends with the estimate of the rate of convergence on balls in a Hilbert space. When summands have finite third moment, the estimate has the order $O(n^{-1/2})$ and is proved by the method of characteristic functions which was substantially improved by F. Götze. At present this method is successfully used to obtain the estimates of convergence rates on smooth sets.

In Section 5.2 the estimates in classical Banach spaces $\mathbf{C}(S)$ and $\mathbf{c}_0$ are considered. Special attention is given to the space $\mathbf{C}[0, 1]$, which is of importance to both the theory of random processes and the theory of Banach spaces. In the case of the limit Wiener process, in particular, the estimate of convergence rate on the balls with centre at zero is of the order $O(n^{-1/6}\ln n)$.

In Section 5.3 the convergence rate in the Prokhorov metric and bounded Lipschitz metric is considered. The estimates of the rate of convergence are reduced to the problems of constructing smooth functions, and to finding the best approximations of functions by smooth functions. Examples confirm that in the general case the order of decrease with respect to the number of summands in the estimates obtained is not improved.

List of Notations

$A^c, \overline{A}, \dot{A}$ A complement, a closure, and an interior of a set A.
A_ε A set of points the distance from which to A is less than ε.
$\mathbf{B}$ A normed or separable Banach space.
$\mathbf{B}^*$ A topologically conjugate space to $\mathbf{B}$.
$\mathrm{BL}_\alpha(\mathbf{B})$ A set of bounded real functions on $\mathbf{B}$, satisfying the Hölder condition with the exponent α a Banach space with respect to the norm

$$\|f\|_{\mathrm{BL}_\alpha} = \sup_{x \in \mathbf{B}} |f(x)| + \sup_{x != y} |f(x) - f(y)| \cdot \|x - y\|^{-\alpha}.$$

$\mathrm{BL}(\mathbf{B}) = \mathrm{BL}_1(\mathbf{B})$.
$\mathscr{B}(\mathbf{B})$ A Borel σ-algebra of a space $\mathbf{B}$.
$\mathbf{C}^0(U, \mathbf{F})$ A class of continuous functions from U in $\mathbf{F}$.
$\mathbf{C}^0(\mathbf{B}) = \mathbf{C}^0(\mathbf{B}, \mathbf{R})$.
$\mathbf{C}_0^b(\mathbf{B})$ A class of continuous bounded real functions on $\mathbf{B}$.

$$\|f\|_\infty = \sup_{x \in \mathbf{B}} |f(x)|.$$

$\mathbf{C}^k(U, \mathbf{F})$ A class of k times continuously Fréchet differentiable functions from U in $\mathbf{F}$ (p. 12).
$\mathbf{C}_u^k(U, \mathbf{F})$ (p. 18).
$\mathbf{C}^\infty(U, \mathbf{F})$ A class of infinitely Fréchet differentiable functions from U in $\mathbf{F}$.
$\mathbf{C}^\infty = \mathbf{C}^\infty(\mathbf{R}) = \mathbf{C}^\infty(\mathbf{R}, \mathbf{R})$.
$\mathbf{C}(S)$ A Banach space of real continuous functions on the metric compact S, $\|x\|_{\mathbf{C}(S)} = \sup_{s \in \mathbf{S}} |x(s)|$.
$\operatorname{cov} X$ Covariance of r.e. X (p. 3).
$\mathbf{c}_0 = \{x = (x_i, i \geq 1) : \lim_{i \to \infty} x_i = 0\}$ A Banach space with respect to the norm $\|x\| = \sup_{i \geq 1} |x_i|$.
$\mathbf{c}_0^+ = \{x \in \mathbf{c}_0 : x_i \geq 0, i \geq 1\}$.
$\mathbf{c}_{0,\lambda} = \{x \in \mathbf{c}_0 : \lim_{i \to \infty} x_i \lambda_i^{-1} = 0\} (\lambda = (\lambda_i) \in \mathbf{c}_0^+)$ A Banach space with respect to the norm $\|x\| = \sup_{i \geq 1} |x_i| \lambda_i^{-1}$.
$C(\cdot, \cdot, \cdot)$ A constant dependent on the parameters in the parentheses.
$\mathbf{D}^k(U, \mathbf{F})$ A class of k times Fréchet differentiable functions from U to $\mathbf{F}$ (p. 12).
$\mathbf{D}_v^k(U, \mathbf{F})$ (p. 18).
$\mathscr{D}(\eta)$ The domain of normal attraction of the Gaussian r.e. η (p. 80).
δ_s The Dirac measure defined by the point s; the element of space $\mathbf{C}^*(S)$.
∂A The boundary of the set A.
$\Delta_n(\xi, \eta; \mathscr{A}) = \sup_{A \in \mathscr{A}} |P\{S_n(\xi) \in A\} - P\{\eta \in A\}|$.
$\Delta_n(X, Y; \mathscr{A})$ (p. 83).
$\{\varepsilon_i, i \geq 1\}$ Rademacher sequence (p. 9).
$F(A, A_1)$ A special class of functions (p. 22).
$\mathbf{F}$ A Banach space.
$\mathbf{H}$ A separable Hilbert space with the norm $|\cdot|$ and the scalar product $(\cdot, \cdot)$.
$H(S, \rho, x)$ A metric entropy (p. 6).
$I_\mathbf{B}$ An identical mapping of space $\mathbf{B}$.
$\mathbf{K}_{\alpha,n} = \mathbf{K}_{\alpha,n}(\varepsilon) = (\varepsilon^{-1/\alpha}, \ldots, \varepsilon^{-n/\alpha})$ (p. 25).
$\mathbf{L}_0(\mathbf{F}) = \mathbf{L}_0(\Omega, \mathscr{F}, P, \mathbf{F})$ A space of all $\mathbf{F}$-r.e.

$\mathbf{L}_p(\mathbf{F}) = \mathbf{L}_p(\Omega, \mathscr{F}, P, \mathbf{F}) = \{X \in \mathbf{L}_0(\mathbf{F}) : E\|X\|^p < \infty\}(0 < p < \infty)$.
$\mathbf{L}_p^0(\mathbf{F}) = \{X \in \mathbf{L}_p(\mathbf{F}) : EX = 0\}(1 \leq p < \infty)$.
$\mathbf{L}_p(S, \nu)$ A Banach space of equivalence classes of real measurable functions on S such that

$$\|f\| = \left(\int_S |f(x)|^p \nu(\mathrm{d}x)\right)^{1/p} < \infty \ (1 \leq p < \infty).$$

$\mathbf{L}^n(\mathbf{B}, \mathbf{F})$ A Banach space of n linear bounded mappings from **B** in **F**.
$\mathbf{L}(\mathbf{B}, \mathbf{F}) = \mathbf{L}^1(\mathbf{B}, \mathbf{F})$.
$L(\xi)$ A distribution of r.e. ξ.
$\mathrm{Lip}_\rho(\mathbf{F}) = \{f \in \mathbf{C}^0(\mathbf{F}) : \|f\|_\rho := \sup_{x \neq y} |f(x) - f(y)| \rho^{-1}(x, y) < \infty\}$.
$\mathrm{Lip}_\alpha(\mathbf{F}) = \mathrm{Lip}_\rho(\mathbf{F})$ Where $\rho(x, y) = \|x - y\|^\alpha$.
$\mathrm{Lip}_\rho(S)$ A Banach space with respect to norm $\|f\|_{\mathrm{Lip}_\rho(S)} = \|f\|_\infty + \|f\|_\rho$, if S is a metric compact.
$\mathbf{l}_p = \{x = (x_i, i \geq 1) : \|x\| = (\Sigma_{i=1}^\infty |x_i|^p)^{1/p} < \infty\}(1 \leq p < \infty)$.
$\mathbf{l}_\infty = \{x = (x_i, i \geq 1) : \|x\| = \sup_{i \geq 1} |x_i| < \infty\}$.
$\mathbf{l}_{\infty,\lambda} = \{x \in \mathbf{l}_\infty : \|x\|_{\mathbf{l}_{\infty,\lambda}} := \sup_{i \geq 1} |x_i| \lambda_i^{-1} < \infty\}(\lambda \in \mathbf{c}_0^+)$.
$\mathrm{lip}_\rho(S) = \{f \in \mathrm{Lip}_\rho(S) : \lim_{\rho(s,t) \to 0} |f(s) - f(t)| \rho^{-1}(s, t) = 0\}$ A Banach space with respect to the norm $\|f\|_{\mathrm{Lip}_\rho(S)} = \|f\|_\infty + \|f\|_\rho$, if S is a metric compact.
$\mathrm{lip}_\alpha[0, 1] = \mathrm{Lip}_\rho(S)$, where $S = [0, 1]$, $\rho(s, t) = |s - t|^\alpha$.
$\mu \circ f^{-1}$ A measure induced from μ by mapping f.
$\nu_3(\xi, \eta)$ A pseudomoment (p. 80).
$P(\mathbf{B})$ A set of probability measures on **B**.
$\pi_\gamma(\mathbf{H}, \mathbf{B})$ A set of γ-radonifying operators from **H** to **B** (p. 7).
$\pi(\mu, \nu)$ Prokhorov metric (p. 7).
$\pi(\xi, \eta) = \pi(L(\xi), L(\eta))$.
$\mathbf{R}_k$ k-dimensional Euclidean space.
$\mathbf{R}^+ = [0, \infty)$.
$\rho_{\mathrm{BL}}(\mu, \nu)$ Bounded Lipschitzian metric (p. 7).
$\rho_{\mathrm{BL}}(\xi, \eta) = \rho_{\mathrm{BL}}(L(\xi), L(\eta))$.
$\mathbf{Q}_u^{k,\alpha}$, $\mathbf{Q}_u^k$, $\mathbf{Q}_u^\infty$ Special classes of functions (p. 22).
$\mathbf{U}_2(\mathbf{B}, \mathbf{F})$ A set of operators of type 2 from **B** in **F** (p. 45).
$S_n(\xi) = n^{-1/2}(\xi_1 + \cdots + \xi_n)$, where ξ, $\xi_1, \ldots, \xi_n$ are independent identically distributed random elements.
T_η Covariance operator of r.e. η.
$V_r(a) = \{x : \|x - a\| < r\}$.
$V_r = V_r(0)$.
$W_t(f) = \{x : f(x) < t\}$.
$\mathbf{W}_a(\mathbf{B})$ Class of balls of the space **B** with centre in a (p. 34).
$\chi_A(\cdot)$ Indicator of the set A.
$\|\cdot\|_{\mathbf{B}}$ Norm of the space **B**.
$:=$ The sign to denote 'equals by definition'.

CHAPTER 1

General Questions of the Distribution Theory in Banach Spaces

This is an auxiliary chapter in which the main facts concerning random elements in infinite-dimensional Banach spaces are collected. All the results are presented without proofs, providing the reader with references to the books [3], [31], [32], [38], [66], [84], [102], [106], [213], [215]. The theory of distributions in Banach (and more general linear topological) spaces is considered in detail in the monograph [215].

1.1. Random Elements in Banach Spaces

Let $(\Omega, \mathscr{A}, P)$ be some probability space, and let **B** be a real Banach space, and $\mathfrak{M}$ some σ-algebra of the subsets of the space **B**. The mapping $f:\Omega \to \mathbf{B}$ is called $\mathfrak{M}$-measurable (measurable, if it is obvious which σ-algebra is meant) if for any set $\mathscr{A} \in \mathfrak{M}$ the inverse image $f^{-1}(A) \in \mathscr{A}$. In probability theory, the most important σ-algebras are the Borel σ-algebra B generated by open sets of the space **B**, and the σ-algebra of cylindrical sets K, defined as follows. The set is called cylindrical if it is of the form

$$\{x \in \mathbf{B} : (f_1(x), \ldots, f_n(x)) \in \mathscr{D}\}, \tag{1.1}$$

where $n \geq 1$, $f_1, \ldots, f_n \in \mathbf{B}^*$, and $\mathscr{D}$ is some Borel set in n-dimensional Euclidean space, $\mathbf{R}_n$. The least σ-algebra containing all the sets of type (1.1) is called the σ-algebra of cylindrical sets. Sometimes, when we wish to indicate the space in which σ-algebra is considered, we shall denote this by $\mathscr{B}(\mathbf{B})$ or $\mathscr{K}(\mathbf{B})$. It is generally accepted that a $\mathscr{B}$-measurable mapping is called strongly measurable and a $\mathscr{K}$-measurable one, weakly measurable.

In the case of a non-separable Banach space **B**, σ-algebras $\mathscr{B}$ and $\mathscr{K}$ do not generally coincide ($\mathscr{K} \subset \mathscr{B}$). We restrict ourselves mainly to the case of separable Banach spaces, and for these the following proposition is true.

PROPOSITION 1.1. *In a separable Banach space $\mathscr{B} = \mathscr{K}$. Hence, the notions of strong and weak measurability coincide.*

The proof of this proposition may be found in [215].

Another reason why we confine ourselves to the case of separable spaces is as follows. If **E** is some linear space with a chosen σ-algebra $\mathscr{E}$, then $(\mathbf{E}, \mathscr{E})$ is called a measurable linear space if the mapping $\mathbf{E} \times \mathbf{E} \times \mathbf{R} \to \mathbf{E}$, defined by the formula $(x, y, \lambda) \to \lambda x + y$, is measurable (in the space $\mathbf{E} \times \mathbf{E} \times \mathbf{R}$ the product of the corresponding σ-algebras is considered). It is known that $(\mathbf{B}, \mathscr{K})$ is always a measurable linear space, and $(\mathbf{B}, \mathscr{B})$, in the case of a non-separable space **B**, may be no longer measurable. So in the following text, unless otherwise indicated, the Banach spaces under consideration are supposed to be separable, and in those few cases where we have to deal with non-separable spaces, they will be provided with σ-algebras of cylindrical sets.

DEFINITION 1.2. A measurable mapping $X: \Omega \to \mathbf{B}$ is called a random element in a Banach space $\mathbf{B}$($\mathbf{B}$-r.e.). The distribution $L(X)$ of $\mathbf{B}$-r.e. X is called the measure on $(\mathbf{B}, \mathscr{B})$, defined by the formula

$$L(X)(A) = P\{\omega \in \Omega : X(\omega) \in A\}.$$

A random element X is called symmetric if $L(X) = L(-X)$. The symmetrization of r.e. X is r.e. $\widetilde{X} = X^1 - X^2$, where X^1, X^2 are independent and $L(X^1) = L(X^2) = L(X)$.

By analogy with the numerical characteristics of real random variables (mean, dispersion), the notions of the mathematical expectation and covariance operator of $\mathbf{B}$-r.e. are defined. Since the integral of a vector-valued function may be defined in several ways, there are also several definitions for mathematical expectation. The two most widely known notions of mathematical expectation are defined below.

DEFINITION 1.3. Let $X: \Omega \to \mathbf{B}$ be a weakly measurable mapping such that for all $f \in \mathbf{B}^*$

$$\int_\Omega |f(X(\omega))| \, P(d\omega) < \infty. \tag{1.2}$$

The Gelfand–Pettis integral of the mapping X is called an element $m \in \mathbf{B}$ (if it exists) such that for all $f \in \mathbf{B}^*$ we have the equality

$$f(m) = \int_\Omega f(X(\omega)) P(d\omega).$$

The element m is called the mathematical expectation of $\mathbf{B}$-r.e. X in the sense of Gelfand–Pettis and is usually denoted by EX.

PROPOSITION 1.4. (a) *If EX exists, then it is unique.*

(b) *If EX_1 and EX_2 exists, then for any real c_1 and c_2, $E(c_1X_1 + c_2X_2)$ exists and $E(c_1X_1 + c_2X_2) = c_1EX_1 + c_2EX_2$.*

(c) *If $P\{\omega \in \Omega: X(\omega) = x_0\} = 1$, then $EX = x_0$.*

(d) *If $\mathbf{B} = \mathbf{R}_1$, then the Gelfand–Pettis integral coincides with the Lebesgue integral.*

(e) *If $A: \mathbf{B} \to \mathbf{B}_1$ is a linear bounded operator, and $\mathbf{B}_1$ is some other Banach space, then from the existence of EX it follows that there exist $E(AX)$ and $E(AX) = AEX$.*

Condition (1.2) is necessary for the existence of mathematical expectation in any space, and it is also sufficient in a reflexive Banach space. In a general case, it is not sufficient (the corresponding simple example is given in [215]).

PROPOSITION 1.5. (a) *If $E\|X\| < \infty$, then EX exists and $\|EX\| \le E\|X\|$.*

(b) *Let $X: \Omega \to \mathbf{B}$ be a measurable mapping satisfying condition (1.2) for all $f \in \mathbf{B}^*$. If there exists a number $\varepsilon > 0$ such that*

$$\inf_{f \in \mathbf{B}^*} P\left\{\omega : |f(X(\omega))| > \varepsilon \left| \int_\Omega f(X(\omega)) P(d\omega) \right| \right\} > 0,$$

then EX in the sense of Gelfand–Pettis exists.

The reader can find the proof of Propositions 1.4 and 1.5 in [213], [215].

We shall now present a definition of mathematical expectation in the sense of Bochner. It should be remembered that the function $Y: \Omega \to \mathbf{B}$ is called simple if $Y(\omega) = \sum_{k=1}^n x_k \chi_{A_k}(\omega)$, where $x_k \in \mathbf{B}$, $A_k \in \mathscr{B}$, $\bigcup_{k=1}^n A_k = \Omega$ and $A_k \cap A_l = \emptyset$ if $k \neq l$. For a simple function the Bochner integral is defined as

$$\int_\Omega Y(\omega)P(d\omega) = \sum_{k=1}^{n} x_k P(A_k).$$

It is easy to verify that for simple functions the Bochner integral coincides with the Gelfand–Pettis integral.

DEFINITION 1.6. $\mathbf{B}$-r.e. X is integrable in the sense of Bochner if there exists a sequence of simple functions $X_n: \Omega \to \mathbf{B}$ such that $X_n \to X$ a.s. and $\int_\Omega \|X_n(\omega) - X(\omega)\| P(d\omega) \to 0$ as $n \to \infty$. In this case mathematical expectation in the sense of Bochner is defined as the limit of integrals of simple functions

$$\int_\Omega X(\omega)P(d\omega) := \lim_{n\to\infty} \int_\Omega X_n(\omega)P(d\omega).$$

We shall denote mathematical expectation in the sense of Bochner by the same symbol EX, since our main concern in this book deals with random elements integrable in the sense of Bochner. It should be noted that the Bochner integral has all the properties mentioned in Proposition 1.4; moreover, the necessary and sufficient conditions of its existence are formulated very simple.

PROPOSITION 1.7. $\mathbf{B}$*-r.e.* X *has a mathematical expectation in the sense of Bochner if and only if* $E\|X\|$ *exists. In this case* (1.3) *holds.*

The proof of this Proposition is available in all standard textbooks on functional analysis (see [85]).

It follows from Propositions 1.5 and 1.7 that the existence of mathematical expectation in the sense of Bochner implies its existence in the sense of Gelfand–Pettis, and both integrals coincide. The opposite statement is not true (see [213]).

The Bochner integral of the function $X: \Omega \to \mathbf{B}$ may be obviously defined over any set $A \in \mathcal{A}$, and from the integrability of the function over the whole space Ω, integrability over any set $A \in \mathcal{A}$ follows. The Gelfand–Pettis integral may be also defined over any set $A \in \mathcal{A}$, but in the case of non-reflexive Banach spaces from the existence of the integral over all the space Ω, its existence over the sets $A \in \mathcal{A}$ does not always follow. The corresponding example can be found by the reader in [213, p. 132].

Now we turn to the notion of the covariance operator of a $\mathbf{B}$-r.e. which is a natural generalization of a covariance matrix of multidimensional random vectors. Let X be a $\mathbf{B}$-r.e. with the mean EX in the sense of Gelfand–Pettis and such that for all $f \in \mathbf{B}^*$.

$$Ef^2(X - EX) < \infty. \tag{1.4}$$

Then in the space $\mathbf{B}^*$ we can define the bilinear form $\operatorname{cov} X$ as

$$\operatorname{cov} X(f, g) := Ef(X - EX)g(X - EX),$$

which is called the covariance of $\mathbf{B}$-r.e. X.

PROPOSITION 1.8. *The covariance of* $\mathbf{B}$*-r.e.* X *is a symmetrical, positive, semidefinite and continuous bilinear form.*

For the proof, see [215].

As usual, to a bounded bilinear form, the bounded linear operator $R: \mathbf{B}^* \to \mathbf{B}^{**}$, representing this form, may be compared and is defined by the formula

$(Rf)(g) = \operatorname{cov} X(f, g)$.

It is this operator that is called the covariance operator of $\mathbf{B}$-r.v. X. Sometimes we denote this by R_X to point out the random element. It is easy to make sure of the truth of the following properties of the covariance operator: if X and Y are independent $\mathbf{B}$-r.e., then

$$R_{X+Y} = R_X + R_Y,$$

and if $A \in \mathbf{L}(\mathbf{B}, \mathbf{F})$, then $R_{AX} = A^{**}R_XA^*$. If $R_X(\mathbf{B}^*) \subset \mathbf{B}$ (here $\mathbf{B}$ is regarded as a subspace of the space $\mathbf{B}^{**}$ at a natural inclusion map $\mathbf{B} \subset \mathbf{B}^{**}$), then we may observe that R_X: $\mathbf{B}^* \to \mathbf{B}$. This is always the case in separable Banach spaces (see [215]).

Two things are to be noted that are useful when computing covariance operators in particular spaces. If $\mathbf{B}$ is a certain Banach space of numeric sequences (for instance, $\mathbf{l}_p$, $\mathbf{c}_0$) then, the random element in these spaces has the form $X = (X_1, X_2, ...)$ and the infinite matrix $A = \{a_{ij}\}_{i,j=1}^{\infty}$, where $a_{ij} = E(X_i - EX_i)(X_j - EX_j)$ corresponds to the covariance operator (if it exists). If $\mathbf{B}$ is a functional space ($\mathbf{C}(S)$, for instance), then a random element in this space is a random process $X = X(t)$, $t \in S$. In this case, an important part is played by the linear functionals δ_t, $t \in T$, defined by the equality $\delta_t(x) = x(t)$, $x \in \mathbf{B}$. If the linear combinations of such functionals are *-weakly dense in $\mathbf{B}^*$, then it is sufficient to know the values of the bilinear form $\operatorname{cov} X$ not for all $f \in \mathbf{B}^*$, but only for functionals of the form δ_t – i.e. we should know the function of two arguments

$$R(t, s) = E\delta_t(X)\delta_s(X) = EX(t)X(s)$$

(assuming that $EX(t) = 0$ and that $EX^2(t) < \infty$ for all $t \in S$). This function is called the covariance function of the process X.

The characteristic functional is another major notion in the theory of distributions in infinite dimensional spaces.

DEFINITION 1.9. By the characteristic functional (ch.f.) of the measure μ (or $\mathbf{B}$-r.e. X with distribution $\mu = L(X)$) we mean the complex-valued functional $\hat{\mu}$: $\mathbf{B}^* \to \mathbf{C}$, defined by the formula

$$\hat{\mu}(f) = E \exp\{if(X)\} = \int_\Omega \exp\{if(X(\omega))\}P(d\omega)$$
$$= \int_{\mathbf{B}} \exp\{if(x)\}\mu(dx).$$

The most essential properties of the characteristic functional are collected in the following proposition.

PROPOSITION 1.10. *Let X be $\mathbf{B}$-r.e., $\mu = L(X)$, then:*

(a) $\hat{\mu}(0) = 1$ *and the functional $\hat{\mu}$ is positively defined (i.e. for any finite collection $\alpha_i \in \mathbf{C}$ and $f_i \in \mathbf{B}^*$, $i = 1, ..., n$,*

$$\sum_{j,i=1}^{n} \hat{\mu}(f_i - f_j)\alpha_i\overline{\alpha_j} \geq 0.$$

where $\overline{\alpha}$ denotes a conjugate complex number);

(b) *$\hat{\mu}$ is uniformly continuous with respect to the norm of the space $\mathbf{B}^*$;*

(c) *if X and Y are independent $\mathbf{B}$-r.e. $\mu_1 = L(X)$, $\mu_2 = L(Y)$, $\mu = L(X+Y)$, then $\hat{\mu}(f) = \hat{\mu}_1(f)\hat{\mu}_2(f)$;*

(d) *for each $n \geq 1$ and any collection $f_i \in \mathbf{B}^*$, $i = 1, ..., n$, the complex-valued function of n arguments*

$$\phi(t_1, ..., t_n) = \hat{\mu}(t_1 f_1 + \cdots + t_n f_n)$$

is a characteristic function of the random n-dimensional vector $(f_1(X), ..., f_n(X))$;

(e) *the distribution is uniquely defined by its characteristic functional, i.e. if* $\hat{\mu}_1(f) = \hat{\mu}_2(f)$ *for all* $f \in \mathbf{B}^*$, *then* $\mu_1 = \mu_2$ *(if* $\mathbf{B}$ *is non-separable, then* μ_1 *and* μ_2 *coincide on the* σ*-algebra of cylindrical sets).*

The proof of these and other properties may be found by the reader in, for instance, [215].

1.2. Main Facts on Gaussian Measures in Banach Spaces

Since the book is intended for questions dealing with the approximation of distribution of sums of Banach-valued elements by Gaussian measures, it is natural that some properties of these measures are required. In this section we present the major properties of Gaussian measures in separable Banach spaces.

DEFINITION 2.1. $\mathbf{B}$-r.e. Y is called Gaussian if, for any $f \in \mathbf{B}^*$, the random variable $f(Y)$ has a Gaussian distribution.

It is easy to see that for Gaussian measures the condition of Proposition 1.8 with $\varepsilon = 1$ is fulfilled, therefore there exists EY for any Gaussian $\mathbf{B}$-r.e. Y. If $EY = 0$, the Gaussian r.e. is called centred, and we shall consider such Gaussian elements later. It can be shown (see, for instance, [106]) that the probability measure μ on $\mathbf{B}$ is centred Gaussian if and only if for the two independent $\mathbf{B}$-r.e. Y_1 and Y_2 with $L(Y_1) = L(Y_2) = \mu$ and the real s, t such that $s^2 + t^2 = 1$, the random elements $sY_1 + tY_2$ and $tY_1 - sY_2$ are independent and have a distribution μ. By the use of this definition X. Fernique succeeded, by a relatively elementary method, in proving the following profound result (it must be noted that in fact he has obtained this result in more general toplogical spaces).

THEOREM 2.2. *Let* $\mathbf{B}$ *be a separable Banach space, and* Y *a centred Gaussian* $\mathbf{B}$*-r.e. There exists* $\alpha > 0$ *such that*

$$E \exp\{\alpha\|Y\|^2\} < \infty.$$

The proof is available in [63] or [106].

The following estimate connecting moments of different orders is often useful.

PROPOSITION 2.3. *Let* Y *be a centred Gaussian* $\mathbf{B}$*-r.e. For any positive* α *and* β *there exists such a constant* $C(\alpha, \beta)$ *independent of* $\mathbf{B}$ *and* $L(Y)$ *that*

$$(E\|Y\|^\alpha)^{1/\alpha} \le C(\alpha, \beta)(E\|Y\|^\beta)^{1/\beta}. \tag{2.1}$$

For the proof see [215].

From Theorem 2.2 it follows that a Gaussian $\mathbf{B}$-r.e. has moments of all orders, in particular $E\|Y\|^2 < \infty$. Therefore, Y has a covariance operator R_Y: $\mathbf{B}^* \to \mathbf{B}$. Since the ch.f. of any Gaussian measure μ in $\mathbf{B}$ has the form

$$\hat{\mu}(f) = \exp\left\{if(EY) - \frac{1}{2}R_Y(f)(f)\right\},$$

the Gaussian measures in $\mathbf{B}$, as in a one-dimensional case, are completely defined by mathematical expectation and the covariance operator.

Further, by $\mathscr{R}(\mathbf{B})$ we denote the class of all covariance operators of probability measures on $\mathbf{B}$, and by $\mathscr{R}_1(\mathbf{B})$ we shall denote the class of Gaussian covariance operators (i.e. $R \in \mathscr{R}_1(\mathbf{B})$ if the Gaussian $\mathbf{B}$−r.e. Y with $R_Y = R$ exists). $\mathbf{B}$−r.e. X (or the corresponding distribution $L(X)$) is called pre-Gaussian if $R_X \in \mathscr{R}_1(\mathbf{B})$. From the above expression of the ch.f. of Gaussian elements, it follows that the problem of describing Gaussian measures in a given Banach space $\mathbf{B}$ is equivalent to a description of the class $\mathscr{R}_1(\mathbf{B})$. Unfortunately, this problem has been completely solved for all but a few spaces. We shall formulate the results concerning the spaces $\mathbf{l}_p$, $\mathbf{c}_0$ and $\mathbf{C}(S)$. In considering the spaces of numerical sequences $\mathbf{l}_p$ and $\mathbf{c}_0$, the standard basis $\{e_n, n \geq 1\}$ will be used. The covariance operator of r.e. with values in these spaces is defined by the symmetric, positive semidefinite infinite matrix $T = \{t_{ij}\}_{i,j=1}^{\infty}$, where $t_{ij} = E\xi_i\xi_j$, ξ_i, $i \geq 1$, are coordinates of the Gaussian vector ξ in the basis $\{e_n, n \geq 1\}$.

THEOREM 2.4. *Operator* $T \in \mathscr{R}_1(\mathbf{l}_p)$, $1 \leq p < \infty$, *if and only if*

$$\sum_{i=1}^{\infty} t_{ii}^{p/2} < \infty.$$

For the proof see [213].

THEOREM 2.5. *If for all* $\varepsilon > 0$

$$\sum_{i=1}^{\infty} \exp\{-\varepsilon t_{ii}^{-1}\} < \infty \tag{2.2}$$

then $T \in \mathscr{R}_1(\mathbf{c}_0)$. *If* $t_{i,j} = 0$ *for* $|i - j| \geq m$, *where* m *is some positive number, then from* $T \in \mathscr{R}_1(\mathbf{c}_0)$ *condition* (2.2) *follows.*

For the proof see [213] and [153].

Further studies of Gaussian measures in $\mathbf{c}_0$ may be found in [39], [40], [196].

In the case of the functional space $\mathbf{C}(S)$ where (S, d) is a metric compact, instead of a covariance operator a covariance function

$$R(t, s) = E\xi(t)\xi(s)$$

is used, as mentioned above.

(Without loss of generality we assume that $E\xi(t) = 0$, $t \in S$.) Let us define a pseudometric τ on S, by setting

$$\tau(s, t) = \tau_\xi(s, t) = E^{1/2}(\xi(t) - \xi(s))^2 = (R(s, s) - 2R(s, t) + R(t, t))^{1/2}.$$

By $N(S, \rho, x)$ we denote the minimal number of balls of radius x (with respect to ρ) covering S. The metric entropy $H(S, \rho, x) = \ln N(S, \rho, x)$.

THEOREM 2.6. *Let* (S, d) *be a metric compact, and* $\xi(t)$ *be a centred Gaussian process on* S *continuous in probability. If for some* $h > 0$

$$\int_0^h H^{1/2}(S, \tau, x)\, dx < \infty,$$

then ξ *with probability* 1 *has continuous trajectories, i.e.* ξ *is a Gaussian* $\mathbf{C}(S)$−*r.e.*

The proof is found in [3].

We shall present another notion related to Gaussian covariances.

DEFINITION 2.7. The operator $u \in \mathbf{L}(\mathbf{H}, \mathbf{B})$ is called γ-radonifying if $u \circ u^* \in \mathscr{R}_1(\mathbf{B})$.

The class of all γ-radonifying operators $u \in \mathbf{L}(\mathbf{H}, \mathbf{B})$ will be denoted by $\pi_\gamma(\mathbf{H}, \mathbf{B})$. Some information on this class can be found in [108], [117], [8].

1.3. Weak Convergence of Probability Measures and Some Inequalities

In spite of the fact that the theory of weak convergence of measures is usually constructed in metric spaces, we confine ourselves to Banach spaces.

Let $\mathbf{B}$ be a separable Banach space,

$$\mathbf{C}_b^0(\mathbf{B}) := \{f \in \mathbf{C}^0(\mathbf{B}) : \|f\|_\infty := \sup_{x \in \mathbf{B}} |f(x)| < \infty\}.$$

If A is a Borel set in $\mathbf{B}$, then by $\mathring{A}$ we denote the set of interior points, and by $\overline{A}$ we denote the closure of the set A, $\partial A = \overline{A} \setminus \mathring{A}$. The set of all probability measures on $\mathbf{B}$ will be denoted by $\mathscr{P}(\mathbf{B})$. Let $\mu \in \mathscr{P}(\mathbf{B})$. The set A is called a μ-continuity set if $\mu(\partial A) = 0$.

DEFINITION 3.1. Let $\mu_n \in \mathscr{P}(\mathbf{B})$, $n \geq 1$, and $\mu \in \mathscr{P}(\mathbf{B})$. The sequence μ_n, $n \geq 1$, weakly converges to the measure μ ($\mu_n \Rightarrow \mu$), if for all $f \in \mathbf{C}_b^0(\mathbf{B})$

$$\lim_{n \to \infty} \int_{\mathbf{B}} f(x) \mu_n(dx) = \int_{\mathbf{B}} f(x) \mu(dx).$$

The measurable function $f : \mathbf{B} \to \mathbf{R}$ is μ-continuous if the set of points of discontinuity of the function f has a zero μ-measure.

THEOREM 3.2. *The following statements are equivalent:*

(a) $\mu_n \Rightarrow \mu$;

(b) $\lim_{n \to \infty} \int_{\mathbf{B}} f(x) \mu_n)(dx) = \int_{\mathbf{B}} f(x) \mu(dx)$ *for all* $f \in \mathrm{BL}(\mathbf{B})$;

(c) $\lim_{n \to \infty} \mu_n(D) \leq \mu(D)$ *for all closed sets* D;

(d) $\lim_{n \to \infty} \mu_n(G) \geq \mu(G)$ *for all open sets* G;

(e) $\lim_{n \to \infty} \mu_n(A) = \mu(A)$ *for all μ-continuous sets* A;

(f) $\lim_{n \to \infty} \int_{\mathbf{B}} f(x) \mu_n(dx) = \int_{\mathbf{B}} f(x) \mu(dx)$ *for all measurable bounded μ-continuous functions* f.

In $\mathscr{P}(\mathbf{B})$ topology may be introduced convergence which coincides with a weak convergence. Traditionally this topology is called a weak topology (from the positions of functional analysis this topology should be called *-weak topology since $\mathscr{P}(\mathbf{B}) \subset (\mathbf{C}_b^0(\mathbf{B}))^*$). The topological space $\mathscr{P}(\mathbf{B})$ is metrized by the Prokhorov metric or the bounded Lipschitzian metric. We remember that the Prokhorov metric π and the bounded Lipschitzian metric ρ_{BL} are defined as follows:

$$\pi(\mu, \nu) := \inf\{\varepsilon \geq 0 : \mu(F) \leq \nu(F_\varepsilon) + \varepsilon \text{ for all closed sets } F \subset \mathbf{B}\};$$

$$\rho_{\mathrm{BL}}(\mu, \nu) := \sup\left\{\left|\int_{\mathbf{B}} f(x)(\mu - \nu)(dx)\right| : \|f\|_{\mathrm{BL}} \leq 1\right\},$$

where

$$\|f\|_{\mathrm{BL}} := \sup_{x \in \mathbf{B}} |f(x)| + \sup_{x \neq y} |f(x) - f(y)| \, \|x - y\|^{-1}.$$

The uniformity classes which play an important part among the questions concerning rates of convergence of probability measures are defined below.

DEFINITION 3.3. The class $\mathcal{F}$ of bounded real measurable functions on $\mathbf{B}$ is said to be μ-uniform if for any sequence μ_n, $n \geq 1$, which weakly converges to μ, we have

$$\lim_{n\to\infty} \sup\left\{ \left| \int_{\mathbf{B}} f(x)(\mu_n - \mu)(dx) \right| : f \in \mathcal{F} \right\} = 0.$$

Correspondingly, the class $\mathcal{A}$ of Borel sets is μ-uniform if

$$\lim_{n\to\infty} \sup_{A \in \mathcal{A}} |\mu_n(A) - \mu(A)| = 0$$

for any sequence $\mu_n \Rightarrow \mu$.

If f is a real function on $\mathbf{B}$ and $\varepsilon > 0$, then the oscillation of the function f on the ball $V_\varepsilon(x)$ and on the whole space $\mathbf{B}$ is defined as follows:

$$\omega(f, \varepsilon, x) := \sup\{ |f(z) - f(y)| : z, y \in V_\varepsilon(x)\},$$

$$\omega(f, \mathbf{B}) := \sup\{ |f(z) - f(y)| : z, y \in \mathbf{B}\}.$$

THEOREM 3.4. *A class of measurable bounded functions $\mathcal{F}$ is μ-uniform if and only if*

(a) $\sup_{f \in \mathcal{F}} \omega(f, \mathbf{B}) < \infty$;

(b) $\lim_{\varepsilon\to 0} \sup_{f \in \mathcal{F}} \int_{\mathbf{B}} \omega(f, \varepsilon, x)\mu(dx) = 0.$

The class A is μ-uniform if and only if

$$\lim_{\varepsilon\to 0} \sup_{A \in \mathcal{A}} \mu((\partial A)_\varepsilon) = 0. \tag{3.1}$$

In the theory of weak convergence of measures an important role is played by the Prokhorov theorem for the formulation of which the following definition is required.

DEFINITION 3.5. A set $\Gamma \subset \mathcal{P}(\mathbf{B})$ is called relatively compact if any sequence from Γ contains a weakly converging subsequence.

THEOREM 3.6. *A set $\Gamma \subset \mathcal{P}(\mathbf{B})$ is relatively compact if and only if it is tight, i.e. for each $\varepsilon > 0$ there exists a compact $K \subset \mathbf{B}$ such that*

$$\sup_{\mu \in \Gamma} \mu(K) > 1 - \varepsilon.$$

We shall present some facts on relationships between different types of convergence of distributions of sums of independent $\mathbf{B}$-r.e. In the case of $p \geq 1$ we denote

$$\mathbf{L}_p^0(\mathbf{B}) := \{X \in \mathbf{L}_p(\mathbf{B}) : EX = 0\},$$

where the mean EX is understood in the sense of Bochner.

THEOREM 3.7 (see [38], [86]). *Let X_j, $j \geq 1$, be a sequence of independent $\mathbf{B}$-r.e., $S_n = \Sigma_{j=1}^n X_j$. The following statements are equivalent:*

(a) *S_n converges in $\mathbf{B}$ a.s.*

(b) *S_n converges in $\mathbf{B}$ in probability.*

(c) *$\mu_n = L(S_n)$ converges weakly.*

If we assume in addition that the X_j are symmetric, $j \geq 1$, then (a)–(c) *are equivalent to the following statements:*

(d) *there exists $\mathbf{B}$-r.e. S such that $f(S_n)$ converges in probability to r.v. $f(S)$ for all $f \in \mathbf{B}^*$;*

(e) *there exists a measure μ on $\mathbf{B}$ such that $\mu_n \cdot f^{-1} \Rightarrow \mu \cdot f^{-1}$ for all $f \in \mathbf{B}^*$.*

THEOREM 3.8. (see [86]). *Let the number* $p \in (0, \infty)$. *If there exists* $\mathbf{B}$*-r.e. such that* $S_n \to S$ *a.s., then the following statements are equivalent:*

(a) S_n *converges in* $\mathbf{L}_p(\mathbf{B})$;

(b) $\sup_{n\geq 1} \|S_n\| \in \mathbf{L}_p$;

(c) $\sup_{n\geq 1} \|X_n\| \in \mathbf{L}_p$.

In conclusion, we present some probabilistic inequalities for the sums of independent random elements.

THEOREM 3.9 (Levy inequality, see [38]). *Let* X_i, $i = 1, 2, \cdots$ *be independent symmetric* $\mathbf{B}$*-r.e. For any* $t > 0$ *the following inequalities take place*

$$P\{\max_{1\leq k\leq n} \|S_k\| > t\} \leq 2P\{\|S_n\| > t\}, \tag{3.2}$$

$$P\{\max_{1\leq k\leq n} \|X_k\| > t\} \leq 2P\{\|S_n\| > t\}. \tag{3.3}$$

If the series $\Sigma_{k=1}^{\infty} X_k$ *converges a.s. to* $\mathbf{B}$*-r.e.* S, *then we have the following inequality:*

$$P\{\max_{n\geq 1} \|S_n\| > t\} \leq 2P\{\|S\| > t\}.$$

By the term 'Rademacher sequence' we shall mean the sequence $\{\varepsilon_i, i \geq 1\}$ of independent identically distributed random variables ε_i with the distribution

$$P\{\varepsilon_i = 1\} = P\{\varepsilon_i = -1\} = \frac{1}{2}.$$

THEOREM 3.10. *Let* $\{\varepsilon_i, i \geq 1\}$ *be a Rademacher sequence. For any numbers* $1 \leq q \leq p < \infty$ *there exists a constant* $C(p, q)$ *such that for any finite collection of elements* $x_i \in \mathbf{B}$, $i = 1, \ldots, n$, *the following inequality holds:*

$$\left(E\|\sum_{i=1}^{n} \varepsilon_i x_i\|^p\right)^{1/p} \leq C(p, q)\left(E\|\sum_{i=1}^{n} \varepsilon_i x_i\|^q\right)^{1/q} \tag{3.4}$$

PROPOSITION 3.11. *Let* X *and* Y *be independent* $\mathbf{B}$*-r.e.* $EY = 0$ *(in the sense of Bochner). If* $\psi: \mathbf{B} \to \mathbf{R}$ *is a convex function, then*

$$E\psi(X) \leq E\psi(X + Y).$$

In particular,

$$E\|X\|^p \leq E\|X + Y\|^p, \; p \geq 1. \tag{3.5}$$

THEOREM 3.12. (see [92]). *Let* $X_1, \ldots, X_n$ *be independent* $\mathbf{B}$*-r.e. and let the number* $L > 0$ *be such that for all* $i = 1, \ldots, n$ *and all* $m \geq 2$ *the following conditions is satisfied:*

$$E\|X_i\|^m \leq \left(\frac{m!}{2}\right) E\|X_i\|^2 L^{m-2}.$$

Then for all $r > 0$ *the following inequality takes place:*

$$P\left\{\|\sum_{i=1}^{n} X_i\| \geq r\right\} \leq$$

$$\leq \exp\left\{-\left[\frac{r^2}{8}-\frac{r^2}{4}E\,\|\sum_{i=1}^{n}X_i\|\right]\Big/\left(\sum_{i=1}^{n}E\,\|X_i\|^2+\frac{rL}{2}\right)\right\}.$$

THEOREM 3.13 (see [3]). *Let* $0 < p < \infty$. *We assume that r.e.* $X_1, \cdots X_n, \cdots \in \mathbf{L}_p^0(\mathbf{B})$ *and r.v.* $\xi_1, ..., \xi_n, \cdots \in \mathbf{L}_p^0(\mathbf{R})$, $\eta_1, ..., \eta_n, \cdots \in \mathbf{L}_p^0(\mathbf{R})$ *are independent in total. If*

$$N := \sup_{j\geq 1} |\eta_j| \in \mathbf{L}_p(\mathbf{R}) \quad \text{and} \quad \alpha := \inf_{j\geq 1} E\,|\xi_j| > 0,$$

then

$$E\,\|\sum_{j=1}^{n}\eta_j X_j\|^p \leq (8\alpha^{-1})^p E N^p E\,\|\sum_{j=1}^{n}\xi_j X_j\|^p.$$

CHAPTER 2

Some Questions of Non-Linear Analysis

In this chapter the approximation of indicator functions of sets in Banach spaces by differentiable functions is considered. The results of this chapter are extensively used in Chapter 5 in constructing the estimates of convergence rate in the CLT. It is for the convenience of the readers that the necessary items on differential calculus in normed spaces are presented first.*

2.1. Differentiation in Normed Spaces

The Fréchet and Gateaux derivatives of vector functions. Let us consider the normed spaces **E** and **F**, the open set $U \subset \mathbf{E}$ and the function $f: U \to \mathbf{F}$.

DEFINITION 1.1. The function $f: U \to \mathbf{F}$ is called differentiable in the sense of Fréchet at the point $x \in U$ if there exists a mapping $f'(x) \in \mathbf{L}(\mathbf{E}, \mathbf{F})$ such that in some neighbourhood of the point x the presentation is true

$$f(x+h) = f(x) + f'(x)(h) + \alpha(h)\|h\|, \tag{1.1}$$

where $\alpha(h) \in \mathbf{F}$ and $\lim_{\|h\|\to 0}\alpha(h) = \alpha(0) = 0$.

The operator $f'(x)$ is called the Fréchet derivative of the function f at the point $x \in U$. It follows from (1.1) that the existence of the derivative $f'(x)$ implies continuity of the function f at the point x. Besides, rewriting (1.1) in the form

$$\|f(x+h) - f(x) - f'(x)(h)\| = o(\|h\|)$$

as $h \to 0$, we obtain that the derivative $f'(x)$ is defined uniquely since the equality $\|T_1(h) - T_2(h)\| = o(\|h\|)$ for $T_1, T_2 \in \mathbf{L}(\mathbf{E}, \mathbf{F})$ is possible only if $T_1 = T_2$. When defining derivatives of higher order, we shall make use of the following lemma.

LEMMA 1.2. *The Banach spaces* $\mathbf{L}^n(\mathbf{E}, \mathbf{L}^m(\mathbf{E}, \mathbf{F}))$ *and* $\mathbf{L}^{n+m}(\mathbf{E}, \mathbf{F})$ *are isometrically isomorphic.*

Proof. If $T \in \mathbf{L}^n(\mathbf{E}, \mathbf{L}^m(\mathbf{E}, \mathbf{F}))$ then $T(x_1, \ldots, x_n) \in \mathbf{L}^m(\mathbf{E}, \mathbf{F})$ and the equality

$$P(x_1, \ldots, x_n, x_{n+1}, \ldots, x_{n+m}) = T(x_1, \ldots, x_n)(x_{n+1}, \ldots, x_{n+m}) \tag{1.2}$$

defines $(n+m)$-linear mapping P of the space $\mathbf{E}^{n+m}$ into the space **F**. On the other hand, any

* More information on differential calculus in normed spaces can be found in [45] and [100].

$(n+m)$-linear mapping P by means of equality (1.2) defines the n-linear mapping $T: \mathbf{E}^n \to \mathbf{L}^m(\mathbf{E}, \mathbf{F})$. It is also clear that $\|P\| = \|T\|$. □

Now let us assume that the function $f: U \to \mathbf{F}$ is differentiable at every point of some open set $V \subset U$. In this case the relation $x \to f'(x)$ defines the function $f': V \to \mathbf{L}(\mathbf{E}, \mathbf{F})$. If this function is differentiable at the point $x \in V$, then its derivative, which is denoted by $f''(x)$ and called the second Fréchet derivative of the function f at the point x is an element of the space $\mathbf{L}(\mathbf{E}, \mathbf{L}(\mathbf{E}, \mathbf{F}))$. In accordance with Lemma 1.2, $f''(x)$ is identified with an element of the space $\mathbf{L}^2(\mathbf{E}, \mathbf{F})$. By induction, derivatives of higher order are defined.

DEFINITION 1.3. The function $f: U \to \mathbf{F}$ is n times differentiable in the sense of Fréchet at the point $x \in U$, if $f'(y), \ldots, f^{(n-1)}(y)$ exists for all y from some neighbourhood $U_x \subset U$ of the point x and the function $f^{(n-1)}: U_x \to \mathbf{L}^{n-1}(\mathbf{E}, \mathbf{F})$ is differentiable in the sense of Fréchet at the point x. In this case $(f^{(n-1)})'(x)$ is identified with an element of the space $\mathbf{L}^n(\mathbf{E}, \mathbf{F})$, is denoted by $f^{(n)}(x)$ and is called the nth Fréchet derivative of the function f at the point x.

The function $f: U \to \mathbf{F}$ is called n times differentiable in the sense of Fréchet if $f^{(n)}(x)$ exists at every point $x \in U$.

Later, we shall denote by $\mathbf{C}^0(U, \mathbf{F})$ the class of continuous functions $f: U \to \mathbf{F}$ and by $\mathbf{D}^n(U, \mathbf{F})$ the class of n times differentiable in the sense of Fréchet functions $f: U \to \mathbf{F}$. We shall also denote

$$\mathbf{C}^n(U, \mathbf{F}) := \{f \in \mathbf{D}^n(U, \mathbf{F}): f^{(n)} \in \mathbf{C}^0(U, \mathbf{L}^n(\mathbf{E}, \mathbf{F}))\}.$$

PROPOSITION 1.4. *The set* $\mathbf{D}^n(U, \mathbf{F})$ *is a linear space and for every fixed* $x \in U$ *the operator* $f \to f^{(n)}(x)$ *linearly maps this space into the space* $\mathbf{L}^n(\mathbf{E}, \mathbf{F})$.

The proof follows directly from Definitions 1.2 and 1.3. □

In most cases, the calculation of Fréchet derivative of a vector function is found to be a complex task. Sometimes this task is simplified if we use the relation of differentiation in the sense of Fréchet with differentiation in the sense of Gateaux.

DEFINITION 1.5. The function $f: U \to \mathbf{F}$ is called n times differentiable in the sense of Gateaux (or n times weakly differentiable) at the point $x \in U$ if

$$\frac{\partial^n}{\partial t_1 \cdots \partial t_n} f(x + t_1 h_1 + \cdots + t_n h_n) \Big|_{t_1 = \ldots = t_n = 0} := d_{h_1} \cdots d_{h_n} f(x)$$

exists for all $h_1, \ldots, h_n \in \mathbf{E}$ and the mapping $(h_1, \ldots, h_n) \to d_{h_1} \cdots d_{h_n} f(x)$, denoted by $f_c^{(n)}(x)$, is an element of the space $\mathbf{L}^n(\mathbf{E}, \mathbf{F})$.

For $x, y \in \mathbf{E}$ we denote

$$[x, y] = \{z \in \mathbf{E}: z = \tau x + (1-\tau) y,\ \tau \in [0, 1]\}.$$

The following result is usually called the mean value theorem.

THEOREM 1.6. *Let* $[x, x+h] \subset U$. *If the function* $f: U \to \mathbf{F}$ *is differentiable in the sense of Gateaux at every point* $y \in [x, x+h]$, *then*

$$\|f(x+h) - f(x)\| \le \sup\{\|f_c'(z)\|: z \in [x, x+h]\}\, \|h\|.$$

Proof. For the arbitrary element $T \in \mathbf{F}^*$ the function $\phi(t) = \langle T, f(x+th)\rangle$, $t \in [0, 1]$ is differentiable at least at the interval $]0, 1[$. By the well-known Lagrange theorem, $|\phi(1) - \phi(0)| \le \sup_{0<t<1} |\phi'(t)|$. Take the functional $T \in \mathbf{F}^*$ such that $\|T\| = 1$ and

$$\langle T, f(x) - f(x+h)\rangle = \|f(x) - f(x+h)\|$$

(this is possible according to the Hahn–Banach theorem). Then

$$\|f(x) - f(x+h)\| = \phi(0) - \phi(1) \leq$$
$$\leq \sup_{0<t<1} \langle | T, f_c'(x+th)(h)\rangle |$$
$$\leq \sup\{\|f_c'(z)\| : z \in [x, x+h]\}\|h\|. \square$$

COROLLARY 1.7. *Let $x \in U$, $h \in \mathbf{E}$, $[x, x+h] \subset U$. If the function $f: U \to \mathbf{F}$ is weakly differentiable at every point of the interval $[x, x+h]$, then for any operator $\Lambda \in \mathbf{L}(\mathbf{E}, \mathbf{F})$ the inequality is true:*

$$\|f(x+h) - f(x) - \Lambda(h)\| \leq \sup\{\|f_c'(y) - \Lambda\| : y \in [x, x+h]\}\|h\|.$$

Proof. It is sufficient to apply Theorem 1.6 for the function $g = f - \Lambda$. □

It can easily be seen that the existence of the Fréchet derivative $f'(x)$ of the function $f: U \to \mathbf{F}$ implies the existence of the Gateaux derivatives $f_c'(x)$ and $f'(x) = f_c'(x)$. The following result shows an inverse relation.

THEOREM 1.8. *If $f_c^{(n)}(x)$ exists in some neighbourhood U_{x_0} of the point x_0 and is continuous at this point, then there exists $f^{(n)}(x_0)$ and $f^{(n)}(x_0) = f_c^{(n)}(x_0)$.*

Proof. Obviously it will be enough to consider the case $n = 1$. By the given $\varepsilon > 0$ we find $\delta > 0$ such that $\|h\| < \delta$ implies $[x_0, x_0 + h] \subset U_{x_0}$ and $\|f_c'(x_0) - f_c'(x_0 + h)\| < \varepsilon$. By the use of Collorary 1.7, where $\Lambda = f_c'(x_0)$, we obtain

$$\|f(x_0 + h) - f(x_0) - f_c'(x_0)(h)\| \leq \sup\{\|f_c'(y) - f_c'(x_0)\| : y \in [x_0, x_0 + h]\}\|h\|.$$

Hence, since for every $y \in [x_0, x_0 + h]$ the relation $\|h\| < \delta$ implies $\|y - x_0\| < \delta$, we obtain

$$\|f(x_0 + h) - f(x_0) - f_c'(x_0)(h)\| = o(\|h\|).$$

This relation completes the proof. □

Thus, if the nth derivative of Gateaux is continuous in the domain U, there exists a continuous derivative in the sense of Fréchet coinciding with the derivative of Gateaux. This gives us a rather simple method of computing the Fréchet derivatives.

From Collorary 1.7, by applying the mean value theorem again, we obtain the following result.

PROPOSITION 1.9. *If the function $f: U \to \mathbf{F}$ is twice differentiable in the sense of Gateaux and* $\sup\{\|f_c'(x)\| : x \in U\} < \infty$, *then $f \in \mathbf{C}^1(U, \mathbf{F})$.*

It should be remembered that the mapping $P \in \mathbf{L}^n(\mathbf{E}, \mathbf{F})$ is called symmetric if $P(x_1, \ldots, x_n) = P(x_{\pi(1)}, \ldots, x_{\pi(n)})$ for any permutation π of the numbers $1, \ldots, n$.

THEOREM 1.10. *If the function $f: U \to \mathbf{F}$ is n times Fréchet differentiable at the point $x_0 \in U$, then the mapping $f^{(n)}(x_0) \in \mathbf{L}^n(\mathbf{E}, \mathbf{F})$ is symmetric.*

Proof. First, consider the case $n = 2$. By the definition of the second derivative

$$f'(y) - f'(x_0) = (f')'(x_0)(y - x_0) + \alpha(y)\,\|y - x_0\|, \tag{1.3}$$

where $\alpha(y) \in \mathbf{L}(\mathbf{E}, \mathbf{F})$ and

$$\lim_{y \to x_0} \alpha(y) = \alpha(x_0) = 0. \tag{1.4}$$

Let x_1, x_2 be arbitrary fixed elements of the space $\mathbf{E}$. Consider the function $\phi(z) = f(z + tx_1)$, which is correctly defined for sufficiently small $t \in \mathbf{R}$ and z, close to x_0. By the use of (1.3), it is easy to check that

$$\phi'(z) = t(f')'(x_0)(x_1) + \alpha(z + tx_1)\,\|z + tx_1 - x_0\| - \alpha(z)\,\|z - x_0\|.$$

In particular,

$$\phi'(x_0) = t(f')'(x_0)(x_1) + \alpha(x_0 + tx_1)\,|t|\,\|x_1\| \tag{1.5}$$

and therefore

$$\phi'(z) - \phi'(x_0)$$
$$= \alpha(z + tx_1)\|z + tx_1 - x_0\| - \alpha(z)\|z - x_0\| - |t|\,\alpha(x_0 + tx_1)\|x_1\|.$$

Hence, taking into account relation (1.4), for every $\varepsilon > 0$ we find $\delta > 0$ such that the conditions $\|z - x_0\| < \delta/2$ and $\|x_1\|\;|t| < \delta/2$ imply the estimate

$$\|\phi'(z) - \phi'(x_0)\| \le 2\varepsilon(\|z - x_0\| + |t|\;\|x_1\|). \tag{1.6}$$

With sufficiently small $t \in \mathbf{R}$ the second difference

$$\Delta_t(x_1, x_2) = f(x_0 + tx_1 + tx_2) - f(x_0 + tx_1) - f(x_0 + tx_2) + f(x_0)$$

is determined correctly. By the use of relations (1.5), (1.6) and Corollary 1.7, we obtain: for

$$|t|\;\|x_1\| < \frac{\delta}{2}, \quad |t|\;\|x_2\| < \frac{\delta}{2},$$

it follows that

$$\|\Delta_t(x_1, x_2) - t^2 f''(x_0)(x_1, x_2)\|$$
$$= \|\phi(x_0 + tx_2) - \phi(x_0) - t^2(f')'(x_0)(x_1)(x_2)\|$$
$$= \|\phi(x_0 + tx_2) - \phi(x_0) - t\phi'(x_0(x_2) + t^2\|x_1\|\alpha(x_0 + tx_1)(x_2)\|$$
$$\le \sup\{\|\phi'(x) - \phi'(x_0)\|\;\|x_2\|\;|t|;\; x \in [x_0, x_0 + tx_2]\} +$$
$$+ \alpha(x_0 + tx_1)t^2\|x_1\|\;\|x_2\|$$
$$\le 3\varepsilon t^2(\|x_1\| + \|x_2\|)\|x_2\|. \tag{1.7}$$

Reversing x_1 and x_2 we also obtain

$$\|\Delta_t(x_2, x_1) - t^2 f''(x_0)(x_2, x_1)\| \le 3\varepsilon t^2(\|x_1\| + \|x_2\|)\|x_1\|. \tag{1.8}$$

Since $\Delta_t(x_1, x_2) = \Delta_t(x_2, x_1)$, it is easy to derive from (1.7) and (1.8) that

$$f''(x_0)(x_1, x_2) = f''(x_0)(x_2, x_1).$$

A further proof will be carried out by means of mathematical induction making use of the following equality

$$(f^{(n-1)})'(x_0)(x_1)(x_2, \ldots, x_n) = f^{(n)}(x_0)(x_1, \ldots, x_n)$$

$$= (f^{(n-2)})''(x_0)(x_1, x_2)(x_3, \dots, x_n).$$

An inductive assumption implies the symmetry of $f^{(n)}(x_0)(x_1, \dots, x_n)$ with respect to the arguments $x_2, \dots, x_n$ and from the proved case $n = 2$ we obtain symmetry with respect of the permutation of x_1 and x_2. This is sufficient for the proof of symmetry. □

Differentiation of a composite function. Let **E**, **F**, **G** be normed spaces, and let U and V be open sets in the spaces **E** and **F** respectively.

THEOREM 1.11. *If the function $f : U \to V$ is n times Fréchet differentiable at the point $x \in U$ and the funtion $g : V \to G$ is n times Fréchet differentiable at the point $y = f(x) \in V$, then the composition $\psi = g \cdot f$ is n times differentiable in the sense of Fréchet at the point x, then we obtain the equality*

$$\psi^{(n)}(x)(h_1, \dots, h_n) = \sum_{k=1} \Sigma\, g^{(k)}(y)[f^{l_1}(x)(h_{i_1^1}, \dots, h_{i_{l_1}^1}), \dots,$$

$$f^{(l_k)}(x)(h_{i_1^k}, \dots, h_{i_{l_k}^k})], \tag{1.9}$$

where the second sum is taken over all the partitions of the set $(h_1, \dots, h_n)$ to k of non-intersecting non-empty sets

$$\{h_{i_1^1}, \dots, h_{i_{l_1}^1}\}, \dots, \{h_{i_1^k}, \dots, h_{i_{l_k}^k}\};$$

besides,

$$l_1 + \cdots + l_k = n,\ l_j > 0,\ i_1^j < \cdots < i_{l_j}^j,\ j = 1, \dots, k,\ i_1^1 < \cdots < i_1^k.$$

Proof. Consider the case $n = 1$. By the assumption

$$f(x + h) = f(x) + f'(x)(h) + \phi_1(h), \tag{1.10}$$

where $\phi_1(h) \in \mathbf{F}$ and $\|\phi_1(h)\| = o(\|h\|)$. By analogy

$$g(y) = g(b) + g'(b)(y - b) + \phi_2(y - b), \tag{1.11}$$

where $\phi_2(y - b) \in \mathbf{G}$ and $\|\phi_2(y - b)\| = o(\|y - b\|)$.

Having applied relations (1.10) and (1.11), for the difference $\psi(x + h) - \psi(x)$ we obtain:

$$\begin{aligned}\psi(x + h) - \psi(x) &= g(f(x + h)) - g(f(x))\\ &= g'(f(x))(f(x + h) - f(x)) + \phi_2(f(x + h) - f(x))\\ &= g'(f(x))(f'(x)(h)) + g'(f(x))(\phi_1(h)) + \phi_2(f(x + h) - f(x))\\ &= g'(f(x))(f'(x)(h)) + \phi_3(h),\end{aligned} \tag{1.12}$$

where

$$\phi_3(h) = g'(f(x))(\phi_1(h)) + \phi_2(f(x + h) - f(x)).$$

Since

$$\|g'(f(x))(\phi_1(h))\| \le \|g'(f(x))\|\ \|\phi_1(h)\|,$$

$$\|\phi_2(f(x + h) - f(x))\| = o(\|f(x + h) - f(x)\|),$$

and

$$\|f(x + h) - f(x)\| \le \|f'(x)\|\ \|h\| + \|\phi_1(h)\|,$$

then

$$\|\phi_2(h)\| = o(\|h\|),$$

and relation (1.12) proves the theorem in the case $n = 1$. A further proof is carried out by induction with respect to n. As these computations are not complicated but rather unwieldy, we let the reader complete the proof. □

The theorem holds if the function f is but weakly n times differentiable at the point $x \in U$. Besides, in formula (1.9) the Fréchet derivatives of the function f must be substituted by the corresponding Gateaux derivatives. However, substituting Fréchet differentiability of the function g by Gateaux differentiability is, generally speaking, impossible. One can be convinced after having considered the following case.

Let

$$\mathbf{E}, \mathbf{F} = \mathbf{R}_2,\ \mathbf{G} = \mathbf{R},\ f(x) = (f_1(x_1, x_2), f_2(x_1, x_2)) = (x_1^2, x_2);$$

$$g(y) = g(y_1, y_2) = \begin{cases} 1 & \text{at } y_1 = y_2^2,\ y_2 > 0, \\ 0 & \text{in the remainder of the cases.} \end{cases}$$

Here the function f is Fréchet differentiable at the point (0, 0), and the function g is Gateaux differentiable at the point (0, 0). However, the function

$$\psi(x) = g \cdot f(x) = g(x_1^2, x_2) = \begin{cases} 1 & \text{at } x_2 = |x_1| > 0, \\ 0 & \text{in the remainder of the cases} \end{cases}$$

is not Gateaux differentiable at the point (0, 0).

The Taylor formula. Let $[x, x + h] \subset U$, the function $f: U \to \mathbf{F}$. On the interval [0, 1] we shall define the function $v: [0, 1] \to \mathbf{F}$, $v(t) = f(x + th)$. By Theorem 1.11, $f \in \mathbf{D}^{n+1}(U, \mathbf{F})$ implies the existence of the derivative $v^{(k)}(t)$, $t \in [0, 1]$, and

$$v^{(k)}(t) = f^{(k)}(x + th)(h)^k, \quad k = 1, \ldots, n + 1. \tag{1.13}$$

Here and elsewhere the image of the vector $(h, h, \ldots, h) \in \mathbf{E}^n$ at the mapping $T: \mathbf{E}^n \to \mathbf{F}$ is denoted by $T(h)^n$ for short.

THEOREM 1.12 (Taylor formula with a remainder term in integral form). *If* $f \in \mathbf{C}^{n+1}(U, \mathbf{F})$ *and* $[x, x + h] \subset U$, *then*

$$f(x + h) = f(x) + f'(x)(h) + \frac{1}{2!} f^2(x)(h)^2 + \cdots + \frac{1}{n!} f^{(n)}(x)(h)^n +$$
$$+ \int_0^1 \frac{(1 - t)^n}{n!} f^{(n+1)}(x + th)(h)^{n+1}\, dt. \tag{1.14}$$

Proof. It is easy to check the validity of the following equality

$$\frac{d}{dt}\left[v(t) + (1 - v)v'(t) + \cdots + \frac{1}{n!}(1 - t)^n v^{(n)}(t)\right] = \frac{1}{n!}(1 - t)^n v^{(n+1)}(t).$$

If the function $g: \mathbf{R} \to \mathbf{F}$ has a continuous derivative g' for $t \in [0, 1]$, then, as is known, (see [100], for instance)

$$g(1) - g(0) = \int_0^1 g'(t)\, dt.$$

Applying this formula to the function

$$g(t) = v(t) + (1 - t)v'(t) + \cdots + \frac{1}{n!}(1 - t)^n v^{(n)}(t)$$

and by use of (1.13), we obtain (1.14). □

THEOREM 1.13. *If* $f \in \mathbf{D}^{n+1}(U, \mathbf{F})$ *and* $[x, x + h] \subset U$, *then*

$$f(x + h) = f(x) + f'(x)(h) + \cdots + (n!)^{-1} f^{(n)}(x)(h)^n + R_n \|h\|^{n+1},$$

where $R_n \in \mathbf{F}$ *and*

$$\|R_n\| \leq ((n + 1)!)^{-1} \sup \{\|f^{(n+1)}(z)\| : z \in [x, x + h]\}.$$

Proof. Consider the function $\phi(t) = \langle T, v(t) \rangle$, $t \in [0, 1]$, where $T \in \mathbf{F}^*$. By the Taylor formula for the real function

$$\phi(1) = \phi(0) + \phi'(0) + \frac{1}{2}\phi''(0) + \cdots + \frac{\phi^{(n)}(0)}{n!} + \frac{\phi^{(n+1)}(\Theta)}{(n + 1)!}, \tag{1.15}$$

where the number $\Theta \in (0, 1)$ in a general case depends on T. Formula (1.15), taking into account (1.13), takes the form

$$\langle T, f(x + h) - f(x) - f'(x)(h) - \cdots - \frac{1}{n!} f^{(n)}(x)(h)^n \rangle$$

$$= \langle T, \frac{1}{(n + 1)!} f^{(n+1)}(x + \Theta h)(h)^{n+1} \rangle.$$

Denoting

$$R_n = \|h\|^{-(n+1)} \left[f(x + h) - f(x) - \cdots - \frac{1}{n!} f^{(n)}(x)(h)^n \right]$$

and choosing T such that $\|T\| = 1$, $\langle T, R_n \rangle = \|R_n\|$, we easily complete the proof. □

The next Taylor formula expresses only an asymptotic property.

THEOREM 1.14. *If the function* $f \in \mathbf{D}^n(U, \mathbf{F})$, $[x, x + h] \subset U$, *then* $\|f(x + h) - f(x) - f'(x)(h) - \cdots - (n!)^{-1} f^{(n)}(x)(h)^n\| = o(\|h\|^n)$.

For the proof, see [45].

An expression of the form $P_n(h) = \Sigma_{k=0}^n T_k(h)^k$, where $T_k \in \mathbf{L}^k(\mathbf{E}, \mathbf{F})$, is called a polynomial of the order n. $(T_0(h)^0$ is understood to be a fixed element of the space $\mathbf{F}$.)

PROPOSITION 1.15. *If for the function* $f : \mathbf{E} \to \mathbf{F}$ *there exists a polynomial* P_n *such that*

$$f(x + h) - f(x) = P_n(h) + \omega(h),$$

where $\omega(h)\|h\|^{-n} \to 0$ *as* $h \to 0$ *uniformly in some neighbourhood of the point* x, *then the function* f *has, at the point* x, *derivatives up to the n-th order and*

$$f^{(k)}(x)(h)^k = k! T_k(h)^k, \quad k = 1, \ldots, n.$$

For the proof, see [122].

Differentiation with respect to the operator. Unlike finite-dimensional spaces, in infinite-dimensional normed spaces the class of Fréchet differentiable functions is not too large. It is known, for instance, that even on a unit ball of a separable Hilbert space there exists a bounded uniformly continuous function, uniformly non-approximable by functions having the first two bounded Fréchet derivatives (see [132]).

Below we shall consider a more general differentiability of functions which allows us to get considerably more results concerning the approximation of functions, and indicator functions in particular.

Let **F**, **B**, **E** be normed spaces, let the operator $u \in \mathbf{L}(\mathbf{F}, \mathbf{B})$, let U be an open set of the space **B**, and let the functions $f: U \to \mathbf{E}$. For $x \in U$ we denote by f_x the functions defined by the equality $f_x(y) = f(x + y)$. The function $f_x \cdot u$ is defined on the open set $u^{-1}(U - x) \subset \mathbf{F}$ and takes the values in **E**:

$$f_x \cdot u(h) = f(x + u(h)).$$

DEFINITION 1.16. The function $f: U \to \mathbf{E}$ is called n times u-differentiable at the point $x \in U$ if the function $f_x \cdot u: u^{-1}(U - x) \to \mathbf{E}$ is n times differentiable in the sense of Fréchet at the point $h = 0$.

The Fréchet derivative $(f_x \cdot u)^{(n)}(0)$ is denoted by $f_u^{(n)}(x)$ and is called the nth u-derivative of the function f at the point x. If $f_u^{(n)}(x)$ exists at every point $x \in U$, then the function f is called n times u-differentiable.

Further on we shall confine ourselves to the following notation: if $\mathbf{S}(V, \mathbf{E})$ is some class of the functions $f: V \to \mathbf{E}$ defined on an arbitrary open set V of the space **F**, then by $\mathbf{S}_u(U, \mathbf{E})$ we denote the class of all functions $f; U \to \mathbf{E}$ such that for each $x \in U$ the function $f_x \cdot u \in \mathbf{S}(u^{-1}(U - x), \mathbf{E})$. The class $\mathbf{C}_u^0(U, \mathbf{E})$, for instance, consists of the functions $f: U \to \mathbf{E}$ such that for each $x \in U$ the function $f_x \cdot u: u^{-1}(U - x) \to \mathbf{E}$ is continuous.

From the definition of the u-derivative we easily obtain the following proposition.

PROPOSITION 1.17. *Let the function $f: U \to \mathbf{E}$ be n times u-differentiable. Then $f \in \mathbf{D}_u^n(U, \mathbf{E})$. Besides, for each $x \in U$ and each $h \in u^{-1}(U - x)$ there holds $(f_x \cdot u)^{(n)}(h) = f_u^{(n)}(x + u(h))$.*

This relation permits us to apply theorems that hold for functions that are differentiable in the sense of Fréchet, as well as for u-differentiable functions. Besides, it is necessary to introduce some changes in the notations. In particular, the Taylor formula takes the following form: if $f \in \mathbf{C}_u^n(U, \mathbf{E})$ and $[x, x + u(h)] \subset U, x \in U, h \in \mathbf{F}$, then

$$f(x + u(h)) = f(x) + f_u'(x)(h) + \frac{1}{2!} f_u''(x)(h)^2 + \cdots + \frac{1}{n!} f_u^{(n)}(x)(h)^n +$$
$$+ \int_0^1 \frac{(1-t)^n}{n!} f_u^{(n+1)}(x + tu(h))(h)^{n+1}\, dt. \tag{1.16}$$

It is easy to check that for any Banach space **F** and any operator $u \in \mathbf{L}(\mathbf{F}, \mathbf{B})$ the relation $f \in \mathbf{D}^n(U, \mathbf{E})$ implies

$$f \in \mathbf{D}_u^n(U, \mathbf{E}),\ f_u^{(n)}(x)(h_1, \ldots, h_n) = f^{(n)}(x)(u(h_1), \ldots, u(h_n)),\ h_1, \ldots, h_n \in \mathbf{F}.$$

If $\mathbf{F} = \mathbf{B}$ and $u = I_\mathbf{B}$, then the class $\mathbf{D}_{I_\mathbf{B}}^n(U, \mathbf{E})$ coincides with the class $\mathbf{D}^n(U, \mathbf{E})$. If the operator u is identical inclusion of the space **F** into the space **B**, then, instead of the term 'u-differentiability' we shall use the term '**F**-differentiability' (the term 'differentiability along the subspace **F**' is also often met). Correspondingly, we change the notations: instead of the operator u, we shall write the subset with respect to which differentiability is considered. For example, $\mathbf{C}_\mathbf{F}^n(\mathbf{B}, \mathbf{R})$ will denote the class of the functions $f: \mathbf{B} \to \mathbf{R}$ which are n times continuously **F**-differentiable and $f_\mathbf{F}^{(n)}(x)$ will denote the nth derivative along the space **F**. For an arbitrary space **F** and an arbitrary operator $u \in \mathbf{L}(\mathbf{F}, \mathbf{B})$, u-differentiability of the function

$f: U \to \mathbf{E}$ is, in fact, equivalent to $\mathbf{B}_u$-differentiability for some space $\mathbf{B}_u \subset \mathbf{B}$. We shall prove this equivalence by the use of the following two lemmas, the proofs of which may be found in [107].

The kernel of the operator $u \in \mathbf{L}(\mathbf{F}, \mathbf{B})$ will be denoted by $\ker u$, and by $\mathbf{F}/\ker u$ we denote the corresponding factor-space provided with the norm $\|\pi(h)\| = \inf\{\|h + y\|, y \in \ker u\}$, where π is a factor mapping $\mathbf{F} \to \mathbf{F}/\ker u$, and $N(u)$ is a set of values of the operator u.

LEMMA 1.18. *For each operator $u \in \mathbf{L}(\mathbf{F}, \mathbf{B})$ there exists an isomorphism $\tilde{u}: \mathbf{F}/\ker u \to N(u)$ such that $u = i \cdot \tilde{u} \cdot \pi$, where i is an operator of identical embedding.*

LEMMA 1.19. *If $u_1 \in \mathbf{L}(\mathbf{F}, \mathbf{B})$, $u_2 \in \mathbf{L}(\mathbf{F}, \mathbf{E})$, then the operator $u \in \mathbf{L}(\mathbf{B}, \mathbf{E})$, such that $u_2 = u \cdot u_1$, exists if and only if* $\ker u_1 \subset \ker u_2$.

Let $u \in \mathbf{L}(\mathbf{F}, \mathbf{B})$. We give the space $N(u)$ the norm $\|z\|_u := \|\tilde{u}^{-1}(z)\|$ and denote the space $(N(u), \|\cdot\|_u)$ by $\mathbf{B}_u$. It is easy to check that $\mathbf{B}_u$ is a Banach space.

THEOREM 1.20. *The function $f \in \mathbf{D}_u^n(U, \mathbf{E})$ if and only if f is n times $\mathbf{B}_u$-differentiable. Besides, for each $x \in U$,*

$$f_u^{(n)}(x) = f_{\mathbf{B}_u}^{(n)}(x) \cdot (\tilde{u} \cdot \pi)^n,$$

where

$$(\tilde{u} \cdot \pi)^n(h_1, \dots, h_n) = (\tilde{u} \cdot \pi(h_1), \dots, \tilde{u} \cdot \pi(h_2)),\ h_1, \dots, h_n \in \mathbf{F}.$$

Proof. Let $n = 1$. it is clear that $f \in \mathbf{D}_{\mathbf{B}_u}^1(U, \mathbf{E})$ implies $f \in \mathbf{D}_u^1(U, \mathbf{E})$. Thus we shall prove the inverse. It is easy to see that $\ker f_u'(x) \supset \ker u$. By Lemma 1.19, we shall find the operator $T_x \in \mathbf{L}(\mathbf{B}_u, \mathbf{B})$ such that $f_n'(x) = T_x \cdot \tilde{u} \cdot \pi$. Since $\{\tilde{h} \in \mathbf{F} / \ker u : \|\tilde{h}\| < r\} = \{\pi(h) : \|h\| < r\}$, we have

$$\sup_{\|z\|_u < r} \|f_x \cdot i(z) - f_x \cdot i(0) - T_x(z)\|$$

$$= \sup_{\|\tilde{h}\| < r} \|f_x \cdot i \cdot \tilde{u}(\tilde{h}) - f_x \cdot i \cdot \tilde{u}(0) - T_x \cdot \tilde{u}(\tilde{h})\|$$

$$= \sup_{\|h\| < r} \|f_x \cdot u(h) - f_x \cdot u(0) - f_u'(x)(h)\| = o(r).$$

where i denotes the embedding operator of the space $\mathbf{B}_u$ into the space $\mathbf{B}$. Hence, it follows that $f \in \mathbf{D}_{\mathbf{B}_u}^1(U, \mathbf{E})$. In a general case the proof is carried out by induction with respect to $k = 1, \dots, n$, making use of the fact that

$$[f_u^{(n-1)}(x)(h_1, \dots, h_{n-1})]_u'(x)(h_n) = f_u^{(n)}(x)(h_1, \dots, h_n),$$

and repeating the reasonings used in the case $n = 1$. □

REMARK 1.21. At the composition of the operator $u \in \mathbf{L}(\mathbf{F}, \mathbf{B})$ on the right, the corresponding class of differentiable functions $f: \mathbf{B} \to \mathbf{E}$ does not decrease, i.e. if $u_1 \in \mathbf{L}(\mathbf{F}_1, \mathbf{F})$, then, $\mathbf{D}_u^k(\mathbf{B}, \mathbf{E}) \subset \mathbf{D}_{u \cdot u_1}^k(\mathbf{B}, \mathbf{E})$.

This section will be completed by constructing real functions on a Banach space $\mathbf{B}$ differentiable along a Hilbert space $\mathbf{H}$ using Gaussian measures for this aim. To simplify the reasonings, we shall assume separability of the spaces $\mathbf{H}$ and $\mathbf{B}$.

Let ν_1, ν_2 be two probability measures on $(\mathbf{B}, \mathscr{B}(\mathbf{B}))$. We recall (see [84], [195]) that the measure ν_1 is absolutely continuous with respect to the measure ν_2 (denoted $\nu_1 < \nu_2$) if from $\nu_2(A) = 0, A \in \mathscr{B}(\mathbf{B})$ it follows that $\nu_1(A) = 0$. This definition is equivalent to the existence of the measurable function

$$\frac{d\nu_1}{d\nu_2}(x) = g(x)$$

(Radon–Nikodym derivative) such that

$$\nu_1(A) = \int_A g(x)\nu_2(dx)$$

for any $A \in \mathscr{B}(\mathbf{B})$. The measures ν_1 and ν_2 are equivalent if $\nu_1 < \nu_2$ and $\nu_2 < \nu_1$. In the case of the operator $u \in \pi_\gamma(\mathbf{H}, \mathbf{B})$ (see Definition 1.2.7), the Gaussian measure with zero mean and covariance operator $u \cdot u^*$ will be denoted by γ_u. For the number $t > 0$ and the element $h \in \mathbf{B}$ the measures γ_u^t and $\gamma_u^t(h, \cdot)$ are defined by the formulas $\gamma_u^t(A) = \gamma_u(t^{-1}A)$ and $\gamma_u^t(h, A) = \gamma_u^t(A + h)$.

First, we shall consider the case when $\mathbf{H}$ is related to $\mathbf{B}$ by the continuous embedding $i: \mathbf{H} \to \mathbf{B}$ which is a γ-radonifying operator. Besides, instead of γ_i we shall write only γ. Here the following terminology can be also used: γ is a Gaussian measure on $\mathbf{B}$ with a zero mean and a covariance operator T, $\mathbf{H}$ is a reproducing Hilbert space; or it may be said that the triplet $(i, \mathbf{H}, \mathbf{B})$ form an abstract Wiener space (see [106]). The proof of the following result can be found in [53] or [183], for instance.

LEMMA 1.22. *If $h \in \mathbf{H}$, then the measure $\gamma^t(h, \cdot)$ is equivalent to the measure γ^t and the Radon–Nikodym derivative is defined by the formula*

$$\frac{d\gamma^t(h, \cdot)}{d\gamma^t}(x) = \exp\left\{-\frac{1}{2t}|h|^2 - \frac{1}{t}\phi_h(x)\right\}, \quad x \in \mathbf{B},$$

where the random functional $\phi_h(\cdot)$ is defined γ^t-almost everywhere on $\mathbf{B}$ and has a normal distribution with a zero mean and with the variance $t|h|^2$. Besides, the mapping $h \to \phi_h$ is linear.

For the measurable bounded function $f: \mathbf{B} \to \mathbf{R}$ we assume

$$\gamma^t f(x) = \int_{\mathbf{B}} f(x + y)\gamma^t(dy).$$

THEOREM 1.23. *The function $\gamma^t f \in \mathbf{C}^\infty_{\mathbf{H}}(\mathbf{B}, \mathbf{R})$, and for all $k \geq 1$ the estimate*

$$\sup_{x \in \mathbf{B}} \sup_{|k| \leq 1} |(\gamma^t f)^{(k)}_{\mathbf{H}}(x)(h)^k| \leq \sqrt{k!}\,\|f\|_\infty t^{-k/2}, \tag{1.17}$$

where $\|f\|_\infty = \sup_{x \in \mathbf{B}} |f(x)|$ is valid.

Proof. Consider the function $g: \mathbf{H} \to \mathbf{R}$, $g(h) = (\gamma^t f)(x + h)$. By the use of Lemma 1.22 we write

$$g(h) = \int_{\mathbf{B}} f(x + y)\psi_y(h)\gamma^t(dy),$$

where

$$\psi_y(h) = \exp\left\{-\frac{1}{2t}|h|^2 + \frac{1}{t}\phi_k(h)\right\}.$$

For all $h \in \mathbf{B}^*$, $y \in \mathbf{B}$ we have $\phi_h(y) = <h, y>$ (see [106]). Consequently, the mapping $h \to \phi_h(y)$ of the space $\mathbf{B}^*$ to the space $\mathbf{R}$ is infinitely differentiable for each $y \in \mathbf{B}$. Since $|\cdot|^2 \in \mathbf{C}^\infty(\mathbf{H}, \mathbf{R})$, it follows from Theorem 1.11 that $\psi_y \in \mathbf{C}^\infty(\mathbf{B}^*, \mathbf{R})$ for each $y \in \mathbf{B}$. Theorem 1.22 guarantees the integrability of the quantities

$$\psi_y^{(k)}(h)(h_1, ..., h_k),\ h, h_1, ..., h_k \in \mathbf{B}^*,\ k \geq 1$$

with respect to $\gamma(dy)$. Consequently, $\gamma f \in \mathbf{C}^\infty_{\mathbf{i}^* \circ i}(\mathbf{B}, \mathbf{R})$. Hence, taking into account the fact that the embedding of the space $\mathbf{B}^*$ into the space $\mathbf{H}$ is dense, it is easy to show that $\gamma f \in \mathbf{C}^\infty_{\mathbf{H}}(\mathbf{B}, \mathbf{R})$. To prove estimate (1.17) we shall use Hermite polynomials.

$$H_0(x) \equiv 1,$$

$$H_k(x) = \exp\left\{\frac{x^2}{2}\right\}\frac{d^k}{dx^k}\exp\left\{-\frac{x^2}{2}\right\},\ k \geq 1.$$

By differentiating the expression

$$\exp\left\{-\frac{x^2}{2}\right\} = (2\pi)^{-1/2}\int_{-\infty}^{+\infty}\exp\left\{-\frac{y^2}{2} + iyx\right\}dy,$$

it is easy to check that

$$H_k(x) = \exp\left\{\frac{x^2}{2}\right\}(2\pi)^{-1/2}\int_{-\infty}^{+\infty}(iy)^k\exp\left\{-\frac{y^2}{2} + iyx\right\}dy.$$

Hence, after elementary transformations, we obtain

$$\sum_{k=0}^{\infty}(-1)^k(k!)^{-1}s^kH_k(y) = \exp\left\{-\frac{s^2}{2} + sy\right\}. \tag{1.18}$$

It will be remembered that the polynomials $(-1)^k(k!)^{-1/2}H_k(x)$, $k = 0, 1, 2, \cdots$, form an orthonormal basis in the space $\mathbf{L}_2(\mathbf{R}, N(0, 1))$, where $N(0, 1)$ is a standard normal distribution. Therefore

$$\int_{\mathbf{B}} H_k^2\frac{\phi_h(y)}{|h|\sqrt{t}}\gamma(dy) = \int_{-\infty}^{\infty} H_k^2(x)N(0, 1)(dx) = k!. \tag{1.19}$$

Let $h \in \mathbf{H}$. By the use of Lemma 1.22 and relations (1.18), (1.19), we find

$$\left|g(h) - g(0) - \int_{\mathbf{B}} f(x+y)\sum_{j=1}^{k}(-1)^j(j!)^{-1}H_j\left(\frac{\phi_k(y)}{|h|\sqrt{t}}\right)|h|^j\gamma(dy)\right|$$

$$= \left|\int_{\mathbf{B}} f(x+y)\sum_{j=k+1}^{\infty}(-1)^j(j!)^{-1}H_j\left(\frac{\phi_h(y)}{|h|\sqrt{t}}\right)|h|^j\gamma(dy)\right|$$

$$\leq \sum_{j=k+1}^{\infty}\sup_{x \in \mathbf{B}}|f(x)|\left(\int_{\mathbf{B}} H_j^2\left(\frac{\phi_h(y)}{|h|\sqrt{t}}\right)|h|^j\gamma(dy)\right)^{1/2}(j!)^{-1}|h|^j$$

$$= \|f\|_\infty\sum_{j=k+1}^{\infty}(j!)^{-1/2}|h|^j t^{-j/2} = o(|h|^k). \tag{1.20}$$

In accordance with the Taylor formula (Theorem 1.14),

$$\left| g(h) - g(0) - \sum_{j=1}^{k} (j!)^{-1} g^{(j)}(0)(h)^j \right| = o(|h|^k). \tag{1.21}$$

From relations (1.20) and (1.21) it is easy to obtain

$$g^{(j)}(0)(h)^j = \int_{\mathbf{B}} f(x+y)(-1)^j H_j\left(\frac{\phi_k(y)}{|h|\sqrt{t}}\right)|h|^j \gamma(dy),\ j \geq 1.$$

There remains to make use of inequality (1.19). □

For the measurable function $f: \mathbf{B} \to \mathbf{R}$ we shall denote

$$\gamma_u^t f(x) = \int_{\mathbf{B}} f(x+y)\gamma_u^t(dy).$$

COROLLARY 1.24. *If the operator $u \in \pi_\gamma(\mathbf{H}, \mathbf{B})$, then for each bounded measurable function $f: \mathbf{B} \to \mathbf{R}$ the function $\gamma_u^t f \in \mathbf{C}_u^\infty(\mathbf{B}, \mathbf{R})$, and for each $k \geq 1$ it holds that*

$$\sup_{x \in \mathbf{B}} \sup_{|h| \leq 1} |(\gamma_u^t f)_u^{(k)}(x)(h)^k| \leq \|f\|_\infty \sqrt{k!}\, t^{-k/2}.$$

Proof. If $u \in \pi_\gamma(\mathbf{H}, \mathbf{B})$, then the space $\mathbf{H}_u = (N(u), \|\cdot\|_u)$ (see Theorem 1.20) is isometric to the Hilbert space $\mathbf{H}_0 \subset \mathbf{B}$ (see [213]). Besides, the operator of identical inclusion of the space $\mathbf{H}$ to the space $\mathbf{B}$ is γ-radonifying. The proof is completed by the use of Theorems 1.20 and 1.23. □

2.2. Smooth Approximation in a Banach Space

Existence of smooth functions with a bounded support. Let $\mathbf{F}$, $\mathbf{B}$ be Banach spaces (not necessarily separable, the operator $u \in \mathbf{L}(\mathbf{F}, \mathbf{B})$, and the integer number $k \geq 1$. In the space $\mathbf{C}_u^k(\mathbf{B}, \mathbf{R})$ we shall consider the following seminorms:

$$\|f\|_i := \sup_{x \in \mathbf{B}} \sup_{\|h\| \leq 1} |f_u^{(i)}(x)(h)^i|,\ i = 1, \dots, k;$$

$$\|f\|_{k,\alpha} := \sup_{h \in \mathbf{F}, h \neq 0} \sup_{\|z\| \leq 1} \|h\|^{-\alpha} |f_u^{(k)}(x + u(h))(z)^k - f_u^{(k)}(x)(z)^k|,$$

where the number $0 < \alpha \leq 1$. It is convenient to assume that $\|\cdot\|_{k,0} := \|\cdot\|_k$. By $\mathbf{Q}_u^{k,\alpha} \equiv \mathbf{Q}_u^{k,\alpha}(\mathbf{B}, \mathbf{R})$, $0 \leq \alpha \leq 1$ ($\mathbf{Q}_u^k$ instead of $\mathbf{Q}_u^{k,0}$) we shall denote the space of the functions $f \in \mathbf{C}_u^k(\mathbf{B}, \mathbf{R})$ satisfying

$$\sum_{j=1}^{k-1} \|f\|_i + \|f\|_{k,\alpha} < \infty.$$

We set

$$\mathbf{Q}_u^\infty := \mathbf{Q}_u^\infty(\mathbf{B}, \mathbf{R}) := \bigcap_{k \geq 1} \mathbf{Q}_u^k.$$

Let $\mathbf{K} = (K_1, \dots, K_k)$ be a collection of non-negative numbers. Let us define the set

$$(\mathbf{Q}_u^{k,\alpha}; \mathbf{K}) := \{f \in \mathbf{Q}_u^{k,\alpha} : \|f\|_i \leq K_i,\ i = 1, \dots, k-1;\ \|f\|_{k,\alpha} \leq K_k\},\ 0 \leq \alpha \leq 1.$$

In the case of $\mathbf{F} = \mathbf{B}$ and $u = I_{\mathbf{B}}$, the index u in the above notations will be omitted.

Let A, A_1 be some sets in the space $\mathbf{B}$, $A \subset A_1 \neq \emptyset$. By $\mathcal{F}(A, A_1)$ we denote the class of functions $f: \mathbf{B} \to \mathbf{R}$, satisfying the conditions $N(f) = [0, 1]$ (here and elsewhere $N(f)$ is a set of values of the function f), $f(x) = 1$, if $x \in A$, $f(x) = 0$ if $x \in \mathbf{B} \setminus A_1$. If $A = \emptyset$, then $\mathcal{F}(A, A_1)$ includes all functions $f: \mathbf{B} \to [0, 1]$ with the support A_1. It will be noted that for each $\varepsilon > 0$ the class $\mathscr{F}(A, A_\varepsilon)$ includes a uniformly continuous function $f(x) = d(\mathbf{B}\ A_\varepsilon, x) \times (d(A, x) + d(\mathbf{B}\backslash A_\varepsilon, x))^{-1}$. However, when constructing the estimates of the rate of convergence in the central limit theorem, it is useful to have information on such sets $A \subset \mathbf{B}$ that the class $\mathscr{F}(A, A_\varepsilon)$ for each $\varepsilon > 0$ has functions with greater smoothness. The presence of functions with greater smoothness depends on the space $\mathbf{B}$, the set A and, naturally, on the smoothness required. In some spaces there may be no bounded set A such that $\mathscr{F}(A, A_\varepsilon)$ would have a Fréchet differentiable function. Therefore, the following question is quite natural: *In which spaces* $\mathbf{B}$ *and for what kind of operator* $u \in \mathbf{L}(\mathbf{F}, \mathbf{B})$ *does the bounded set* $A \subset \mathbf{B}$ *exist, such that* $\mathscr{F}(A, A_\varepsilon) \cap \mathbf{Q}_u^{k,\alpha} = \emptyset$ *for each* $\varepsilon > 0$? Of course, at present an exhaustive answer to this general question is not known. Below we shall present one sufficient condition for the existence of such a set which is useful at proving negative results, and in the following items of this section, as well as in supplements, positive results will be given. But first we shall present the following elementary proposition.

PROPOSITION 2.1. *A bounded set* $A \subset \mathbf{B}$ *such that* $\mathscr{F}(A, A_\varepsilon) \cap \mathbf{Q}_u^{k,\alpha} \neq \emptyset$ *for all* $\varepsilon > 0$ *exists if and only if there exists a non-trivial function* $f \in \mathbf{Q}_u^{k,\alpha}(\mathbf{B}, \mathbf{R})$ *with a bounded support.*

Proof. Let $f \in \mathbf{Q}_u^{k,\alpha}$, $f \not\equiv 0$, and the support of the function f be a bounded set. Let $x_0 \in \mathbf{B}$ be such an element that $f(x_0) \neq 0$, and $\tilde{f}(x) = f(x + x_0)(f(x_0))^{-1}$. Since f has a bounded support, then there exists such a number $r > 0$ that $\tilde{f}(x) = 0$ at $\|x\| > r$. Take the function $\phi \in \mathbf{C}^\infty(\mathbf{R})$ with the properties $N(\phi) = [0, 1]$, $\phi(0) = 1$, $\phi(1) = 0$. Then it is easy to check that the function $g_\varepsilon(x) = 1 - \phi(\tilde{f}(r\varepsilon^{-1}x))$ belongs to the class $F(A, A_\varepsilon) \cap \mathbf{C}_u^{k,\alpha}$, where $A = \{0\}$. The necessity of a condition is obvious. □

THEOREM 2.2. *Let* $u \in \mathbf{L}(\mathbf{F}, \mathbf{B})$ *and* $0 < \alpha \leq 1$. *If the class* $\mathbf{Q}_u^{1,\alpha}$ *contains a non-trivial function with a bounded support then the convergence of the series*

$$\sum_{i=1}^{\infty} \|x_i\|^{1+\alpha}, \quad x_i \in \mathbf{F}, \ i \geq 1,$$

implies convergence in probability of the series

$$\sum_i \varepsilon_i u(x_i).$$

Proof. According to Proposition 2.1, for each $\varepsilon > 0$ there exists a function $g_\varepsilon \in \mathbf{Q}_u^{1,\alpha}$ such that $N(g_\varepsilon) = [0, 1]$, $g_\varepsilon(0) = 1$, $g_\varepsilon(x) = 0$, if $\|x\| > \varepsilon$. We denote $f(x) = 1 - g_\varepsilon(x)$ and

$$I_{m,n} := \int_{\mathbf{B}} f(x) L\left(\sum_{i=m}^{n} \varepsilon_i u(x_i)\right)(dx), \quad n > m \geq 1.$$

By the use of the fact that $E\varepsilon_i = 0$, and by the Taylor formula (1.14) we have:

$$I_{m,n} = \int_{\mathbf{B}} \int_{\mathbf{F}} f(x + u(y)) L(\varepsilon_m x_m)(dy) L\left(\sum_{i=m+1}^{n} \varepsilon_i u(x_i)\right)(dx)$$

$$= \int_{\mathbf{B}} \int_{\mathbf{F}} [f(x + u(y)) - f(x) - f_u'(x)(y)] \times$$

$$\times L(\varepsilon_m x_m)(dy) L\left(\sum_{i=m+1}^{n} \varepsilon_i u(x_i)\right)(dx) + I_{m+1,n}$$

$$= I_{m+1,n} + \int_B \int_F \int_0^1 [f'_u(x + \tau u(y))(y) - f'_u(x)(y)]\, d\tau \times$$

$$\times L(\varepsilon_m x_m)(dy) L\left(\sum_{i=m+1}^{n} \varepsilon_i u(x_i)\right)(dx)$$

$$\leq \frac{\|x_m\|^{1+\alpha} \|g_\varepsilon\|_{1,\alpha}}{1+\alpha} + I_{m+1,n}. \tag{2.1}$$

Since $f(0) = 0$, these arguments can be applied for estimating the quantity $I_{n,n}$. Thus we get

$$I_{n,n} \leq (\alpha + 1)^{-1} \|x_n\|^{1+\alpha} \|g_\varepsilon\|_{1,\alpha}. \tag{2.2}$$

Taking into account that

$$P\left\{\|\sum_{i=m}^{n} \varepsilon_i u(x_i)\| > \varepsilon\right\} \leq I_{m,n},$$

from relations (2.1), (2.2) we obtain

$$P\left\{\|\sum_{i=m}^{n} \varepsilon_i u(x_i)\| > \varepsilon\right\} \leq (\alpha + 1)^{-1} \|g_\varepsilon\|_{1,\alpha} \sum_{i=m}^{n} \|x_i\|^{1+\alpha}.$$

Hence the proposition of the theorem follows. □

REMARK 2.3. It must be noted that by the use of Proposition 3.1.1, Theorem 2.2 may be reformulated as follows: if there exists a non-trivial function $f \in \mathbf{Q}_u^{1,\alpha}(\mathbf{B}, \mathbf{R})$ with a bounded support, then the operator u is an operator of the type $1 + \alpha$.

The following result may serve as an example of Theorem 2.2.

PROPOSITION 2.4. *Let $p \in [1, 2)$, $\mathbf{L}_p(S, \nu)$ be an infinite-dimensional space where (S, ν) is a space with σ-finite measure ν. If $p - 1 < \alpha \leq 1$, then the space $\mathbf{Q}^{1,\alpha}(\mathbf{L}_p(S, \nu), \mathbf{R})$ does not contain a non-trivial function with a bounded support.*

Proof. It is sufficient to construct a sequence $x_i \in L_p(S, \nu)$, $i \geq 1$, such that

$$\sum_{i=1}^{\infty} \|x_i\|^{1+\alpha} < \infty,$$

but the series

$$\sum_{i=1}^{\infty} \varepsilon_i u(x_i)$$

diverges. Since the measure ν is σ-finite, then there will be a sequence of pairwise non-intersecting sets A_i, $i \geq 1$, such that $\nu(A_i) < \infty$ and $S = \bigcup_{i \geq 1} A_i$. We shall define the function $x_i(t)$ assuming it to be equal to 0, if $t \in A_i$ and $x_i(t) = \lambda_i(\nu(A_i))^{-1/p}$, if $t \overline{\in} A_i$. Here $\lambda_i > 0$, $i \geq 1$. Then

$$\sum_{i=1}^{n} \|x_i\|^{1+\alpha} = \sum_{i=1}^{n} \lambda_i^{1+\alpha}, \quad E\|\sum_{i=1}^{n} \varepsilon_i x_i\| = \left(\sum_{i=1}^{n} \lambda_i^p\right)^{1/p}.$$

Thus, if the sequence $\{\lambda_i, i \geq 1\}$ is such that

$$\sum_{i=1}^{\infty} \lambda_i^{1+\alpha} < \infty \text{ and } \sum_{i=1}^{\infty} \lambda_i^{p} = \infty,$$

thus

$$E \| \sum_{i=1}^{n} \varepsilon_i x_i \| \to \infty$$

at $n \to \infty$, and the series

$$\sum_{i=1}^{\infty} \varepsilon_i x_i$$

cannot converge in probability. (This follows from Theorems 1.3.7 and 1.3.8.) □

Relation between the uniform classes and approximation of functions. Further along Section 2.2, by $\mathscr{W} = \{W_t, t \in \mathbf{R}\}$ we shall denote the class of sets $W_t \subset \mathbf{B}$, such that $t' < t$ implies $W_{t'} \subseteq W_t$. In the case of the function $f: \mathbf{B} \to \mathbf{R}$ the class $\mathscr{W}_f = \{W_t(f), t \in \mathbf{R}\}$ consists of the sets $W_t(f) = \{x \in \mathbf{R}: f(x) \leq t\}$, and if $\mathscr{F}$ is some class of functions $f: \mathbf{B} \to \mathbf{R}$, then $\mathscr{W}_{\mathscr{F}} = \{\mathscr{W}_f, f \in \mathscr{F}\}$.

Let $\mathbf{K} = \mathbf{K}(\varepsilon) = \mathbf{K}(\varepsilon, k) = (K_1(\varepsilon), ..., K_k(\varepsilon))$ be a collection of non-negative functions $K_1(\varepsilon), ..., K_k(\varepsilon)$ of the positive argument $\varepsilon > 0$. Most frequently we shall make use of the functions $K_i(\varepsilon) = \varepsilon^{-1/\alpha}$, wnere $\alpha \in (0, 1]$, so we introduce special notations for such a collection $\mathbf{K}_{\alpha,k}(\varepsilon)$, i.e. $\mathbf{K}_{\alpha,k}(\varepsilon) = (\varepsilon^{-1/\alpha}, ..., \varepsilon^{-k/\alpha})$.

DEFINITION 2.5. The class $\mathscr{W} = \{W_t, t \in \mathbf{R}\}$ will be called $(\mathbf{Q}_u^{k,\alpha}; \mathbf{K})$-uniform if, for all $\varepsilon > 0$ and $t \in \mathbf{R}$, the relation $\mathscr{F}(W_t, W_{t+\varepsilon}) \cap (\mathbf{Q}_u^{k,\alpha}; \mathbf{K}(\varepsilon)) \neq \emptyset$ is satisfied.

The class $\mathscr{W}_{\mathscr{F}}$ will be called $(\mathbf{Q}_u^{k,\alpha}; \mathbf{K})$-uniform if this property is possessed by the subclass $\mathscr{W}_f$ for any function $f \in \mathscr{F}$. In the cases when there is no necessarity to know the type of the function $\mathbf{K}(\varepsilon)$ we shall use the term '$\mathbf{Q}_u^{k,\alpha}$-uniform' instead of '$(\mathbf{Q}_u^{k,\alpha}; \mathbf{K})$-uniform'.

REMARK 2.6. For an arbitrary operator $v \in \mathbf{L}(\mathbf{F}_1, \mathbf{F})$, $(\mathbf{Q}_u^{k,\alpha}, \mathbf{K})$-uniformity of the class W implies $(\mathbf{Q}_{u \cdot v}^{k,\alpha}, \overline{\mathbf{K}})$-uniformity of the same class, where

$$\overline{\mathbf{K}} = (\|v\| K_1, ..., \|v\|^{k+\alpha} K_k)$$

DEFINITION 2.7. The function $f: \mathbf{B} \to \mathbf{R}$ is called uniformly $(\mathbf{Q}_u^{k,\alpha}; \mathbf{K})$-approximable if for each $\varepsilon > 0$ there exists a function $g_\varepsilon \in (\mathbf{Q}_u^{k,\alpha}, \mathbf{K}(\varepsilon))$ such that

$$\sup_{x \in \mathbf{B}} | f(x) - g_\varepsilon(x) | \leq \varepsilon.$$

The results below establish a relation between the uniformity of the class $\mathscr{W}_f$ and the uniform approximation of the function f.

THEOREM 2.8. *Let the function $f \in \underline{\mathbf{C}}^0(\mathbf{B}, \mathbf{R})$ be such that the class $\mathscr{W}_f$ is $(\mathbf{Q}_u^{k,\alpha}, \mathbf{K})$-uniform. Then the function f is uniformly $(\mathbf{Q}_u^{k,\alpha}, \overline{\mathbf{K}})$-approximable where*

$$\overline{\mathbf{K}}(\varepsilon) = (\overline{K}_1(\varepsilon), ..., \overline{K}_k(\varepsilon)), \ \overline{K}_i(\varepsilon) = 2\varepsilon K_i(\varepsilon), \ i = 1, ..., k.$$

Proof. Let the number $\varepsilon > 0$ and the functions

$$g_m \equiv g_{m,\varepsilon} \in \mathscr{F}(W_{m\varepsilon}(f), W_{m\varepsilon+\varepsilon}(f)) \cap (\mathbf{Q}_u^{k,\alpha}, \mathbf{K}(\varepsilon)),$$

$$m = 0, \pm 1, \pm 2, \cdots$$

Let us assume $\phi_m(x) = \varepsilon(1 - g_m(x))$, if $m = 0, 1, 2, \dots$; $\phi_m(x) = -\varepsilon g_m(x)$, if $m = -1, -2, \cdots$. We shall define the function $\phi \equiv \phi_\varepsilon : \mathbf{B} \to \mathbf{R}$, assuming

$$\phi(x) = \sum_{m=-\infty}^{\infty} \phi_m(x). \tag{2.3}$$

It will be noted that the series in (2.3) converges since, at a fixed $x \in \mathbf{B}$, it is but the finite number of the non-vanishing summands of the series. Now we shall estimate $\sup_{x \in \mathbf{B}} |f(x) - \phi(x)|$. For each $x \in \mathbf{B}$ there exists such a least number $m = m_x$, that $x \in W_{m\varepsilon+\varepsilon}(f)$, i.e. $f(x) < m\varepsilon + \varepsilon$ and $f(x) > m\varepsilon$. Suppose that $m_x \geq 0$ (the case $m_x < 0$ is considered by analogy). Then

$$|f(x) - \phi(x)| = \left| f(x) - \sum_{i=0}^{m_x} \phi_i(x) \right| = |f(x) - (m_x + 1)\varepsilon| \leq \varepsilon.$$

There remains to show that $\phi \in (\mathbf{Q}_u^{k,\alpha}, \overline{\mathbf{K}}(\varepsilon))$. Since $m_x\varepsilon \leq f(x) < (m_x + 1)\varepsilon$, then by continuity of the function f the point x, together with some neighbourhood U_x, is contained in the set

$$(m_x - 1)\varepsilon < f(x) < (m_x + 1)\varepsilon.$$

All functions ϕ_i for $i \geq m_x + 1$ or $i < m_x - 2$ are constant on the set

$$\{y \in \mathbf{B}: (m_x - 1)\varepsilon < f(y) < (m_x + 1)\varepsilon\}$$

and, moreover, on the neighbourhood U_x. Therefore, by differentiating the series (2.3) term by term we have

$$\begin{aligned}\phi_u'(x)(h) &= (\phi_{m_x-1})_u'(x)(h) - (\phi_{m_x})_u'(x)(h) \\ &= -\varepsilon[(g_{m_x-1})_u'(x)(h) + (g_{m_x})_u'(x)(h)],\ h \in \mathbf{F}.\end{aligned}$$

Since $\|g_m\|_1 \leq K_1(\varepsilon)$, then $\|\phi\|_1 \leq 2\varepsilon K_k(\varepsilon)$. By analogy we obtain the estimates

$$\|\phi_i\|_i \leq 2\varepsilon K_i(\varepsilon),\ i = 1, \dots, k - 1, \text{ and } \|\phi\|_{k,\alpha} \leq 2\varepsilon K_k(\varepsilon).\square$$

THEOREM 2.9. *Let the number $\alpha \in [0, 1]$, the integer number $k \geq 1$, the function $\mathbf{K} = \mathbf{K}(\varepsilon) = (K_1(\varepsilon), \dots, K_k(\varepsilon))$, $\varepsilon > 0$. If the function $f: \mathbf{B} \to \mathbf{R}$ is uniformly $(\mathbf{Q}_u^{k,\alpha}, \mathbf{K})$-approximable, then the class $\mathcal{W}_f$ is $(\mathbf{Q}_u^{k,\alpha}, \mathbf{K})$-uniform, where*

$$\hat{\mathbf{K}} = \hat{\mathbf{K}}(\varepsilon) = (\hat{K}_1(\varepsilon), \dots, \hat{K}_k(\varepsilon)).$$

Besides,

$$\hat{K}_i(\varepsilon) = C(i)\, \Sigma_i' \varepsilon^{-s} K_{j_1}\left(\frac{\varepsilon}{8}\right), \dots, K_{j_s}\left(\frac{\varepsilon}{8}\right),\ 1 \leq i \leq k - 1, \tag{2.4}$$

where Σ_i' denotes that the summation is over all possible collections $(j_1, \dots, j_s)$ of such integer indices $j_1, \dots, j_s \geq 1$, that $j_1 + \cdots + j_s = i$ (s can take values from 1 to i);

$$\begin{aligned}\hat{K}_k(\varepsilon) = C(k)\Bigg[&\varepsilon^{-1} K_k K_k\left(\frac{\varepsilon}{8}\right) + \Sigma_k' \varepsilon^{-s-\alpha} K_1^\alpha\left(\frac{\varepsilon}{8}\right) K_{j_1}\left(\frac{\varepsilon}{8}\right) \cdots K_{j_s}\left(\frac{\varepsilon}{8}\right) + \\ &+ \Sigma_k'' \varepsilon^{-s} K_{j_1}^{1-\alpha}\left(\frac{\varepsilon}{8}\right) K_{j_1+1}^\alpha\left(\frac{\varepsilon}{8}\right) K_{j_2}\left(\frac{\varepsilon}{8}\right) \cdots K_{j_s}\left(\frac{\varepsilon}{8}\right)\Bigg],\end{aligned} \tag{2.5}$$

where Σ_k'' denotes that the summation is over all collections $(j_1, \dots, j_s)$ of integer indices $j_1, \dots, j_s \geq 1$, $j_1 + \cdots + j_s = k$ and $j_1 < k$. If the number $\alpha = 0$, then (2.4) is preserved at $i = k$ also.

Proof. For each $\varepsilon > 0$ there exists a function $g = g_\varepsilon \in (\mathbf{Q}_u^{k,\alpha}; \mathbf{K}(\varepsilon/8))$, satisfying the condition

$$\sup_{x \in \mathbf{B}} |f(x) - g(x)| \leq \frac{\varepsilon}{8}.$$

Let the function $\phi \in \mathbf{C}^\infty(\mathbf{R})$ be such that $N(\phi) = [0, 1]$; $\phi(t) = 1$, if $t < 1/8$; $\phi(t) = 0$, if $t > 7/8$. Suppose $\psi(x) = \phi(\varepsilon^{-1}(g(x) - t))$. We have

$$\left| \frac{f(x) - t}{\varepsilon} - \frac{g(x) - t}{\varepsilon} \right| \leq \frac{1}{8}. \tag{2.6}$$

If $x \in W_t(f)$, then $(f(x) - t)\varepsilon^{-1} < 0$ and by (2.6), $(g(x) - t)\varepsilon^{-1} < 1/8$. Hence, according to the properties of the function ϕ, it follows that $\psi(x) = 1$. By analogy, $\psi(x) = 0$, is checked if $x \in \mathbf{B} \backslash W_{t+\varepsilon}(f)$. There remains to evaluate the derivatives of the function ψ. Let $z \in \mathbf{F}$, $\|z\| \leq 1$. By the formula of differentiation of composite functions (Theorem 1.11) we have

$$\psi_u^{(i)}(x)(z)^i = \Sigma_i' \varepsilon^{-l} \phi^{(l)} \left(\frac{g(x) - t}{\varepsilon} \right) g_u^{(j_1)}(x)(z)^{j_1} \cdots g_u^{(j_l)}(x)(z)^{j_l}. \tag{2.7}$$

Taking into account that $g \in (\mathbf{Q}_u^{k,\alpha}; \mathbf{K}(\varepsilon/8))$, from (2.7) we easily obtain (2.4). We shall prove (2.5). Let $h \in \mathbf{F}$. To simplify the writing, we introduce the following:

$$U_m(x) := U_m(x, z) := g^{j_m}(x)(z)^{j_m},$$

$$\delta_m(x, h) := U_m(x + u(h)) - U_m(x), \quad m = 1, \ldots, k;$$

$$\delta_0(x, h, \varepsilon) := \phi^{(l)} \left(\frac{g(x + u(h)) - t}{\varepsilon} \right) - \phi^{(l)} \left(\frac{g(x) - t}{\varepsilon} \right).$$

It is easy to check that

$$|(\psi_u^{(k)}(x + u(h)) - \psi_u^{(k)}(x))(z)^k| = I_1 + I_2, \tag{2.8}$$

where

$$I_1 := |\Sigma_k' \varepsilon^{-l} \delta_0(x, h, \varepsilon) U_1(x + h) \cdots U_l(x + h)|$$

$$I_2 := \left| \Sigma_k' \varepsilon^{-l} \phi^{(s)} \left(\frac{g(x) - t}{\varepsilon} \right) \sum_{s=1}^{l} U_1(x) \cdots U_{s-1}(x) \delta_s(x, h) U_{s+1}(x + h) \cdots U_l(x + h) \right|.$$

In accordance with the conditions of the theorem we have

$$|U_m(x)| \leq K_{j_m} \left(\frac{\varepsilon}{8} \right), \quad m = 1, \ldots, l; \tag{2.9}$$

$$|\delta_m(x, h)| \leq \begin{cases} K_{j_m+1} \left(\frac{\varepsilon}{8} \right) \|h\|, & j_m \neq k; \\ K_k \left(\frac{\varepsilon}{8} \right) \|h\|^\alpha, & j_m = k. \end{cases} \tag{2.10}$$

Let us estimate $\delta_0(x, h, \varepsilon)$. Since $|g(x + h) - g(x)| \leq K_1 \|h\|$, and the auxiliary function ϕ has bounded derivatives satisfying the Lipschitz condition, then

$$|\delta_0(x, h, \varepsilon)| \leq C(l); \tag{2.11}$$

$$|\delta_0(x, h, \varepsilon)| \le C(l)K_1\left(\frac{\varepsilon}{8}\right)\|h\|\varepsilon^{-1}. \tag{2.12}$$

Here the number $C(l)$ depends on the auxiliary functions ϕ. Having raised (2.11) and (2.12) to the powers $1-\alpha$ and α respectively, and multiplying the inequalities obtained, we have

$$|\delta_0(x, h, \varepsilon)| \le C(l)K_1^{\alpha}\left(\frac{\varepsilon}{8}\right)\|h\|^{\alpha}\varepsilon^{-\alpha}. \tag{2.13}$$

By analogy, from (2.9) and (2.10) we obtain the estimate

$$|\delta_m(x, h)| \le C(m)K_{j_m}^{1-\alpha}\left(\frac{\varepsilon}{8}\right)K_{j_m+1}^{\alpha}\left(\frac{\varepsilon}{8}\right)\|h\|^{\alpha}. \tag{2.14}$$

From (2.13), (2.14), (2.8) we easily obtain (2.5). □

COROLLARY 2.10. *Let the function* $f \in \mathbf{C}^0(\mathbf{B}, \mathbf{R})$, *and the number* $k \ge 1$. *The following statements are equivalent:*

(1) *There exists a constant* $C > 0$ *such that the class* $\mathcal{W}_f$ *is* $(\mathbf{Q}_u^k, C\mathbf{K}_{1,k})$*-uniform.*

(2) *There exists a* $C > 0$ *such that the function f is uniformly* $(\mathbf{Q}_u^k, C\mathbf{K})$*-approximable where*

$\mathbf{K} = \mathbf{K}(\varepsilon) = \varepsilon K_{1,k}(\varepsilon)$.

It must be noted that in a general case for $(\mathbf{C}_u^k, \mathbf{K})$-uniform class $\mathcal{W}_f$ the order of the function $K_i(\varepsilon)$ cannot be better than $O(\varepsilon^{-i})$. In fact, let us consider the class $\mathcal{W}_{\|\cdot\|}$ and assume that this call is $(\mathbf{C}^k, \mathbf{K})$-uniform, $\mathbf{K} = \mathbf{K}(\varepsilon) = (K_1(\varepsilon), \ldots, K_k(\varepsilon))$. For each function $g_\varepsilon \in \mathcal{F}(W_t, W_{t+\varepsilon}) \cap (\mathbf{Q}^k, \mathbf{K}(\varepsilon))$ we have

$$\sup_{\|h\|\le 1} |g_\varepsilon^{(i)}(x)(h)^i| \le C_iK_i(\varepsilon)\chi_{(t\le\|x\|\le t+\varepsilon)}(x).$$

If $x \in \mathbf{B}$, $t \le \|x\| \le t + \varepsilon$, then we shall find the element $y \in \mathbf{B}$ such that $\|x - y\| \le \varepsilon$ and $y \in \{x: t \le \|x\| \le t + \varepsilon\}$. From the mean value theorem we obtain

$$\sup_{\|h\|\le 1} |g_\varepsilon^{(i-1)}(x)(h)^{i-1}| \le \sup_{\|h\|\le 1} |g_\varepsilon^{(i-1)}(x) - g_\varepsilon^{(i-1)}(y))(h)^{i-1}| \le C_iK_i(\varepsilon)\varepsilon.$$

By proceeding with this process we have

$$\sup_{\|h\|\le 1} |g_\varepsilon'(x)(h)| \le C_iK_i(\varepsilon)\varepsilon^{i-1}.$$

Furthermore, we have

$$1 = \sup_{x\in\mathbf{B}} |g_\varepsilon(x)| = \max\left(1, \sup_{t\le\|x\|\le t+\varepsilon} |g_\varepsilon(x)|\right).$$

If $x \in \mathbf{B}$ such that $t \le \|x\| \le t + \varepsilon$, then we shall find $y \in \mathbf{B}$ such that $\|x - y\| = \varepsilon$, $\|y\| > t + \varepsilon$. Consequently, $g_\varepsilon(y) = 0$ and again, from the mean value theorem we have $1 \le \max(C_iK_i(\varepsilon)\varepsilon^i; 1)$.

The concept of uniformity of a class of sets has also been used to construct the estimates of rates of convergence. From this point of view, the following two problems are the most essential:

1. To describe, for a given operator $u \in \mathbf{L}(\mathbf{F}, \mathbf{B})$, the set of $\mathbf{Q}_u^k$-uniform classes $\mathcal{W} = (W_t, t \in \mathbf{R})$ and to define for each such class the function $\mathbf{K}(\varepsilon, k)$, with which it is $(\mathbf{Q}_u^k, \mathbf{K})$-uniform.

2. To describe, for the given class of sets $\mathcal{W} = (W_t, t \in \mathbf{R})$, the operators $u \in \mathbf{L}(\mathbf{F}, \mathbf{B})$ or the subspace $\mathbf{F} \subset \mathbf{B}$ such that $\mathcal{W}$ is $(\mathbf{Q}_u^k, \mathbf{K})$-uniform or $(\mathbf{Q}_{\mathbf{F}}^k, \mathbf{K})$-uniform.

Of course, both problems are rather general and at present only some partial results may be presented.

Uniform classes with respect to γ-radonifying operators A most comprehensive description of $\mathbf{Q}_u^k$-uniform classes $\mathscr{W}$ is obtained in the case of the operator $u \in \pi_\gamma(\mathbf{H}, \mathbf{B})$, as well as taking into account Remark 2.6, in the case of the operator $u \in \mathbf{L}(\mathbf{F}, \mathbf{B})$, admitting factorization through a Hilbert space by means of a radonifying operator, i.e. in the case

$$u \in \mathbf{L}(\mathbf{F}, \mathbf{H}) \cdot \pi_\gamma(\mathbf{H}, \mathbf{B}).$$

We shall define for an operator $u \in \pi_\gamma(\mathbf{H}, \mathbf{B})$ and a measurable function $f : \mathbf{B} \to \mathbf{R}$

$$\sigma_f(t) := \sup_{x \in \mathbf{B}} |f(x) - \gamma_u^t f(x)|, \quad t > 0.$$

LEMMA 2.11. *If $u \in \pi_\gamma(\mathbf{H}, \mathbf{B})$, and the function $f : \mathbf{B} \to \mathbf{R}$ is uniformly continuous and bounded, then $\sigma_f(t) \to 0$ as $t \to 0$.*

Proof. For an arbitrary $\varepsilon > 0$, there may be found a $\delta > 0$ such that $\|x - y\| < \delta$ implies $|f(x) - f(y)| < \varepsilon$. Therefore,

$$|f(x) - \gamma_u^t f(x)| \leq \int_{\|y\| < \delta} \varepsilon \gamma_u^t(dy) + \|f\|_\infty \gamma_u^t(y : \|y\| > \delta)$$
$$\leq \varepsilon + 2\|f\|_\infty \gamma_u(y : \|y\| > \delta t^{-1/2}).$$

There remains to note that $\gamma_u(y : \|y\| > \delta t^{-1/2}) \to 0$ when $t \to 0$. □

THEOREM 2.12. *If $u \in \pi_\gamma(\mathbf{H}, \mathbf{B})$, then each uniformly continuous function $f : \mathbf{B} \to \mathbf{R}$ is uniformly $(\mathbf{Q}_u^k, \mathbf{K})$-approximable for any $k \geq 1$, where*

$$K_i(\varepsilon) = i!\,\|f\|_\infty (\sup(t > 0 : \sigma_f(t) \leq \varepsilon))^{-i/2}, \quad i = 1, \ldots, k.$$

Proof. It is sufficient to apply Theorem 1.23 and Lemma 2.11. □

COROLLARY 2.13. *If $u \in \pi_\gamma(\mathbf{H}, \mathbf{B})$, then each function f from the class $\mathrm{BL}_\alpha(\mathbf{B})$ is uniformly $(\mathbf{Q}_u^k, \mathbf{K})$-approximable for each $k \geq 1$, where*

$$K_i(\varepsilon) = i!\,\|f\|_\infty \|f\|_{\mathrm{BL}_\alpha}^{i/\alpha} \left(\int_{\mathbf{B}} \|y\|^\alpha \gamma_u(dy) \right)^{1/\alpha} \varepsilon^{-i/\alpha}, \quad i = 1, \ldots, k.$$

Proof. If $f \in \mathrm{BL}_\alpha(\mathbf{B})$, then

$$\sigma_f(t) \leq \int_{\mathbf{B}} \|y\|^\alpha \gamma_u^t(dy) \|f\|_{\mathrm{BL}_\alpha} = t^{\alpha/2} \|f\|_{\mathrm{BL}_\alpha} \left(\int_{\mathbf{B}} \|y\|^\alpha \gamma_u(dy) \right).$$

Therefore,

$$\sup\{t > 0 : \sigma_f(t) \leq \varepsilon\} \geq \varepsilon^{2/\alpha} \left(\int_{\mathbf{B}} \|y\|^\alpha \gamma_u(dy)\ \|f\|_{\mathrm{BL}_\alpha} \right)^{-2/\alpha}. \square$$

With increasing smoothness of the function $f : \mathbf{B} \to \mathbf{R}$ the quality of approximation is improving in the sense of improving the function $\mathbf{K}(\varepsilon)$.

COROLLARY 2.14. *If $u \in \pi_\gamma(\mathbf{H}, \mathbf{B})$, then each bounded function f from the class $\mathbf{Q}^{1,\alpha}(\mathbf{B}, \mathbf{R})$, $0 < \alpha \leq 1$, is uniformly $(\mathbf{Q}_u^k, \mathbf{K})$-approximable for each $k \geq 1$, where*

$$K_i(\varepsilon) = i!\|f\|_\infty(1+\alpha)^{-i/1+\alpha}\|f\|_{1,\alpha}^{i/1+\alpha}\left(\int_{\mathbf{B}}\|y\|^{1+\alpha}\gamma_u(dy)\right)^{i/1+\alpha}\varepsilon^{-i/1+\alpha},\ i=1,\ldots,k.$$

Proof. Applying the Taylor formula and taking into account that the mean of the measure γ_u^t is equal to zero, we have

$$|\gamma_u^t f(x) - f(x)| = \left|\int_{\mathbf{B}}(f(x+y) - f(x) - f'(x)(y))\gamma_u^t(dy)\right|$$

$$= \left|\int_{\mathbf{B}}\int_0^1 (f'(x \mid \tau y)(y) - f'(x)(y))\, d\tau\gamma_u^t(dy)\right|$$

$$\leq (1+\alpha)^{-1}\int_{\mathbf{B}}\|y\|^{1+\alpha}\gamma_u^t(dy)\|f\|_{1,\alpha}$$

$$= (1+\alpha)^{-1}t^{(1+\alpha)/2}\|f\|_{1,\alpha}\int_{\mathbf{B}}\|y\|^{1+\alpha}\gamma_u(dy).$$

Hence

$$\sup\{t>0: \sigma_f(t)\leq\varepsilon\}$$

$$\geq\left(\varepsilon(1+\alpha)\left(\|f\|_{1,\alpha}\int_{\mathbf{B}}\|y\|^{\alpha+1}\gamma_u(dy)\right)^{-1}\right)^{2/1+\alpha}. \square$$

In the case of the measurable function $f: \mathbf{B}\to\mathbf{R}$ and the operator $u\in\pi_\gamma(\mathbf{H},\mathbf{B})$

$$\tilde{\sigma}_f(t) := \sup_{x\in\mathbf{B}}\int_{\mathbf{B}}|f(x) - f(x-y)|\gamma_u^t(dy).$$

THEOREM 2.15. *If $u\in\pi_\gamma(\mathbf{H},\mathbf{B})$, then for each uniformly continuous function $f: \mathbf{B}\to\mathbf{R}$ the class $\mathcal{W}_f$ is $(\mathbf{Q}_u^k, \mathbf{K})$-uniform for each $k\geq 1$, where*

$$K_i(\varepsilon) = C(i)(\sup\{t>0: \tilde{\sigma}_f(t)<\varepsilon\})^{-i/2},\ i=1,\ldots,k.$$

Proof. Let $\varepsilon>0$, s, $t\in\mathbf{R}$, the continuous function $\phi: \mathbf{R}\to[0,1]$ be linear on the interval $[0,1]$ and $\phi(\tau)=0$ if $\tau\leq 0$, $\phi(\tau)=1$, if $\tau\geq 1$. Assume that $f_{\varepsilon,s}(x) = \phi(\varepsilon^{-1}(f(x)-s))$. Just as in the proof of Lemma 2.11, we are convinced that $\tilde{\sigma}_f(t)\to 0$ at $t\to 0$. Since the function ϕ satisfies the condition $|\phi(\tau)-\phi(s)|\leq|\tau-s|$, consequently,

$$\tilde{\sigma}_{f_{\varepsilon,s}}(t) = \sup_x\left|\int_{\mathbf{B}}(f_{\varepsilon,s}(x \mid y) - f_{\varepsilon,s}(x))\gamma_u^t(dy)\right|$$

$$\leq\varepsilon^{-1}\sup_x\int_{\mathbf{B}}|f(x+y)-f(x)|\gamma_u^t(dy) = \varepsilon^{-1}\tilde{\sigma}_f(t).$$

By Theorem 2.12, the function $f_{\varepsilon,s}$ is $(\mathbf{Q}_u^k, \tilde{\mathbf{K}}(\delta))$-approximable, where

$$\tilde{K}_i(\delta)\equiv\tilde{K}_i(\delta,\varepsilon) = i!(\sup\{t>0; \tilde{\sigma}_f(t)\varepsilon^{-1}\leq\delta\})^{-i/2},\ i=1,\ldots,k.$$

Theorem 2.9, in its turn, implies $(\mathbf{Q}_u^k, \hat{\mathbf{K}}(\delta,\varepsilon))$-uniformity of the class $\mathcal{W}_{f_{\varepsilon,s}}$, where

$$\hat{K}_j(\delta,\varepsilon) = C_j\Sigma\delta^{-1}\tilde{K}_{i_1}\left(\frac{\delta}{8},\varepsilon\right)\cdots\tilde{K}_{i_l}\left(\frac{\delta}{8},\varepsilon\right),\ j=1,\ldots,k.$$

Assuming that $\delta=1$, we shall find the function

$$\psi_\varepsilon \in \mathscr{F}(W_{f_{\varepsilon,s}}(0), W_{f_{\varepsilon,s}}(1)) \cap (\mathbf{Q}_u^k, \widehat{\mathbf{K}}(1, \varepsilon)).$$

Since the conditions $f_{\varepsilon,s} \leq 0$, $f_{\varepsilon,s} \geq 1$ by the definition of the function ϕ are equivalent to the conditions $f(x) \leq s$, $f(x) \geq s + \varepsilon$, then

$$\psi_\varepsilon \in \mathscr{F}(W_f(s), W_f(s + \varepsilon)) \cap (\mathbf{Q}_u^k, \widehat{\mathbf{K}}(1, \varepsilon)).$$

It remains to note that

$$\widehat{K}_i(1, \varepsilon) \leq C(i)(\sup\{t > 0: \widetilde{\sigma}_f(t) < \varepsilon\})^{-i/2}. \square$$

As usual, $\mathrm{Lip}_\alpha(\mathbf{B})$ stands for the set of functions $f : \mathbf{B} \to \mathbf{R}$ satisfying the Hölder condition with the exponent $\alpha \in (0, 1]$:

$$\|f\|_\alpha := \sup_{x \neq y} |f(x) - f(y)| \, \|x - y\|^{-\alpha} < \infty.$$

THEOREM 2.16. *If $u \in \pi_\gamma(\mathbf{H}, \mathbf{B})$ then for each $k \geq 1$ there exists a constant $C = C(u, k)$ such that the class $\mathscr{W}_F = \{\mathscr{W}_f, \|f\|_\alpha \leq 1\}$ is $(\mathbf{Q}_u^k, \mathbf{K}_\alpha)$-uniform.*

Proof. It is analogous to Corollary 2.13. □

This result, together with Theorem 2.8, implies the following refinement of Corollary 2.13.

THEOREM 2.17. *If $u \in \pi_\gamma(\mathbf{H}, \mathbf{B})$, then for each $k \geq 1$ there exists a constant $C = C(u, k)$ such, that every function $f \in \mathrm{Lip}_\alpha(\mathbf{B})$ is uniformly $(\mathbf{Q}_u^k, C\mathbf{K}_{\alpha,k})$-approximable.*

Proof. The assertion of the theorem follows from Theorem 2.16 and Corollary 2.10. □

It should be noted that in every Banach space $\mathbf{B}$ there exists a subspace $\mathbf{H} = \mathbf{H_B}$ such that the operator of the identical inclusion $\mathbf{H_B} \subset \mathbf{B}$ is γ-radonifying (see [106]). Consequently, for each Banach space $\mathbf{B}$ the class $\{W_f, \|f\|_{\mathrm{BL}} \leq 1\}$ is $\mathbf{Q}_{\mathbf{H_B}}^\infty$-uniform and, from Remark 2.5, $\mathbf{Q}_{\mathbf{F}}^\infty$-uniform for the arbitrary space $\mathbf{F}$ such that $\mathbf{F} \subset \mathbf{H_B}$ and this inclusion is continuous. As an example, the corresponding results in the case of the spaces $\mathbf{c}_0$ and $\mathbf{C}[0, 1]$ may be presented.

For the sequence $\sigma = (\sigma_1, \sigma_2, \ldots) \in \mathbf{c}_0^+$ we shall define the space

$$\mathbf{l}_2(\sigma) := \{x = (x_1, x_2, \ldots): \|x\|_\sigma := \sum_i x_i^2 \sigma_i^{-1} < \infty\}.$$

In the case of the space $\mathbf{c}_0$ we have the following two results.

THEOREM 2.18. *If for each $\varepsilon > 0$ the series $\Sigma_i \exp\{-\varepsilon \sigma_i^{-2}\}$ converges, then the class $\{W_f, f \in \mathrm{Lip}_\alpha(\mathbf{c}_0), \|f\|_{\mathrm{Lip}_\alpha} \leq 1\}$ is $(\mathbf{Q}_{\mathbf{l}_2(\sigma)}^k, C\mathbf{K}_{\alpha,k})$-uniform.*

Proof. The operator of the inclusion $\mathbf{l}_2(\sigma) \subset \mathbf{c}_0$ is γ-radonifying. □

THEOREM 2.19. *If $\lambda = (\lambda_1, \lambda_2, \ldots)$, $\lambda_i = i^{-\beta}$, $\beta > 1/2$, then for each $k \geq 1$ there exists the constant $C = C(k, \beta)$ such that the class $\{W_f, f \in \mathrm{Lip}_\alpha(\mathbf{c}_0), \|f\|_\alpha \leq 1\}$ is $(\mathbf{Q}_{\mathbf{l}_{\infty,\lambda}}^k, C\mathbf{K}_{\alpha,k})$-uniform.*

Proof. It is easy to check that under the condition $\beta > 1/2$ there exists the Hilbert space $\mathbf{H}$ such that the operator of inclusion $i_\lambda: \mathbf{l}_{\infty,\lambda} \subset \mathbf{c}_0$ is an element of the set $\mathbf{L}(\mathbf{l}_{\infty,\lambda}, \mathbf{H}) \cdot \pi_\gamma(\mathbf{H}, \mathbf{c}_0)$. □

In the space C[0, 1] we have the following example of a $\mathbf{Q}^k_{\mathrm{Lip}_\alpha[0,1]}$-uniform class.

THEOREM 2.20. *If the number* $\alpha \in (1/2, 1]$, *then for each* $k \geq 1$ *and* $\delta \in (0, 1]$ *there exists a constant* $C = C(k, \alpha, \delta)$ *such that the class* $\{\mathcal{W}_f, f \in \mathrm{Lip}_\delta(\mathbf{C}[0, 1]), \|f\|_\delta \leq 1\}$ *is* $(\mathbf{Q}^k_{\mathrm{Lip}_\alpha[0,1]}, C\mathbf{K}_{\delta,k})$*-uniform.*

Proof. By i_α we denote the operator of inclusion $\mathrm{Lip}_\alpha[0, 1] \subset \mathbf{C}[0, 1]$. The theorem will be proved if we show that there exists a Hilbert space $\mathbf{H}_\alpha$ such that $i_\alpha \in \mathbf{L}(\mathrm{Lip}_\alpha[0, 1], \mathbf{H}_\alpha) \cdot \pi_\gamma(\mathbf{H}_\alpha, \mathbf{C}[0, 1])$. For convenience, instead of the interval [0, 1] let us consider the interval $[-\pi, \pi]$. The set of functions

$$\{e_n(t), n = 0, \pm 1, \pm 2, \cdots\}, \; e_0(t) = (2\pi)^{-1/2},$$

$$e_n(t) = \frac{\cos nt}{\sqrt{2\pi}}, \; e_{-n}(t) = \frac{\sin nt}{\sqrt{2\pi}}$$

form a complete orthonormal system in the Hilbert space $\mathbf{L}_2[-\pi, \pi]$. Suppose

$$\|\phi\|^2_\beta := |a_0|^2 + \sum_{n=-\infty}^{\infty} n^{2\beta} |a_n|^2, \; \beta > 0.$$

It is clear that the set of functions ϕ with a finite form $\|\phi\|_\beta$ form a Hilbert space which is denoted by $\mathbf{W}_\beta$. Besides, the embedding $\mathbf{W}_\beta \subset \mathbf{L}_2[-\pi, \pi]$ is continuous. Since

$$\|\phi\|_\infty := \sup_{x \in [-\pi,\pi]} |\phi(x)| \leq \sum_{n=-\infty}^{\infty} |a_n(\phi)| \leq \left(1 + \sum_{n=1}^{\infty} n^{-2\beta}\right)^{1/2} \|\phi\|_\beta,$$

then the embedding $\mathbf{W}_\beta \subset \mathbf{C}[-\pi, \pi]$ is continuous if $\beta > 1/2$. Let $\alpha = 1/2 + \varepsilon$, $\varepsilon > 0$. Suppose $\beta = (1 + \varepsilon)2^{-1}$ and $\mathbf{H}_\alpha = \mathbf{W}_\beta$. Let $\gamma_0, \gamma_1, \gamma_{-1}, \cdots$ be independent standard Gaussian random variables,

$$Y(t) = \sum_{n=1}^{\infty} n^{-\beta}(\gamma_n e_n(t) + \gamma_{-n} e_{-n}(t)) + \gamma_0 e_0(t).$$

In [99, Ch. XIII] it is shown that for $\beta > 1/2$ the realizations of the random process $Y(t)$ with probability 1 are continuous. Since the functions $e_0(t), e_1(t), e_{-1}(t), \cdots$ form an orthonormalized basis of the space $\mathbf{W}_\beta$, this means that the operator of the embedding $\mathbf{W}_\beta \subset \mathbf{C}([-\pi, \pi])$ is γ-radonifying. We shall prove the continuity of the embedding $\mathrm{Lip}_\alpha[-\pi, \pi]$ into the space $\mathbf{H}_\alpha$. For the function $\phi \in \mathrm{Lip}_\alpha[-\pi, \pi]$ assume $\psi(t) = \phi(t) - a_0(\phi)$. Let the number $\nu \in (0, 1]$. By $D^\nu\psi$ we denote the 'fractional derivative' of the order ν of the function ψ,

$$D^\nu\psi(t) := \sum_{n=0}^{\infty} n^\nu[a_n(\psi)e_n(t) + a_{-n}(\psi)e_{-n}(t)].$$

It is well known (see [231]) that if $\phi \in \mathrm{Lip}_\alpha[-\pi, \pi]$, then for $\nu < \alpha$ the function $D^\nu\psi \in \mathrm{Lip}_{\alpha-\nu}[-\pi, \pi]$. By carrying out more precise computations of constants appearing in the proof of this fact, the following inequality can be obtained:

$$\|D^\nu\psi\|_{\mathrm{Lip}_{\alpha-\nu}[-\pi,\pi]} \leq C(\alpha, \nu)\|\psi\|_{\mathrm{Lip}_\alpha[-\pi,\pi]} \tag{2.15}$$

with some constant $C(\alpha, \nu) > 0$. Therefore $D^\beta\psi \in \mathrm{Lip}_{\varepsilon/2}[-\pi, \pi]$ and we have

$$\|\phi\|^2_{\mathbf{W}_\beta} = |a_0(\phi)|^2 + \|D^\beta\psi\|_{L_2[-\pi,\pi]}.$$

Let

$$g(t) = \int_{-\pi}^{t} (D^{\beta}\psi)(s)ds.$$

We have

$$\|\phi\|^2_{W_\beta} = |a_0(\phi)|^2 + \int_{-\pi}^{\pi} |g'(t)|^2 dt \le |a_0(\phi)|^2 +$$

$$+ 2\pi[\sup\{|g(t) - g(\tau)| \; |t - \tau|^{-1} : t, \tau \in [-\pi, \pi]\}]^2. \tag{2.16}$$

By applying (2.15) we obtain

$$|g(t) - g(\tau)| \; |t - \tau|^{-1}$$

$$\le |t - \tau|^{-1} \left| \int_{\tau}^{t} (D^{\beta}\psi)(s)\, ds - (t - \tau)(D^{\beta}\psi)(\tau) \right| +$$

$$+ |D^{\beta}\psi(\tau)| \le |t - \tau|^{-1} \int_{\tau}^{t} |D^{\beta}\psi(s) - D^{\beta}\psi(\tau)| \, ds + |D^{\beta}\psi(\tau)|$$

$$\le \|D^{\beta}\psi\|_{\mathrm{Lip}_{\epsilon/2}[-\pi,\pi]} |t - \tau|^{-1} \int_{\tau}^{t} (s - t)^{\epsilon/2} ds + |D^{\beta}\psi(\tau)|$$

$$\le C(\alpha)\|\psi\|_{\mathrm{Lip}_\alpha[-\pi,\pi]} + |D^{\beta}\psi(\tau)|. \tag{2.17}$$

Since $a_0(D^{\beta}\psi) = 0$, then there exists a number $\tau_0 \in [-\pi, \pi]$ such that $D^{\beta}\psi(\tau_0) = 0$. Therefore, by (2.15),

$$|D^{\beta}\psi(\tau)| = |D^{\beta}\psi(\tau) - D^{\beta}\psi(\tau_0)| \le (4\pi)^{\epsilon/2} \|D^{\beta}\psi\|_{\mathrm{Lip}_{\epsilon/2}[-\pi,\pi]} \le C(\alpha)\|\psi\|_{\mathrm{Lip}_\alpha[-\pi,\pi]}.$$

Taking (2.17) into consideration, from (2.16) we obtain

$$\|\phi\|^2_{W_\beta} \le |a_0(\phi)|^2 + C(\alpha)\|\psi\|^2_{\mathrm{Lip}_\alpha[-\pi,\pi]} \le C(\alpha)(\|\phi\|_{C[-\pi,\pi]} + \|\phi\|_{\mathrm{Lip}_\alpha[-\pi,\pi]}),$$

which means the continuity of the embedding of $\mathrm{Lip}_\alpha[-\pi, \pi]$ into $\mathbf{H}_\alpha$.

Some other examples of uniform classes. It is natural to expect that for the fixed functions $f: \mathbf{B} \to R$ the set of operators $u \in \mathbf{L}(\mathbf{F}, \mathbf{B})$ such that the class $\mathscr{W}_f$ is $\mathbf{Q}^k_u$-uniform (for some $k \ge 3$), is greater than the set $\mathbf{L}(\mathbf{F}, \mathbf{H}) \cdot \pi_\gamma(\mathbf{H}, \mathbf{B})$. If the function f itself is smooth enough, then we can apply Theorem 2.9, and u may be any operator.

THEOREM 2.21. *If the function* $f \in \mathbf{C}^k_u(\mathbf{B}\backslash\{0\}, \mathbf{R})$ *and the following conditions are satisfied*

$$|f(x)| \le C_0\|x\|, \; x \in \mathbf{B}, \tag{2.18}$$

$$\|f^{(i)}_u(x)\| \le C_i\|x\|^{1-i}, \; i = 1, \ldots, k, \; x \in \mathbf{B}\backslash\{0\}, \tag{2.19}$$

then the class $\mathscr{W}_f$ *is* $(\mathbf{Q}^k_u, C\mathbf{K}_{1,k})$*-uniform, where the constant C depends only on the constants* $C_0, C_1, \ldots, C_k$.

Proof. Let the function $\phi \in \mathbf{C}^\infty(\mathbf{R})$ satisfy the conditions $N(\phi) = [0, 1]$, $\phi(t) = 1$, if $|t| \ge 1$, $\phi(t) = 0$, if $|t| \le 1/2$. Suppose $g(x) \equiv g_\epsilon(x) := f(x)\phi(\epsilon^{-1}f(x))$. By denoting $t = \epsilon^{-1}f(x)$, we have

$$g'_u(x)(h) = f'_u(x)(h)(\phi(t) + \phi'(t) \cdot t), \; x \in \mathbf{B}, \; h \in \mathbf{F},$$

$$g_u''(x)(h)^2 = f_u''(x)(h)^2(\phi(t) + t\phi'(t))^2 + \varepsilon^{-1}(f_u'(x)(h))^2(2\phi'(t) + t\phi''(t)).$$

Since $\phi(t) = 0$ for $|t| \le 1/2$, and, moreover, for $\|x\| \le \varepsilon(2C_0)^{-1}$, by the use of estimates (2.18) and (2.19) we obtain

$$|g_u'(x)(h)| \le C\|h\|,$$

$$|g_u''(x)(h)^2| \le C \cdot \varepsilon^{-1}\|h\|^2.$$

By induction on $i \ge 1$ we prove that

$$\|g\|_i \le C\varepsilon^{1-i}, \quad i = 1, ..., k.$$

Further,

$$\sup_x |f(x) - g_\varepsilon(x)| = \sup_x |f(x)|\ |1 - \phi(\varepsilon^{-1}f(x))| \le \varepsilon,$$

since $\phi(\varepsilon^{-1}f(x)) = 1$, if $|f(x)| \ge \varepsilon$. The proof is completed by applying Theorem 2.9. □

The particular case $\mathbf{F} = \mathbf{B}$, $u = I_{\mathbf{B}}$ and $f(\cdot) = \|\cdot\|$ in the above formulated theorem is important. As we shall consider the class $\mathscr{W}_{\|\cdot\|}$ rather often (this is a class of balls with the centre at zero), we denote it by $\mathbf{W}_0$, and to point the space under consideration we shall write $\mathbf{W}_0(\mathbf{B})$. Because of the importance of this particular case, it will be formulated as a theorem.

THEOREM 2.22. *If the norm of the space* $\mathbf{B}\backslash\phi(x) = \|x\|$ *satisfies the conditions* $\phi \in \mathbf{C}^k(\mathbf{B}\backslash\{0\}, \mathbf{R})$,

$$\sup_{\|x\|=1} \|\phi^{(i)}(x)\| \le C_i, \quad i = 1, ..., k, \tag{2.20}$$

then there exists the constant $C = C(C_1, ..., C_k)$ *such that the class* $\mathbf{W}_0$ *is* $(\mathbf{Q}^k, C\mathbf{K}_{1,k})$*-uniform.*

Proof. Taking into account the homogeneity of the norm, we have

$$\|\phi^{(i)}(x)\| \le C_i\|x\|^{1-i}, \quad i = 1, ..., k.$$

There remains to apply Theorem 2.21. □

Let (S, ν) be a space with σ-finite measure ν, $p \ge 1$, the integer r being defined by the inequalities $p - 1 \le r < p$.

PROPOSITION 2.23. *If* $p = 2l$, l *is an integer, then the norm* ϕ *of the space* $\mathbf{L}_p \equiv \mathbf{L}_p(S, \nu)$ *belongs to the class* $\mathbf{C}^\infty(\mathbf{L}_p\backslash\{0\}, \mathbf{R})$ *and* (2.20) *is valid for all* k. *In other cases* $\phi \in \mathbf{C}^r(\mathbf{L}_p\backslash\{0\}, \mathbf{R})$ *and* (2.20) *holds for* $k = r$.

Proof. We denote $\phi_1(x) = (\phi(x))^p$, $\psi: \mathbf{R}_1 \to \mathbf{R}_1$, $\psi(t) = |t|^p$. It is obvious that $\psi \in \mathbf{C}^r(\mathbf{R}_1, \mathbf{R}_1)$ and

$$\psi^{(k)}(t) = p(p-1) \cdots (p-k+1)|t|^{p-k}(\operatorname{sgn} t)^k.$$

By using the Taylor formula, we can write

$$\left|\psi(t+h) - \sum_{k=0}^{r} \psi^{(k)}(t)h^k(k!)^{-1}\right| \le p(p-1) \cdots (p-r+1)\frac{|h|^r}{r!}.$$

Let $A_k(x)$, $k = 1, ..., r$, denote a continuous k-linear form on $\mathbf{L}_p$ defined by the equality

$$A_k(x)(h_1, ..., h_k) = \int_S \psi^{(k)}(x(s))h_1(s) \cdots h_k(s)\nu(ds),$$

where $h_i \in \mathbf{L}_p$, $i = 1, ..., k$. (The boundedness of the form A_k can be proved by induction applying the Hölder inequality.) If we set $A_0(x) = \phi_1(x)$, then

$$|\phi_1(x+h) - \sum_{k=0}^{r} (k!)^{-1}A_k(x)(h)^k|$$

$$\leq \int_S |\psi(x(s)) + h(s)) - \sum_{k=0}^{r} (k!)^{-1}\psi^{(k)}(x(s))(h(s))^k| \nu(ds)$$

$$\leq p(p-1) \cdots (p-r-1)(r!)^{-1}\|h\|^r.$$

Hence, by Proposition 1.15, we obtain that $\phi_1^{(k)}(x) = A_k(x)$, $k = 1, ..., r$. and that $\phi_1 \in \mathbf{C}^r(\mathbf{L}_p, \mathbf{R})$. If $p = 2l$, then $\psi(t) = t^{2l}$ and it is obvious that $\phi_1^{(2l-1)}$ is linear with respect to x; therefore, $\phi_1 \in \mathbf{C}^\infty(\mathbf{L}_{2l}, \mathbf{R})$. Since $\phi(x) = (\phi_1(x))^{1/p}$ we easily get the assertions of the proposition. □

If the function $f : \mathbf{B} \to \mathbf{R}$ is differentiable but the growth of $f(x)$ is more rapid than $C\|x\|$, i.e. conditions (2.18) and (2.19) are not satisfied, then, by the use of Theorems 2.21 and 2.22, we can obtain the following result.

THEOREM 2.24. *Let* $f \in \mathbf{C}^k(\mathbf{B}\backslash\{0\}, \mathbf{R})$ *and the following conditions are satisfied*

$$|f(x)| \leq \overline{C}_0\|x\|^\alpha, \tag{2.21}$$

$$\|f^{(i)}(x)\| \leq \overline{C}_i\|x\|^{\alpha-i}, \ i = 1, ...,, \ \alpha > 0. \tag{2.22}$$

(a) *If the norm of the space* **B** *satisfies the conditions of Theorem* 2.22, *then the class* $\mathcal{W}_g$ *is* $(\mathbf{Q}^k, \overline{C}\mathbf{K}_{1,k})$*-uniform, where* $g(x) = f(x)\|x\|^{1-\alpha}$ *and the constant* $\overline{C}$ *depends on* $\overline{C}_i$, $i = 0, 1, ..., k$ *and* C_i *from* (2.20);

(b) *If in addition to* (2.21) $f(x) \geq \overline{C}_0\|x\|^\alpha$, *then the class* $\mathcal{W}_f$ *is* $(\mathbf{Q}^k, \overline{C}\mathbf{K}_{1,k})$*-uniform where* $\overline{C}$ *depends on* $\overline{C}_0$ *and* C_{i_i}, $i = 0, 1, ..., k$.

The proof of (a) is reduced to checking the conditions of Theorem 2.21 for the function g. To prove (b) it is sufficient to note that $\mathcal{W}_f = \mathcal{W}_{\tilde{f}}$, where $\tilde{f}(x) = (f(x))^{1/\alpha}$, and for the function $\tilde{f}$ the conditions of Theorem 2.21 are satisfied. □

We shall consider two examples of the functions satisfying conditions (2.21) and (2.22). One elementary example is represented by the homogeneous forms of the power $k \geq 2$. It is easy to check the validity of the following proposition.

PROPOSITION 2.25. *Let* $f(x) = \langle P(x, ..., x), x \rangle$, *where* $P \in \mathbf{L}^k(\mathbf{B}, \mathbf{B}^*)$ *and is a symmetric mapping. Then* f *satisfies conditions* (2.21) *and* (2.22) *where* $\alpha = k + 1$.

For example, the functional $f : \mathbf{L}_2(0, 1) \to \mathbf{R}$, defined by the formula

$$f(x) = \int_0^1\int_0^1 K(s, t)x^2(t)x^2(s)\, ds\, dt,$$

where $K(s, t)$ is a continuous bounded function, satisfies the conditions of (2.21) and (2.22) where $\alpha = 4$. In the same way we can construct differentiable functionals in the space $\mathbf{L}_p$.

Another class of differentiable functionals may be obtained by the use of Nemytski and Hammerstein operators thoroughly studied in non-linear functional analysis (see, for instance,

[212]). We recall that the Nemytski operator, acting from one functional space (i.e. a space of real functions) into another, has the form

$$Q(u) = q(u(y), y),$$

and the Hammerstein operator is of the form

$$\Gamma(u) = \int_0^1 K(x, y) q(u(y), y)\, dy.$$

Let us consider the following functional in the space $\mathbf{L}_p \equiv \mathbf{L}_p(0, 1), p > 4$,

$$f(x) = \int_0^1 h(t) \int_0^1 K(t, y) q(x(y), y)\, dy\, dt,$$

$$x \in \mathbf{L}_p, \; h \in \mathbf{L}_q, \; p^{-1} + q^{-1} = 1,$$

which is a composition of a linear continuous functional and the Hammerstein operator, i.e. $f = H \cdot \Gamma$, where

$$\Gamma: \mathbf{L}_p \to \mathbf{L}_p, \; \Gamma(u) = \int_0^1 K(x, y) q(u(y), y)\, dy,$$

$$H: \mathbf{L}_p \to \mathbf{R}, \; H(u) = \int_0^1 h(t) u(t)\, dt.$$

PROPOSITION 2.26. *Let $q(u, x)$ be a function, continuous in u for almost all x and measurable with respect to x. For a fixed u, $|q(u, x)| \leq C|u|^{p-1}$, there exists, with respect to u, a continuous derivative $q_u^{(i)}(u, x)$ and $|q_u^{(i)}(u, x)| < C|u|^{p-i-1}$, $i = 1, 2, 3$, for all u and $x \in [0, 1]$. If*

$$\|K\|_p = \left(\int_0^1 \int_0^1 |K(x, y)|^p\, dx\, dy \right)^{1/p} < \infty, \tag{2.23}$$

then the functional f satisfies the conditions of Theorem 2.24, where $k = 3$ and $\alpha = p - 1$.

Proof. The estimate

$$|f(x)| \leq C \|K\|_p \|h\|_q \|x\|_p^{p-1}$$

immediately follows from the conditions of the propositions and one has to compute only the derivatives of the functional f. Since H is a linear continuous functional, then

$$f^{(i)}(x) = H \cdot \Gamma^{(i)}(x),$$

and it is necessary to show that the Hammerstein operator is thrice continuously Fréchet differentiable. With this aim in view, we shall make use of the following two theorems from [212], which will be given without proof. □

THEOREM 2.27. *In order for the Nemytski operator $Q(u) = q(u(y), y)$, – where $q(u, y)$ is continuous in u at almost every x and is measurable with respect to x for a fixed u – to be a continuous operator acting from $\mathbf{L}_p$ to $\mathbf{L}_{p_1}$ $(p > 0, p_1 > 0)$, it is necessary and sufficient that*

$$|q(u, x)| \leq a(x) + b|u|^r, \; r = pp_1^{-1}, \; a \in \mathbf{L}_{p_1}, \; b > 0. \tag{2.24}$$

THEOREM 2.28. *Let condition (2.23) be satisfied with some $p > 2$ and the measurable with respect to x on $[0, 1]$ function $q(u, x)$ has a partial derivative $q_u'(u, x)$ continuous in $u \in \mathbf{R}$;*

besides, $Q(u) = q_u'(u(x), x)$ *is a continuous operator from* $\mathbf{L}_p$ *to* $\mathbf{L}_r$*, where* $r = p(p-2)^{-1}$. *Then at each point* $u \in \mathbf{L}_p$*, the operator* Γ *has a Fréchet differential*

$$\Gamma'(u)(v) = \int_0^1 K(x, y) q_u'(u(y), y) v(y)\, dy.$$

Since in the conditions of the proposition it is given that $|q_u'(u, x)| < C|u|^{p-2}$, then condition (2.24) is satisfied with $a(x) \equiv 0$ and $r = p - 2$, $p_1 = (p-2)^{-1}$. Hence from Theorem 2.28 the existence of the first derivative of the operator Γ follows. We shall now show that

$$\Gamma''(u)(v, z) = \int_0^1 K(x, y) q_u''(u(y), y) v(y) z(y)\, dy. \tag{2.25}$$

To this end we shall estimate the difference

$$\begin{aligned} |I(x)| &= |\Gamma'(u+z)(v) - \Gamma'(u)(v) - \Gamma''(u)(v, z)| \\ &= \left| \int_0^1 K(x, y)[q_u'(u(y) + z(y), y) - q_u'(u(y), y) - q_u''(u(y), y) z(y)] v(y)\, dy \right|. \end{aligned} \tag{2.26}$$

Applying the Hölder inequality we have

$$|I(x)| \le \left(\int_0^1 K(x, y)^p\, dy \right)^{1/p} \| [q_u'(u + z, \cdot) - q_u'(u, \cdot) - q_u''(u, \cdot) z] v \|_q. \tag{2.27}$$

It is easy to check that

$$\begin{aligned} &\| [q_u'(u + z, \cdot) - q_u'(u, \cdot) - q_u''(u, \cdot) z] v \|_q \\ &\quad = \| [q_u''(u + \Theta z, \cdot) - q_u''(u, \cdot)] z \cdot v \|_q \\ &\quad \le \|z\|_p \|v\|_p \|q_u''(u + \Theta z, \cdot) - q_u''(u, \cdot)\|_s, \end{aligned} \tag{2.28}$$

where $s = p(p-3)^{-1}$. The operator $Q_1(u) = q_u''(u(y), y)$ satisfies condition (2.24), where $r = p - 3$, and is a continuous operator from $\mathbf{L}_p$ to $\mathbf{L}_s$; therefore,

$$\lim_{\|z\|_p \to 0} \|q_u''(u + \Theta z, \cdot) - q_u''(u, \cdot)\|_s = 0. \tag{2.29}$$

From (2.26) – (2.29) it follows that

$$\lim_{\|z\|_p \to 0} \|z\|_p^{-1} \|\Gamma'(u + z)(v) - \Gamma'(u)(v) - \Gamma''(u)(v, z)\|_p = 0$$

and this proves equality (2.25). We shall now estimate the second derivative of the functional f:

$$\begin{aligned} |f''(u)(z_1, z_2)| &= \left| \int_0^1 \int_0^1 K(x, y) q_u''(u(y), y) z_1(y) z_2(y) h(x)\, dx\, dy \right| \\ &\le \|K\|_p \|h\|_q \|z_1\|_p \|z_2\|_p \|q_u''\|_s. \end{aligned}$$

But

$$\|q_u''\|_s = \left(\int_0^1 |q_u''(u(y), y)|^s\, dy \right)^{1/s}$$

$$\leq C\left(\int_0^1 |u(y)|^{s(p-3)}\,dy\right)^{(p-3)/p} = C\|u\|_p^{p-3},$$

therefore, we finally obtain

$$\|f''(u)\| \leq C\|K\|_p\|h\|_q\|u\|_p^{p-3},$$

which coincides with (2.22) at $\alpha = p - 1, i = 2$.

By analogy we may show that

$$\Gamma'''(u)(v, z, w) = \int_0^1 K(x, y)q_u'''(u(y), y)v(y)z(y)w(y)\,dy,$$

and this derivative admits the estimate required. □

To conclude this section we shall consider the class of balls $\mathbf{W}_0(\mathbf{B})$ in the spaces $\mathbf{B}$ with a non-smooth norm, namely, in the spaces $\mathbf{c}_0$ and $\mathbf{l}_1$ whose norms are Fréchet non-differentiable. In these spaces it is necessary to make use of differentiation along the subspaces. The results of this type have already been presented in Theorems 2.18 and 2.19 (in the case of the space $\mathbf{c}_0$) but it should be noted that in the above theorems the subspaces $\mathbf{B}_1 \subset \mathbf{c}_0$, with respect to which the class $\mathbf{W}_0(\mathbf{c}_0)$ is $\mathbf{Q}^k_{\mathbf{B}_1}$-uniform, are relatively small. By the use of direct methods of constructing approximating functions, it is possible to enlarge these spaces.

THEOREM 2.29. *For all $k \geq 1$ and $p \geq 1$ there exists a constant $C = C(k, p)$ such that the class $\mathbf{W}_0(\mathbf{c}_0)$ is $(\mathbf{Q}^k_{\mathbf{l}_p}, C\mathbf{K}_{1,k})$-uniform.*

Proof. One may assume that the number $p \geq 1$ is large enough since, from the $(\mathbf{Q}^k_{\mathbf{l}_p}, C\mathbf{K}_{1,k})$-uniformity, $(\mathbf{Q}^k_{\mathbf{l}_q}, C\mathbf{K}_{1,k})$-uniformity follows if $p \geq q$.

Let the function $\phi \in \mathbf{C}^\infty(\mathbf{R}_1)$, $\phi(t) = 1$, if $t \leq 0$, $\phi(t) = 1 - t^m$, if $0 \leq t \leq \delta$, $1 - \delta^m = 1/2$, $\phi(t) = 0$, if $t \geq 1$; the function decreases monotonically and is $m - 1$ times continuously differentiable (the number $m \geq k + 1$).

Suppose

$$g_\varepsilon(x) = \prod_{i=1}^{\infty} \phi(t_i),$$

where $t_i = \varepsilon^{-1}(x_i - r)$, $r > 0$, $\varepsilon > 0$, $x = (x_1, x_2, \ldots) \in \mathbf{c}_0$. It is clear that $g_\varepsilon(x) = 1$ if $\|x\| \leq r$ and $g_\varepsilon(x) = 0$ if $\|x\| > r + \varepsilon$. Thus, the theorem will be proved if we prove the corresponding estimates for the derivatives of the function g_ε. We shall only consider the case $k = 2$ since in the case of a single, arbitrary k, the notation becomes more complicated. As in the whole proof we shall consider the derivatives along the subspace $\mathbf{l}_p$, so, to simplify the writing, we shall write g_ε'' instead of $(g_\varepsilon)''_{\mathbf{l}_p}$. Let $h, f \in \mathbf{l}_p$, $\|f\| = \|h\| = 1$. Taking into account that $x_i \to 0$ when $i \to \infty$, we have

$$\varepsilon^2 g_\varepsilon''(x)(f, h) = \sum_i f_i h_i b_i \prod_{j \neq i} \phi(t_j) + \sum_{i \neq j} f_i h_j a_i a_j \prod_{l \neq i,j} \phi(t_l), \tag{2.30}$$

where $a_i = \phi'(t_i)$, $b_i = \phi''(t_i)$, and the sums and products in (2.30) are finite. Let the last non-zero term in these sums have the number l. Suppose $\tilde{a}_i = |a_i|$, $\tilde{b}_i = |b_i|$ if $\delta \leq t_i < 1$ and $\tilde{a}_i = \tilde{b}_i = 0$ if $t_i \overline{\in} (0, 1)$. Denoting by s the number of such t_i that $\delta \leq t_i < 1$, $i \leq l$, and observing that $\phi(t_i) \leq 1/2$ for $t_i \geq \delta$, $\phi(t_i) \geq 1/2$ for $t_i < \delta$, and that the functions ϕ' and ϕ'' are bounded by a number depending only on m, from (2.30) we obtain

$$|\varepsilon^2 g_\varepsilon''(x)(f,h)|$$
$$\leq 2^s C(m)\left\{\left(\sum_{i=1}^{l} \tilde{a}_i |f_i|\right)\left(\sum_{i=1}^{l} \tilde{a}_i |h_i|\right) + \sum_{i=1}^{l} \tilde{b}_i |f_i|\ |h_i|\right\} \prod_{0<t_i<\delta} \phi(t_i).$$

Let $p^{-1} + q^{-1} = 1$. Applying the Hölder inequality we find

$$|\varepsilon^2 g_\varepsilon''(x)(f,h)| \leq 2^{-s} C(m)\left\{\left(\sum_{i=1}^{l} \tilde{a}_i^q\right)^{2/q} + \left(\sum_{i=1}^{l} \tilde{b}_i^2\right)^{1/2}\right\} \prod_{0<t_i<s} \phi(t_i).$$

Since $1 - t \leq l^{-t}$ for $t \geq 0$, then, denoting $A = \Sigma_{0<t_i<\delta} t_i^m$, we have

$$\sum_{i=1}^{l} \tilde{a}_i^q \leq s + C(p,m)A, \text{ if } (m-1)q \geq m,$$
$$\sum_{i=1}^{l} \tilde{b}_i^2 \leq s + C(p,m)A, \text{ if } 2(m-2) \geq m.$$

Therefore, by selecting $m = m(p)$ in an appropriate way, we obtain

$$|\varepsilon^2 g_\varepsilon''(x)(f,h)| \leq 2^{-s} C(p) e^{-A}\{(s+A)^{q/2} + (s+A)^{1/2}\} \leq C(p).\square$$

By choosing a smoother function ϕ used at constructing the function g_ε and having carried a more detailed analysis of the derivatives of the function g_ε, we can obtain the following result, whose proof is available in [15].

THEOREM 2.30. *Let the sequence* $\lambda = (\lambda_i, i \geq 1) \in \mathbf{c}_0^+$ *be such that for each* $\varepsilon > 0$.

$$\sum_{i=1}^{\infty} \exp\{-\varepsilon\lambda_i^{-1}(\ln \lambda_i)^{-2}\} < \infty.$$

Then the class $\mathbf{W}_0(\mathbf{c}_0)$ *and* $\mathscr{W} = \{(x \in \mathbf{c}_0 : x_i - a_i < r, i \geq 1), r \in \mathbf{R}\}$, $a = (a_i, i \geq 1) \in \mathbf{c}_0$ *are* $(\mathbf{Q}^n_{\mathbf{c}_{0,\lambda}}, C(\lambda)\mathbf{K}_{1,n})$*-uniform for each* $n \geq 1$.

We shall now consider the space $\mathbf{l}_1$. Let $\lambda = (\lambda_i, i \geq 1) \in \mathbf{c}_0^+$ and $\Sigma_{i=1}^{\infty}\lambda_i^2 < \infty$. By the sequence λ the diagonal operator $u_\lambda \in \mathbf{L}(\mathbf{l}_2, \mathbf{l}_1)$, $u_\lambda(x) = (\lambda_i x_i, i \geq 1)$, $x \in \mathbf{l}_2$ is defined.

THEOREM 2.31. *If the number* $m \geq 2$ *and* $\Sigma_{i=1}^{\infty}\lambda_i^{m/m-1} < \infty$, *then the class* $\mathbf{W}_0(\mathbf{l}_1)$ *is* $(\mathbf{Q}^{m-1,1}_{u_\lambda}, C\mathbf{K}_{1,m})$*-uniform.*

Proof. According to Theorem 2.9, it is sufficient to prove that the norm $\|\cdot\|$ of the space $\mathbf{l}_1$ can be uniformly approximated by a function of the class $(\mathbf{Q}^{m-1,1}_{u_\lambda}, C\mathbf{K})$, where $\mathbf{K} = \mathbf{K}(\varepsilon) = (1, \varepsilon^{-1}, \ldots, \varepsilon^{-m+1})$.

Let us consider the function $\phi \in \mathbf{C}^\infty(\mathbf{R})$ with the properties $N(\phi) = [0,1]$; $\phi(t) = 0$, if $|t| \leq 1/2$; $\phi(t) = 1$, if $|t| \geq 1$ and denote by g the function $g(t) = \phi(|t|)|t|$. Set $\sigma_i = \lambda_i^{m/m-1}$ and for $x \in \mathbf{l}_1$, $\varepsilon > 0$ we define

$$\psi(x) = \psi_\varepsilon(x) = \sum_{i=1}^{\infty} g(x_i(\varepsilon\sigma_i)^{-1})\varepsilon\sigma_i.$$

We have

$$|\psi(x) - \|x\|| = \sum_{i=1}^{\infty} |x_i|(1 - \phi(x_i(\varepsilon\sigma_i)^{-1}))$$

$$\leq \sum_{i:\ |x_i| \leq \varepsilon\sigma_i} |x_i| \leq \varepsilon \sum_{i=1}^{\infty} \sigma_i.$$

Therefore, it is sufficient to prove that $\psi \in \mathbf{D}^m_{u_\lambda}$ and to estimate the derivatives.

Let $x \in \mathbf{l}_1$, $h \in \mathbf{l}_2$, $h^{(1)}, \ldots, h^{(l)} \in \mathbf{L}_2$, $l = 1, \ldots, m$. Let us define

$$A_l(x, h)(h^{(1)}, \ldots, h^{(l)}) = \sum_{i=1}^{\infty} h_i^{(1)} \cdots h_i^{(l)} g^{(i)}\left(\frac{x_i + \lambda_i h_i}{\varepsilon\sigma_i}\right)\lambda_i^l \varepsilon^{1-l}\sigma_i^{1-l}. \tag{2.31}$$

It is easy to show that for all $l = 1, \ldots, m$ the operator $A_l(x, h) \in \mathbf{L}^l(\mathbf{l}_2, \mathbf{R})$. Besides, $A_l(x, h)$ as a function of h, is continuous at $h \to 0$. We shall prove the relationship

$$|I(h)| := \left|\psi_x \cdot u(h) - \psi_x \cdot u(0) - \sum_{i=1}^{m} (l!)^{-1} A_l(x, 0)(h)^l\right| = o(\|h\|^m). \tag{2.32}$$

(We recall that the function $\psi_x \cdot u : \mathbf{l}_2 \to \mathbf{R}$ is defined as follows: $\psi_x \cdot u(h) = \psi(x + u(h))$.)

We have

$$I(h) = \sum_{i=1}^{\infty}\left\{g\left(\frac{x_i + \lambda_i h_i}{\varepsilon\sigma_i}\right)\varepsilon\sigma_i - g\left(\frac{x_i}{\varepsilon\sigma_i}\right)\varepsilon\sigma_i - \sum_{i=1}^{m}(l!)^{-1}A_l(x, 0)(h)^l\right\}$$

$$= \sum_{i=1}^{\infty}\left\{g\left(\frac{x_i + \lambda_i h_i}{\varepsilon\sigma_i}\right) - g\left(\frac{x_i}{\varepsilon\sigma_i}\right) - \sum_{l=1}^{m} g^{(l)}\left(\frac{x_i}{\varepsilon\sigma_i}\right)\left(\frac{h_i\lambda_i}{\varepsilon\sigma_i}\right)^l\right\}\sigma_i\varepsilon.$$

Hence

$$|I(h)| \leq \sum_{i=1}^{\infty} o\left(\left|\frac{\lambda_i h_i}{\varepsilon\sigma_i}\right|^m\right)\varepsilon\sigma_i = \sum_{i=1}^{\infty} o(|h_i|^m)\varepsilon^{1-m}.$$

This relation proves (2.32). According to Proposition 1.15, the function $\psi_x \cdot u$ is Fréchet differentiable at zero and

$$(\psi_x \cdot u)^l(0)(h) = A_l(x, 0)(h)^l.$$

Hence we easily obtain

$$|\psi_u^{(l)}(x)(h)^l| \leq C(l)\sum_{i=1}^{\infty}|h_i|^l\lambda_i^l\sigma_i^{1-l}\varepsilon^{1-l}$$

$$= C(l)\varepsilon^{1-l}\sum_{i=1}^{\infty}|h_i|^l\lambda_i^{(m-l)/(m-1)} \leq C(l)\varepsilon^{1-l}\|h\|^l.$$

REMARK 2.32. If the sequence $\lambda = (\lambda_i, i \geq 1)$ satisfies the condition that there exists a sequence $(\rho_i, i \geq 1) \in \mathbf{c}_0^+$, $\Sigma_{i=1}^{\infty}\rho_i^{m/m-1} < \infty$ such that $\lim_{i\to\infty}\lambda_i\rho_i^{-1} = 0$, then it is easy to show that the class $\mathbf{W}_0(\mathbf{l}_1)$ is $(\mathbf{Q}^m_{u_\lambda}, C\mathbf{K}_{1,m})$-uniform.

REMARK 2.33. If the sequence $\lambda = (\lambda_i, i \geq 1) \in \mathbf{c}_0^+$ and $\Sigma_{i=1}^{\infty}\lambda_i < \infty$, then the operator $u_\lambda \in \pi_\gamma(\mathbf{l}_2, \mathbf{l}_1)$ (see [117]). Hence, in this case the class $\mathbf{W}_0(\mathbf{l}_1)$ is $(\mathbf{Q}^m_{u_\lambda}, \mathbf{K}_{1,m})$-uniform for each $m \geq 1$.

2.3. Supplements

Numerous publications devoted to the questions of existence on Banach space of smooth real functions with bounded supports are available (see [33], [62], [126], [198], [199], [200], [219]). To formulate some results most closely connected with the materials of Section 2.2, the following notations are necessary:

$$\mathbf{D}^n_\alpha(\mathbf{B}, \mathbf{R}) := \{f \in \mathbf{D}^n(\mathbf{B}, \mathbf{R}) : \sup_{x,h \in \mathbf{B},\, h \neq 0} \|f^{(n)}(x+h) - f^{(n)}(x)\| \, \|h\|^{-\alpha} < \infty\},$$

$n \geq 0,\ 0 < \alpha \leq 1;$

$$\mathbf{C}^n_\alpha(\mathbf{B}, \mathbf{R}) := \lim_{h \to 0} \sup_{x \in \mathbf{B}} \|f^{(n)}(x+h) - f^{(n)}(x)\| \, \|h\|^{\alpha} = 0\},$$

$n \geq 1,\ 0 \leq \alpha < 1.$

Let $\mathbf{S}(\mathbf{B}, \mathbf{R})$ be some class of smooth functions $f : \mathbf{B} \to \mathbf{R}$. The space $\mathbf{B}$ is called $\mathbf{S}$-smooth if the class $\mathbf{S}(\mathbf{B}, \mathbf{R})$ contains a non-trivial function with a bounded support.

2.3.1. THEOREM [33]. (a) *Any finite-dimensional Banach space is* $\mathbf{Q}^\infty$*-smooth.*
(b) *Every Banach space is* $\mathbf{D}^0_1$*-smooth.*
(c) $\mathbf{L}_p$ *for* $p = 2k$, $k \geq 1$, *is* $\mathbf{Q}^\infty$*-smooth.*
(d) *If* $p \geq 1$ *is an odd integer, then* $\mathbf{L}_p$ *is* $\mathbf{D}^{p-1}_1$*-smooth, but is not* $\mathbf{C}^{p-1}_1$*-smooth.*
(e) *If* p *is not an integer and* $m = [p]$, *then* $\mathbf{L}_p$ *is* $\mathbf{D}^m_{p-m}$*-smooth, but not* $\mathbf{C}^m_{p-m}$*-smooth.*
(f) *The space* $\mathbf{C}[0, 1]$ *is not* $\mathbf{D}^1$*-smooth.*
(g) *The space* $\mathbf{c}_0$ *has an equivalent norm from the class* $\mathbf{C}^\infty(\mathbf{c}_0 \backslash \{0\}, \mathbf{R})$ *and is* $\mathbf{C}^\infty$*-smooth.*

2.3.2. Let $\mathcal{K}(\mathbf{H})$ be a Banach space (with respect to the operator norm) of all compact operators acting in the Hilbert space $\mathbf{H}$. The space

$$\mathbf{S}_p := \left\{A \in \mathcal{K}(\mathbf{H}) : \|A\|_p := \left(\sum_{k=1}^{\infty} \lambda_k^p\right)^{1/p} < \infty\right\}, \quad 1 \leq p < \infty,$$

where $\lambda_1 \geq \lambda_2 \geq \cdots$ is a sequence of non-zero eigenvalues of the operator $(A^*A)^{1/2}$, is a Banach space with respect to the norm $\|\cdot\|_p$. In particular, $\mathbf{S}_1$ is a space of nuclear operators, $\mathbf{S}_2$ is a space of the Hilber – Schmit operators.

THEOREM [206]. (a) $\mathbf{S}_p$ *for* $p = 2k$, $k \geq 1$ *is* $\mathbf{Q}^\infty$*-smooth.*
(b) *If* $p \geq 1$ *is an odd integer, then* $\mathbf{S}_p$ *is* $\mathbf{D}^{p-1}_1$*-smooth.*
(c) *If* p *is not an integer and* $m = [p]$, *then* $\mathbf{S}_p$ *is* $\mathbf{D}^m_{p-m}$*-smooth.*
(d) *The space* $\mathcal{K}(\mathbf{H})$ *is* $\mathbf{C}^\infty$*-smooth.*

2.3.3. The Banach space $\mathbf{F}$ is finitely represented in $\mathbf{B}$ if for each $\varepsilon > 0$ and each finite-dimensional subspace $\mathbf{F}_1 \subset \mathbf{F}$, there exists a finite-dimensional space $\mathbf{B}_1 \subset \mathbf{B}$ and such an isomorphism $T : \mathbf{F}_1 \to \mathbf{B}_1$, that $\|T\| \, \|T^{-1}\| \leq 1 + \varepsilon$. The space $\mathbf{B}$ is called super-reflexive if each space, finitely represented in $\mathbf{B}$, is reflexive.

THEOREM [199]. *A Banach space* $\mathbf{B}$ *is* $\mathbf{C}^1_0$*-smooth if and only if* $\mathbf{B}$ *is super-reflexive.*

COROLLARY. *The Banach spaces* $\mathbf{c}_0$, $\mathbf{l}_1$, $\mathbf{C}(S)$, *where* S *is a metric compact, are not* $\mathbf{C}^1_0$*-smooth.*

2.3.4. THEOREM [62]. *The Banach space* $\mathbf{B}$ *is* $\mathbf{D}^1_1$*-smooth (respectively* $\mathbf{C}^1_0$*-smooth) if and only if there exists the equivalent norm* $\phi \in \mathbf{C}^1(\mathbf{B}\ \{0\}, \mathbf{R})$ *and*

$$\sup\{\|\phi'(x) - \phi'(y)\| \, \|x - y\|^{-1} : x \neq y,\ \|x\| = \|y\| = 1\} < \infty$$

(respectively, the mapping $\phi' : \mathbf{B}\backslash\{0\} \to \mathbf{L}(\mathbf{B}, \mathbf{R})$ *is uniformly continuous on a unit sphere).*

2.3.5. The module of smoothness $\rho_{\mathbf{B}}$ of the space **B** is defined by means of the equality

$$\rho_{\mathbf{B}}(\tau) = \sup\left\{\frac{\|x + \tau y\| + \|x - \tau y\|}{2} - 1 : \|x\| = \|y\| = 1\right\}, \ \tau > 0.$$

A Banach space **B** is called p-smoothable, $1 < p \leq 2$ (correspondingly, uniformly p-smoothable, $1 \leq p < 2$), if **B** permits renorming such that $\sup_{0 < \tau \leq 2} \rho_{\mathbf{B}}(\tau)\tau^{-p} < \infty$ (correspondingly $\lim_{\tau \to 0} \rho_{\mathbf{B}}(\tau)\tau^{-p} = 0$). By the use of the method of Theorem 2.3.4, it is easy to establish the following relation between the notions of smoothness and smoothability of the space **B**.

THEOREM [175]. *A Banach space is p-smoothable* $1 < p \leq 2$ *(respectively, uniformly p-smoothable,* $1 \leq p < 2$*), if and only if it is* $\mathbf{D}^1_{p-1}$*-smooth (respectively,* $\mathbf{C}^1_{p-1}$*-smooth).*

Hence, taking into account the result of [164] we obtain the following result.

THEOREM. *If a Banach space* **B** *is* $\mathbf{C}^1_\alpha$*-smooth, then there exists such a number* $\varepsilon > 0$ *that* **B** *is* $\mathbf{D}^1_{\alpha+\varepsilon}$*-smooth.*

2.3.6. The concept of smoothness of Banach spaces is naturally transferred to operators.

Let $u \in \mathbf{L}(\mathbf{B}, \mathbf{F})$. The function $f : \mathbf{B} \to \mathbf{R}$ has a u-bounded support if the operator u maps the support of the function f onto a bounded set.

The operator $u \in \mathbf{L}(\mathbf{B}, \mathbf{F})$ is called **S**-smoothing if the class $\mathbf{S}(\mathbf{B}, \mathbf{R})$ contains a non-trivial function with a u-bounded support.

It is shown in [175] that most of the properties of **S**-smoothness of Banach spaces are preserved for **S**-smoothing operators. However, in the general case some properties are not found. In particular, there exists a $\mathbf{C}^1_\alpha$-smoothing, $0 \leq \alpha < 1$, operator $u \in \mathbf{L}(\mathbf{l}_{1+\alpha}, \mathbf{l}_{1+\alpha})$, which is not $\mathbf{D}^1_\beta$-smoothing for any $1 > \beta > \alpha$.

It would be interesting to compare the following two results with Theorems 2.30 and 2.20.

THEOREM [175]. *Let* $(\lambda_i)_{i \geq 1} \in \mathbf{c}_0^+$. *If there exists such a number* $\varepsilon > 0$ *that*

$$\sum_{i=1}^{\infty} \exp\left\{-\frac{\varepsilon}{\lambda_i^q}\right\} < \infty$$

for some $q < 1 + 1/\alpha$*, then the inclusion operator* $i_\lambda : \mathbf{c}_{0,\lambda} \to \mathbf{c}_0$ *is* $\mathbf{D}^1_\alpha$*-smoothing.*

Let (S, d) be a metric compact.

THEOREM [175]. *If there exists a continuous with respect to* α *semimetric* d *on* S *such that*

$$\int_0^1 \log^q N_\rho(S, \varepsilon)\, d\varepsilon < \infty$$

for some $q > \alpha/(\alpha + 1)$*, then the inclusion operator* $i_\rho : \mathrm{Lip}_\rho(S) \to \mathbf{C}(S)$ *is* $\mathbf{D}^1_\alpha$*-smoothing.*

2.3.7. When solving the problem of constructing a smooth function 'well approximating' an indicator function of the given subset of a Banach space, as well as a smooth approximation of functions defined on a Banach space, the theory of differentiable measures can be applied (see [7], [53], [65]).

The measure μ on a Banach space is called differentiable [7] if, for each Borel set $A \in \mathbf{B}$, there exists $\lim_{t \to 0} (\mu(A + th) - \mu(A))/t$, denoted by $d_h\mu(A)$ and called as a derivative of the measure μ in the direction $h \in \mathbf{B}$ on the set A. The iterated derivatives $d_{h_1} \cdots d_{h_k}\mu(A)$ in the directions of $h_1, \ldots, h_k$ – **B** are defined by induction.

It can be shown that Theorem 1.23 is, in fact, equivalent to the following statement: the Gaussian measure μ is infinitely differentiable in all directions from the Hilbert space $\mathbf{H} \subset \mathbf{B}$ associated with the Gaussian measure μ.

It occurs that this result can be considerably strengthened: the Gaussian measure μ is an analytical measure, i.e.

$$\mu(A + h) = \mu(A) + d_h\mu(A) + \cdots + \frac{1}{n!}\underbrace{d_h \cdots d_h}_{n \text{ times}} \mu(A) + \ldots,$$

besides, the series converges in variation uniformly over all Borel sets $A \in \mathbf{B}$ and directions h, covering any pre-set bounded subset of the Hilbert space $\mathbf{H} \subset \mathbf{B}$ (see [11] where one can find somewhat more general and more exact results).

CHAPTER 3

The Central Limit Theorem in Banach Spaces

In this chapter we study the weak convergence of the distributions of sums of independent identically distributed random elements of a Banach space to a Gaussian distribution. First we consider type-2 operators whose relation to the CLT is established in Section 3.2. By means of this relation the CLT is proved in the spaces $\mathbf{C}(S)$ and $\mathbf{c}_0$, in particular.

3.1. Operator of Type p

DEFINITION 1.1. Let **B** and **F** be Banach spaces, not necessarily separable, and let the number $p \in [1, 2]$, $\{\varepsilon_i, i \geq 1\}$ be a Rademacher sequence. The operator $u \in \mathbf{L}(\mathbf{B}, \mathbf{F})$ is called an operator of type p if there exists a constant $C = C(u, p)$ such that for any finite set of elements $x_1, ..., x_n$ of the space **B** the following inequality holds:

$$\left(E\left\|\sum_{i=1}^{n} \varepsilon_i u(x_i)\right\|\right) \leq C\left(\sum_{i=1}^{n} \|x_i\|^p\right)^{1/p}. \tag{1.1}$$

DEFINITION 1.2. A Banach space **B** is called a space of type p, $1 \leq p \leq 2$, if the identical operator $I_{\mathbf{B}}$ is an operator of type p.

Statements equivalent to Definition 1.1, are collected in the next theorem. In this chapter the Banach space **F** is not necessarily separable.

THEOREM 1.3. *Let $u \in \mathbf{L}(\mathbf{F}, \mathbf{B})$ and $1 \leq p \leq 2$. The following statements are equivalent:*

(a) *u is an operator of type p;*

(b) *there exists a constant C such that for any independent r.e. $X_i \in \mathbf{L}_p^0(\mathbf{F})$, $i = 1, ..., n$, the inequality takes place:*

$$E\left\|\sum_{i=1}^{n} u(X_i)\right\|^p \leq C\sum_{i=1}^{n} E\|X_i\|^p; \tag{1.2}$$

(c) *for any sequence $\{x_i, i \geq 1\} \subset \mathbf{F}$ such that $\sum_{i=1}^{\infty} \|x_i\|^p < \infty$ the series $\sum_{i=1}^{\infty} \varepsilon_i u(x_i)$ converges a.s. in* **B**.

Proof. The implication (b) $\Rightarrow$ (a) is trivial. We shall prove (a) $\Rightarrow$ (b). Let $X_1, ..., X_n$ be independent $\mathbf{F}$-r.e. defined on a probability space $(\Omega, \mathcal{A}, P)$ and the Rademacher sequence ε_i, $i \geq 1$, defined on $(\Omega', \mathcal{A}', P')$. It may be considered that r.e. X_i and r.v. ε_i, $i = 1, ..., n$, are given on the product of the spaces $(\Omega \times \Omega', \mathcal{A} \times \mathcal{A}', P \times P')$. By E we shall denote the mean with respect to the measure $P \times P'$ and, by E', to the measure P'. Applying Proposition 1.3.11 we have

$$E\left\|\sum_{i=1}^{n} u(X_i)\right\|^p \leq 2^p E\left\|\sum_{i=1}^{n} u(X_i)\varepsilon_i\right\|^p. \tag{1.3}$$

In accordance with (1.1), we can write

$$E' \left\| \sum_{i=1}^{n} \varepsilon_i u(X_i(\omega)) \right\|^p \le C \sum_{i=1}^{n} \|X_i(\omega)\|^p. \tag{1.4}$$

From (1.3) and (1.4), applying the Fubini theorem, we obtain (1.2). We shall prove the implication (a) $\Rightarrow$ (b). If we have (1.2) and $\Sigma_{i=1}^{\infty} \|x_i\|^p < \infty$, then $\Sigma_{i=1}^{n} \varepsilon_i u(x_i)$, $n \ge 1$, converges in $\mathbf{L}_p(\mathbf{B})$. This implies convergence of this sequence in probability and by Theorem 1.3.7 convergence a.s. (c) $\Rightarrow$ (a). Let us assume that we have (c) but (a) is not true, i.e. one can find sequences $a_n \to \infty$, $x_j^{(n)} \in \mathbf{F}$ and a Rademacher sequence $\varepsilon_j^{(n)}$, $j = 1, ..., k_n$, $n \ge 1$, such that $\Sigma_{n=1}^{\infty} a_n^{-1} < \infty$ and for all $n \ge 1$

$$E \left\| \sum_{j=1}^{k_n} \varepsilon_j^{(n)} u(x_j^{(n)}) \right\|^p \ge a_n \sum_{j=1}^{k_n} \|x_j^{(n)}\|^p.$$

Without loss of generality, we can assume that

$$a_n \sum_{j=1}^{k_n} \|x_j^{(n)}\|^p = 1,$$

otherwise it is possible to take the new elements

$$\tilde{x}_j^{(n)} = \left(a_n \sum_{j=1}^{k_n} \|x_j^{(n)}\|^p \right)^{-1/p} x_j^{(n)}.$$

The latter inequality means that the series $\Sigma_{n=1}^{\infty} \Sigma_{j=1}^{k_n} \varepsilon_j^{(n)} u(x_j^{(n)})$ does not converge in $\mathbf{L}_p(\mathbf{B})$. On the other hand,

$$\sum_{n=1}^{\infty} \sum_{j=1}^{k_n} \|x_j^{(n)}\|^p = \sum_{n=1}^{\infty} a_n^{-1} < \infty,$$

and from (b) there follows the convergence

$$\sum_{n=1}^{\infty} \sum_{j=1}^{k_n} \varepsilon_j^{(n)} u(x_j^{(n)})$$

a.s. and by Theorem 1.3.8 also in $\mathbf{L}_p(\mathbf{B})$. The contradiction obtained proves the implication (c) $\Rightarrow$ (a). □

By $\mathbf{U}_p(\mathbf{F}, \mathbf{B})$ we shall denote the class of operators $u \in \mathbf{L}(\mathbf{F}, \mathbf{B})$, which are operators of type p. For $u \in \mathbf{U}_p(\mathbf{F}, \mathbf{B})$ we shall define the number $\|u\|_p$ as the least constant C with which (1.1) holds. The standard reasonings show that

$$\|u\|_p = \sup_{n \ge 1} \sup \left\{ \left(E \left\| \sum_{i=1}^{n} \varepsilon_i u(x_i) \right\|^p \right)^{1/p} : \sum_{i=1}^{n} \|x_i\|^p \le 1 \right\}$$

and that $\mathbf{U}_p(\mathbf{F}, \mathbf{B})$ becomes a Banach space with respect to the norm $\|\cdot\|_p$. It follows from Definitions 1.1 and 1.2 that $\mathbf{U}_1(\mathbf{F}, \mathbf{B}) = \mathbf{L}(\mathbf{F}, \mathbf{B})$ and for $1 < p \le 2$ $\mathbf{U}_p(\mathbf{F}, \mathbf{B}) = \mathbf{L}(\mathbf{F}, \mathbf{B})$, if at least one of the spaces $\mathbf{F}$ and $\mathbf{B}$ is a space of type p.

Later we shall confine ourselves to a consideration of type-2 operators and spaces whose relation to the CLT will be established in Section 3.2.

To check the fact that a given operator is an operator of type-2, the following theorem is useful.

THEOREM 1.4. *Let* $\mathbf{B}_1$, $\mathbf{B}_2$ *be separable Banach spaces and the operator* $u \in \mathbf{L}(\mathbf{B}_1, \mathbf{B}_2)$. *If for each r.e.* $\xi \in \mathbf{L}_2^0(\mathbf{B}_1)$ $\mathbf{B}_2$*-r.e.* $u(\xi)$ *is pre-Gaussian, then* $u \in \mathbf{U}_2(\mathbf{B}_1, \mathbf{B}_2)$.

Proof. Let the sequence $x_i \in \mathbf{B}_1$, $i \geq 1$, be such that $\Sigma_{i=1}^{\infty} \|x_i\|^2 < \infty$. By Theorem 1.3 it is sufficient to show that the series $\Sigma_{i=1}^{\infty} \varepsilon_i y_i$ converges a.s. in $\mathbf{B}_2$ where $y_i = u(x_i)$. Without loss of generality we shall assume that $x_i \neq 0$ for all $i \geq 1$. Let us dfine the measure μ in $\mathbf{B}_1$ by the following equality:

$$\mu = \sum_{i=1}^{\infty} \frac{\alpha_i \left(\delta_{z_i} + \delta_{-z_i}\right)}{2},$$

where $\alpha_i = \|x_i\|^2$, $z_i = x_i \|x_i\|^{-1}$. It is obvious that the random variable corresponding to the measure μ belongs to $\mathbf{L}_2^0(\mathbf{B}_1)$, so the measure $\mu \cdot u^{-1}$ with the covariance

$$T_{\mu \cdot u^{-1}}(f, g) = \sum_{i=1}^{\infty} <f, y_i><g, y_i>, \ f, g \in \mathbf{B}_2^*.$$

is pre-Gaussian and on $\mathbf{B}_2$ there exists a Gaussian measure ν with the same covariance. If we denote

$$S_n = \sum_{i=1}^{n} \gamma_i y_i ,$$

where γ_i, $i \geq 1$, is a standard Gaussian sequence, then for each $f \in \mathbf{B}_2^*$

$$E \exp\{i<f, S_n>\} \underset{n \to \infty}{\to} \int_{\mathbf{B}_2} \exp\{i <f, x>\} \nu(dx)$$

and from Theorem 1.3.7 the series $\Sigma_{i=1}^{\infty} \gamma_i y_i$ converges a.s. By Theorem 1.3.13 we obtain convergence of the series $\Sigma_{i=1}^{\infty} \varepsilon_i y_i$ a.s. □

Next we consider some examples of spaces and operators of type-2.

PROPOSITION 1.5. *Let (S, ν) be a measurable space with the σ-finite measure ν, $r \geq 2$. Then $\mathbf{L}_r(S, \nu)$ is a Banach space of type-2.*

Proof. Let $x_k \in \mathbf{L}_r(S, \nu)$, $k = 1, ..., n$, and let ε_k, $k = 1, ..., n$ be a Rademacher sequence. Having applied inequality (1.3.4), for any $t \in S$ we obtain

$$\left(E \left| \sum_{i=1}^{n} \varepsilon_i x_i(t) \right|^p \right)^{1/p} \leq C \left(E \left| \sum_{i=1}^{n} \varepsilon_i x_i(t) \right|^2 \right)^{1/2} = C \left(\sum_{i=1}^{n} |x_i(t)|^2 \right)^{1/2} .$$

Whence we have

$$E \left| \sum_{i=1}^{n} \varepsilon_i x_i(t) \right|^r \leq C \left(\sum_{i=1}^{n} |x_i(t)|^2 \right)^{r/2} .$$

By integrating this inequality with respect to the measure ν and by the use of the Minkowski inequality, we confirm the validity of the inequality

$$\left(E \left\| \sum_{i=1}^{n} \varepsilon_i x_i \right\|^r \right)^{2/r} \leq C \left(\int_S \left(\sum_{i=1}^{n} |x_i(t)|^2 \right)^{r/2} \nu(dt) \right)^{2/r}$$

$$\leq C \sum_{i=1}^{n} \left(\int_S |x_i(t)|^r \nu(dt) \right)^{2/r} = C \sum_{i=1}^{n} \|x_i\|^2. \tag{1.5}$$

There remains to apply (in the case $r > 2$) the moment inequality, and from (1.5) we obtain (1.1) with $u = I_{\mathbf{L}_r}$ and $p = 2$. □

Other examples of space of type-2 can be found in the supplement. Here we shall present examples of operators $u \in \mathbf{U}_2(\mathbf{F}, \mathbf{B})$; and as $\mathbf{B}$, we shall consider the spaces $\mathbf{c}_0$ and $\mathbf{C}(S)$. The two spaces are of importance both for the probability theory and that of Banach spaces. The space $\mathbf{C}[0, 1]$ is a universal separable Banach space in the sense that any separable Banach space may be isometrically included into $\mathbf{C}[0, 1]$. The random elements with values in $\mathbf{C}(S)$ are random processes with continuous paths. The space $\mathbf{c}_0$ may be defined as 'boundary', in a sense: there exist many assertions that are valid for any Banach space not containing $\mathbf{c}_0$ as a subspace and already not true in $\mathbf{c}_0$. It should be noted that both spaces are not spaces of type p, $1 < p \leq 2$. This may be confirmed by the use of Example 2.2 in Section 3.2. In this example symmetric i.i.d. $\mathbf{c}_0$ r.e. $X_1, \dots, X_n$ with $P(\|X_i\| \leq 1) = 1$ are given for which

$$P(\|X_1 + \cdots + X_n\| \geq n(\ln n)^{-1}) \to 1$$

at $n \to \infty$. Then for any $p > 1$,

$$E\|X_1 + \cdots + X_n\|^p (nE\|X_1\|^p)^{-1} \to \infty.$$

Let us first consider the space $\mathbf{C}(S)$, where (S, d) is a metric compact. Before passing on the formulation of results, two auxiliary lemmas will be presented.

Let $\psi: \mathbf{R}^+ \to \mathbf{R}^+$ be an increasing convex function $\psi(0) = 0$. As usual, $\mathbf{L}_\psi(\Omega, A, P)$ denotes an Orlicz space defined by the function ψ, i.e.

$$\mathbf{L}_\psi(\Omega, A, P) = \left\{X \in \mathbf{L}_0 : \text{there exists } c > 0 : \int_\Omega \psi\left(\frac{|x(\omega)|}{c}\right) P(d\omega) < \infty\right\}.$$

For $X \in \mathbf{L}_\psi(\Omega, A, P)$ we define

$$\|X\|_\psi = \inf\left\{c > 0 : \int_\Omega \psi\left(\frac{|X(\omega)|}{c}\right) P(d\omega) \leq 1\right\}.$$

LEMMA 1.6. *If* $X_i \in \mathbf{L}_\psi(\Omega, A, P)$, $i = 1, \dots, N$, *then*

$$E \sup_{i \leq N} |X_i| \leq \psi^{-1}(N) \sup_{i \leq N} \|X_i\|_\psi$$

Proof. Without loss of generality we may assume, that $\sup_{i \leq N} \|X_i\|_\psi \leq 1$. Consequently,

$$E\psi(X_i) \leq 1, \quad 1 \leq i \leq N.$$

Since ψ is convex and increasing, then

$$\psi(E \sup_{i \leq N} |X_i|) \leq E\psi(\sup_{i \leq N} |X_i|) = E \sup_{i \leq N} \psi(X_i)$$
$$\leq E \sum_{i=1}^N \psi(X_i) \leq N.$$

Therefore, $E \sup_{i \leq N} |X_i| \leq \psi^{-1}(N)$. □

We denote $\psi_2(t) = \exp\{t^2\} - 1$.

LEMMA 1.7. *For an arbitrary* $a_1, \dots, a_n \in \mathbf{R}$ *the following inequality is valid:*

$$\left\| \sum_{i=1}^n \varepsilon_i a_i \right\|_{\psi_2} \leq 2\left(\sum_{i=1}^n a_i^2\right)^{1/2}.$$

Proof. It is sufficient to make use of a well-known inequality for sub-Gaussian r.v. (see [38] or [99]).

$$P\left(\left|\sum_{i=1}^{n} \varepsilon_i a_i\right| > t\right) \le \exp\left\{-t^2\left(2\sum_{i=1}^{n} a_i^2\right)^{-1}\right\} \tag{1.6}$$

and to carry out some simple calculations. □

Let (S, d) be a metric compact, $u \in \mathbf{L}(\mathbf{F}, \mathbf{C}(S))$. By δ_s, $s \in S$, we denote a linear continuous functional defined by the equality $<\delta_s, x> = x(s)$, $x \in \mathbf{C}(S)$. We shall define the semimetric d_u on S by setting $d_u(s, t) = \|u^*(\delta_s) - u^*(\delta_t)\|$, where $u^* \in \mathbf{L}((\mathbf{C}(S))^*, \mathbf{F}^*)$ is a conjugate operator.

Let ρ be a continuous semimetric with respect to d. We recall that $H(S, \rho, x) = \ln N(S, \rho, x)$, where by $N(S, \rho, x)$ we denote a minimum number of balls (of radius x with respect to semimetric ρ) covering S. By $K(\rho, x)$ we denote the set of the centres of these balls.

THEOREM 1.8. *Let there exist a semimetric $\rho \ge d_u$, continuous with respect to d, satisfying the condition*

$$\int_0^a H^{1/2}(S, \rho, x)\, dx < \infty \tag{1.7}$$

for some $a > 0$. Then $u \in \mathbf{U}_2(\mathbf{B}, \mathbf{C}(S))$ and

$$\|u\|_2 \le C \inf_{k \ge 1}\left\{H^{1/2}(S, \rho, 2^{-k}) \sup_{s \in K(\rho, 2^{-k})} \|u^*(\delta_s)\| + \right.$$

$$\left. + \sum_{j \ge k} 2^{-j} H^{1/2}(S, \rho, 2^{-j})\right\}. \tag{1.8}$$

Proof. Fix the number $K \ge 1$. Let $x_1, ..., x_n \in B$, $Z_n := \Sigma_{j=1}^{n} u(x_i)\varepsilon_i$. Later, to simplify the notations, instead of Z_n, $H(S, \rho, 2^{-k})$ and $N(S, \rho, 2^{-k})$ we shall write Z, H_k, N_k, $\alpha_k = 2^{-k}$. For the number $\tau > 0$ we have

$$P\{\|Z\| > \tau\} \le I_1 + I_2, \tag{1.9}$$

where

$$I_1 = P\left\{\sup_{s \in K(\rho, \alpha_k)} |Z(s)| > \frac{\tau}{2}\right\},$$

$$I_2 = P\left\{\sup_{\rho(s,t) \le \alpha_k} |Z(s) - Z(t)| > \frac{\tau}{2}\right\}.$$

From the definition of the number $N(S, \rho, x)$ it follows that there exist points $t_j^k \in S$, $j = 1, 2, \cdots, N_k$, $k = 1, 2, \cdots$, such that

$$S = \bigcup_{j=1}^{N_k} V_{\alpha_k}(t_j^k).$$

Therefore, it is possible to choose a partition $\{A_j^k, j = 1, 2, ..., N_k\}$ of the set S such that $A_j^k \subset V_{\alpha_k}(t_j^k)$ and the diameter of the set A_j^k would not exceed $2\alpha_k$.

Let us define

$$Z^k(t) = \sum_{i=1}^{N_k} \chi_{A_i^k}(t) Z(t_i^k).$$

It is easy to check that

$$Z(t) = Z^K(t) + \sum_{k\geq K+1} (Z^k(t) - Z^{k-1}(t));$$

also

$$\sup_{s,t:\rho(s,t)<\alpha_k} |Z(s) - Z(t)| \leq 2 \sum_{k\geq K+1} \sup_t |Z^k(t) - Z^{k-1}(t)| \leq$$

$$\leq 2 \sum_{k\geq K+1} \sup_{(i,j)\in\Lambda_k} |Z^k(t_j^k) - Z^k(t_i^k)|, \tag{1.10}$$

where $\Lambda_k \subset \{1, 2, ..., N_k\} \times \{1, 2, ..., N_{k-1}\}$ contains the pairs (i, j) such that $A_j^k \cap A_i^{k-1} \neq \emptyset$. From (1.10) and Lemma 1.6 we obtain

$$I_2 \leq 2r^{-1}E \sup_{\rho(s,t)\leq\alpha_k} |Z(s) - Z(t)| \leq$$

$$\leq 4r^{-1} \sum_{k\geq K+1} \psi_2^{-1}(N_k^2) \sup_{(i,j)\in A_k} \|Z^k(t_j^k) - Z^k(t_i^k)\|_{\psi_2}.$$

By Lemma 1.7, we have

$$\|Z^k(t_j^k) - Z^k(t_i^k)\|_{\psi_2} \leq 8\alpha_k \left(\sum_{i=1}^n \|x_i\|^2\right)^{1/2}$$

Finally, taking into account that $\psi_2^{-1}(N_k^2) \leq 2H_k^{1/2}$, we obtain the estimate

$$I_2 \leq 32r^{-1} \sum_{k\geq K+1} \alpha_k H_k^{1/2} \left(\sum_{i=1}^n \|x_i\|^2\right)^{1/2}. \tag{1.11}$$

To estimate the quantity I_1 we again make use of Lemmas 1.6 and 1.7:

$$I_1 \leq 2r^{-1}E \sup_{s\in K(\rho,\alpha_k)} |Z(s)| \leq 2r^{-1}\psi_2^{-1}(N_k) \sup_{s\in K(\rho,\alpha_k)} \|Z(s)\|_{\psi_2} \leq$$

$$\leq 4r^{-1}H_K^{1/2} \sup_{s\in K(\rho,\alpha_k)} \|u^*(\delta_s)\| \left(\sum_{i=1}^n \|x_i\|^2\right)^{1/2}. \tag{1.12}$$

It is easy to obtain (1.8) from (1.9), (1.11), (1.12), the definition of the norm $\|u\|_2$ and Theorem 1.3.10. □

COROLLARY 1.9. *Let the semimetric ρ, continuous with respect to d, satisfy condition* (1.7). *Then the inclusion operators*

$$i_1 : \mathrm{Lip}_\rho(S) \to \mathrm{C}(S), \; i_2 : \mathrm{lip}_\rho(S) \to \mathrm{C}(S)$$

are operators of type-2.

We shall now consider the case of the space $\mathbf{c}_0$. By Λ_0 (Λ_1, respectively) we denote the class of sequences $\lambda = (\lambda_1, ..., \lambda_n, ...)$ such that for some $\varepsilon > 0$ (for each $\varepsilon > 0$, respectively)

$$\sum_{i=1}^{\infty} \exp\left\{-\frac{\varepsilon}{\lambda_1^2}\right\} < \infty. \tag{1.13}$$

Here and later, in order not to write the condition $\lambda_i \neq 0$ under the sign of the sum, the convention $\exp\{-\varepsilon\lambda_i^{-1}\} = 0$ is accepted if $\lambda_i = 0$.

First, we shall prove two propositions on the classes introduced.

PROPOSITION 1.10. *If $\lambda \in \Lambda_1$, then there exists $a \in \mathbf{c}_0^+$ such that*

$$\frac{\lambda}{a} := \left(\frac{\lambda_1}{a_1}, \ldots, \frac{\lambda_n}{a_n}, \cdots\right) \in \Lambda_1.$$

Proof. Let us choose the sequence of number $b_k > 0, k = 0, 1, 2, \cdots$, such that

$$\sum_{k=1}^{\infty} b_k < \infty, \tag{1.14}$$

and then we shall construct the sequence n_k, $k \geq 1$, such that the following inequalities are valid

$$\sum_{i=n_k+1}^{\infty} \exp\{-b_{k-1}\lambda_i^{-2}\} < b_k, \; k = 1, 2, \cdots \tag{1.15}$$

Suppose $a_i^2 = b_{k-1}^{1/2}$ for

$i \in I_k := \{n_k + 1, n_k + 2, ..., n_{k+1}\}$, $(I_0 = \{1, 2, ..., n_1\})$.

It is obvious that $a \in \mathbf{c}_0^+$. By the use of (1.14) and (1.15) we obtain for each $\varepsilon > 0$:

$$\begin{aligned}\sum_{i=1}^{\infty} \exp\left\{-\frac{\varepsilon a_i^2}{\lambda_i^2}\right\} &= \sum_{k=0}^{\infty} \sum_{i \in I_k} \exp\left\{-\frac{\varepsilon a_i^2}{\lambda_i^2}\right\} \\ &= \sum_{k:\varepsilon > \sqrt{b_{k-1}}} \sum_{i \in I_k} \exp\left\{-\frac{b_{k-1}}{\lambda_i^2}\right\} + \sum_{k:\varepsilon \leq \sqrt{b_{k-1}}} \sum_{i \in I_k} \exp\left\{-\frac{\varepsilon (b_{k-1})^{1/2}}{\lambda_i^2}\right\} \\ &\leq \sum_{k:\varepsilon^2 > b_{k-1}} \sum_{i \in I_k} \exp\left\{-\frac{b_{k-1}}{\lambda_i^2}\right\} + \sum_{k:\varepsilon^2 \leq b_{k-1}} \sum_{i \in I_k} \exp\left\{-\frac{\varepsilon (b_{k-1})^{1/2}}{\lambda_i^2}\right\} \\ &\leq \sum_{k=1}^{\infty} b_k + \sum_{k:\varepsilon^2 \leq b_{k-1}} \sum_{i \in I_k} \exp\left\{-\frac{\varepsilon (b_{k-1})^{1/2}}{\lambda_i^2}\right\} < \infty,\end{aligned} \tag{1.16}$$

since there will be but a finite number of terms in the second sum. □

PROPOSITION 1.11. *If for all* $\varepsilon > 0$

$$\sum_{i=1}^{\infty} \exp\{-\varepsilon \lambda_i^{-2}\} = \infty, \tag{1.17}$$

then there exists such a sequence $a_n > 0$, $a_n \to \infty$, *that for all* $\varepsilon > 0$

$$\sum_{i=1}^{\infty} \exp\{-\varepsilon a_i^2 \lambda_i^{-2}\} = \infty. \tag{1.18}$$

Proof. Choosing a sequence of numbers $b_k \to 0$ such that $\Sigma_{k=1}^{\infty} b_k = \infty$, we construct a sequence n_k, $k \geq 0$ $(n_0 = 0)$, such that the following inequalities are valid

$$\sum_{i=n_k+1}^{n_{k+1}} \exp\{-\lambda_i^{-2} b_{k+1}^{-1}\} > b_{k+1}.$$

Then we easily find that the sequence $a_i^2 = (b_{k+1})^{-1/2}$ for $i \in I_k$ can be taken as the one sought for. □

By $\{e_k, k \geq 1\}$, as usual, we shall denote the standard basis in the space $\mathbf{l}_1 = \mathbf{c}_0^*$ and let $\lambda_i := \|u^*(e_i)\|$.

THEOREM 1.12. *If the sequence* $\lambda = (\lambda_1, ..., \lambda_n, ...) \in \Lambda_0$, *then* $u \in \mathbf{U}_2(\mathbf{B}, \mathbf{c}_0)$ *and*

$$\|u\|_2 \leq C \inf_{\delta > \varepsilon_0} \left(\delta + \delta \sum_{k:\lambda_k^2 \leq \delta} \exp\left\{ -\frac{\delta}{\lambda_k^2} \right\} + \sum_{k:\lambda_k^2 > \delta} \lambda_k^2 \right)^{1/2}, \tag{1.19}$$

where

$$\varepsilon_0 = \inf\left\{ \varepsilon : \sum_{i=1}^{\infty} \exp\{-\varepsilon \lambda_i^{-2}\} < \infty \right\}.$$

Proof. Let the numbers $\delta \geq \varepsilon_0$, $t > 0$. We denote $A = (\Sigma_{i=1}^n \|x_i\|^2)^{1/2}$, where $x_1, ..., x_n \in \mathbf{B}$. First consider the case $A < t(2\delta)^{-1/2}$. We have

$$P\left\{ \left\| \sum_{i=1}^n u(x_i)\varepsilon_i \right\| > t \right\} \leq I_1 + I_2,$$

where

$$I_1 = P\left\{ \sup_{k:\lambda_k^2 \leq \delta} \left| \sum_{i=1}^n \langle x_i, u^*(e_k) \rangle \varepsilon_i \right| > t \right\},$$

$$I_2 = P\left\{ \sup_{k:\lambda_k^2 > \delta} \left| \sum_{i=1}^n \langle x_i, u^*(e_k) \rangle \varepsilon_i \right| > t \right\}.$$

By the use of (1.6) we obtain

$$P\left\{ \left| \sum_{i=1}^n \langle x_i, u^*(e_k) \rangle \varepsilon_i \right| > t \right\} \leq \exp\{-t^2(2A\lambda_k^2)^{-1}\}. \tag{1.20}$$

It should be noted that the function $s^{-2}\exp\{-s^{-2}\}$ increases on the interval (0, 1], therefore for $s \in (0, 1]$, $a \in (0, 1]$ we have

$$\exp\{-(s \cdot a)^{-2}\} \leq a^2 \exp\{-s^{-2}\}.$$

Making use of this inequality with $s = (\delta)^{1/2}\lambda_k^{-1}$ and $a = A(2\delta)^{1/2}t^{-1}$, as well as estimate (1.20) we have

$$I_1 \leq \sum_{k:\lambda_k^2 \leq \delta} P\left\{ \left| \sum_{i=1}^n \langle x_i, u^*(e_k) \rangle \varepsilon_i \right| > t \right\} \leq$$

$$\leq 2\delta t^{-2} A^2 \sum_{k:\lambda_k^2 \leq \delta} \exp\{-\delta \lambda_k^{-2}\}.$$

To estimate the quantity I_2, it is sufficient to apply the Chebyshev inequality. As a result, we shall obtain

$$I_2 \leq A^2 t^{-2} \sum_{k:\lambda_k^2 > \delta} \lambda_k^2.$$

Thus, in the case under consideration

$$P\left\{ \left\| \sum_{i=1}^n u(x_i)\varepsilon_i \right\| > t \right\} \leq \left(2\delta \sum_{k:\lambda_k^2 \leq \delta} \exp\left\{ -\frac{\delta}{\lambda_k^2} \right\} + \sum_{k:\lambda_k^2 > \delta} \lambda_k^2 \right) A^2 t^{-2}. \tag{1.21}$$

If $A > t(2\delta)^{-1/2}$, then

$$P\left\{ \left\| \sum_{i=1}^n u(x_i)\varepsilon_i \right\| > t \right\} \leq 2\delta t^{-2} A^2. \tag{1.22}$$

From (1.21) and (1.22) for any $\delta \geq \varepsilon_0$ we have

$$E\left\|\sum_{i=1}^{n} u(x_i)\varepsilon_i\right\| \leq \sup_{t\geq 0} t\left(P\left\{\left\|\sum_{i=1}^{n} u(x_i)\varepsilon_i\right\| > t\right\}\right)^{1/2} \leq$$

$$\leq \left(2\delta + 2\delta \sum_{k:\lambda_k^2 \leq \delta} \exp\{-\delta\lambda_k^{-2}\} + \sum_{k:\lambda_k^2 > \delta} \lambda_k^2\right)^{1/2} A. \square$$

COROLLARY 1.13. *If* $\lambda \in \Lambda_0$, *then the inclusion operators* $j_1 : \mathbf{c}_{0,\lambda} \to \mathbf{c}_0$ *and* $j_2 : \mathbf{l}_{\infty,\lambda} \to \mathbf{c}_0$ *are operators of type-2.*

The proof of the corollary follows from the easily checked equalities $\|j_i^*(e_k)\| = \lambda_k$, $i = 1, 2$. □

It is interesting to note that the assertion of Corollary 1.13 may be proved in another, rather simple, way by the use of Theorem 1.5. We present this proof in the case of the operator j_1.

Thus, let $\lambda \in \Lambda_0$, and $\xi = (\xi_1, \xi_2, \dots, \xi_n, \dots) \in \mathbf{L}_2^0(\mathbf{c}_{0,\lambda})$, i.e. $|\xi_n\lambda_n^{-1}| \underset{n\to\infty}{\to} 0$ a.s., and $E\sup_{n\geq 1} |\xi_n\lambda_n^{-1}|^2 < \infty$. By Lebesgue's dominated convergence theorem, it follows that $E|\xi_n\lambda_n^{-1}|^2 \to 0$, which means that $\sigma_n^2\lambda_n^{-2} \to 0$, where $\sigma_n = E\xi_n^2$. But then the sequence $\sigma = (\sigma_1, \dots, \sigma_n, \dots) \in \Lambda_1$, and the application of Theorem 1.2.5 enables us to conclude that $\operatorname{cov}\xi \in \mathcal{R}_1(\mathbf{c}_0)$. It follows from Theorem 1.5 that the operator j_1 is an operator of type-2.

The chapter will be concluded by the assertion that, in a particular case of the diagonal operator $u \in \mathbf{L}(\mathbf{c}_0, \mathbf{c}_0)$, the condition of Theorem 1.12 is also necessary.

PROPOSITION 1.14. *Let*

$$\lambda \in \mathbf{c}_0^+,\ u_\lambda : \mathbf{c}_0 \to \mathbf{c}_0,\ u_\lambda(x) = (\lambda_1 x_1, \dots, \lambda_n x_n, \dots).$$

The operator $u_\lambda \in \mathbf{U}_2(\mathbf{c}_0, \mathbf{c}_0)$ *if and only if* $\lambda \in \Lambda_0$.

Proof. Sufficiency obviously follows from Theorem 1.12 since $\|u_\lambda^*(e_k)\| = \lambda_k$. Now we assume that $\lambda \in \Lambda_0$, i.e. for all $\varepsilon > 0$, (1.17) holds. Then by Proposition 1.11 there exists a sequence $a_n \to \infty$ for $n \to \infty$ such that for all $\varepsilon > 0$, (1.18) is satisfied. Suppose $\xi = \sum_{n=1}^{\infty} a_n^{-1}\varepsilon_n e_n$, where $\{e_n, n \geq 1\}$ is a standard basis in $\mathbf{c}_0$. It is evident that $u_\lambda(\xi) \in \mathbf{L}_0(\mathbf{c}_0)$, but $u_\lambda(\xi)$ is not a pre-Gaussian r.e. in $\mathbf{c}_0$. According to Theorem 1.4, $u_\lambda \overline{\in} \mathbf{U}_2(\mathbf{c}_0, \mathbf{c}_0)$.

3.2. The Central Limit Theorem

Let $\xi, \xi_1, \dots, \xi_n, \cdots$ be i.i.d. $\mathbf{B}$-r.e., $E\xi = 0$. Let us assume that for all $f \in \mathbf{B}^*$

$$Ef^2(\xi) < \infty \tag{2.1}$$

and by T we denote the covariance operator of r.e. ξ. We also denote

$$S_n(\xi) = n^{-1/2} \sum_{i=1}^{n} \xi_i,\ \mu_n = L(S_n(\xi)).$$

DEFINITION 2.1. We say that $\mathbf{B}$-r.e. ξ satisfies the central limit theorem (and write $\xi \in \mathrm{CLT}(\mathbf{B})$ or simply $\xi \in \mathrm{CLT}$ if it is clear which space is meant) if μ_n is a weakly convergent sequence.

If by μ we denote a limit measure for the sequence μ_n, then by condition (2.1) and the CLT in a finite-dimensional space, it follows that μ is a Gaussian measure with zero mean and covariance operator T. Hence we may conclude that

$$T \in \mathscr{R}_1(\mathbf{B}) \tag{2.2}$$

is necessary for $\xi \in$ CLT.

We shall draw the reader's attention to the fact that norming under consideration is by means of the sequence $n^{-1/2}$. In this case, as is known, in a finite-dimensional Banach space the condition

$$E\,\|\xi\|^2 < \infty \tag{2.3}$$

is necessary and sufficient for the CLT to hold. Besides, in this case all three conditions ((2.1) – (2.3)) coincide. The situation becomes much more complicated in infinite-dimensional Banach spaces. First, condition (2.3) in a general case is neither sufficient nor necessary. Moreover, even the boundedness of $\mathbf{B}$ – r.e. ξ, a much stronger condition, not allways ensures the validity of the CLT. The above is confirmed by the examples below (see also Supplements 3.3.4).

EXAMPLE 2.2. Let $\xi = (a_1\varsigma_1, \ldots, a_n\varsigma_n, \ldots)$, where ς_i are independent r.v. defined on some probability space $(\Omega, \mathscr{A}, P)$,

$$P\{\varsigma_i = 1\} = P\{\varsigma_i = -1\} = p_i,$$

$$P\{\varsigma_i = 0\} = 1 - 2p_i, \ (a_i) \in \mathbf{c}_0, \ a_i \leq 1.$$

Then $P\{\xi \in \mathbf{c}_0\} = 1$, $P\{\|\xi\| \leq 1\} = 1$, but if we choose a_i and p_i such that $\sigma_i^2 = E\,(a_i\varsigma_i)^2 = 2a_i^2 p_i = (\ln i)^{-1}$, $i \geq 3$, then condition (1.2.2) is not satisfied and this, together with independence of r.v. ς_i, enables us to assert that $T \overline{\in} \mathscr{R}_1(\mathbf{c}_0)$.

EXAMPLE 2.3. Now, in the above example let us set

$$p_n = (\ln(n+7))^{-1}, \ a_n = (\ln\ln\ln(n+15))^{-1}, \ n \geq 1.$$

Since $\sigma_n^2 = 2a_n^2 p_n$, it is easily seen that $P\{\|\xi\| \leq 1\} = 1$, $\operatorname{cov}\xi \in \mathscr{R}_1(\mathbf{c}_0)$. Thus, condition (2.2) is satisfied. Nevertheless, it will be shown that $\xi \overline{\in}$ CLT. Let $\xi_i = (a_1\varsigma_{i1}, \ldots, a_k\varsigma_{ik}, \ldots)$ be independent copies of r.e. ξ, $S_n := n^{-1/2}\Sigma_{i=1}^n \xi_i$. Consider the events $E_{nk} = {}_{i=1}^{n}\{\varsigma_{ik} = 1\}$, $A_n = {}_{k \leq N_n} E_{nk}$, where $N_n = \exp\{\exp\{n\}\}$. Since $P\{E_{nk}\} = (\ln(k+7))^{-n}$, then

$$P\{A_n\} = 1 - P\left\{\bigcap_{k \leq N_n} E_{nk}^c\right\} = 1 - \prod_{k \leq N_n}(1 - (\ln(k+7))^{-n}) \leq 1 - (1 - (\ln(N_n + 7))^{-n})^{N_n}.$$

Simple computations show that

$$P\{A_n\} \to 1 \text{ as } n \to \infty. \tag{2.4}$$

But for all $\omega \in A_n$ we have

$$\|S_n(\omega)\| = n^{-1/2}\sup_{k \geq 1}\left|\sum_{i=1}^{n} a_k\varsigma_{ik}\right| \geq n^{-1/2}\sup_{k \leq N_n}\left|\sum_{i=1}^{n} a_k\varsigma_{ik}\right|$$

$$> n^{-1/2} n a_{N_n} = n^{1/2}(\ln\ln\ln(\exp\{\exp\{n\}\} + 15))^{-1} \to \infty,$$

as $n \to \infty$. It follows from relation (2.4) that $\xi \overline{\in}$ CLT since for $\mathbf{B}$ – r.e. ξ, satisfying the CLT, for any sequence $b_n \to \infty$, the relation $P\{\|S_n(\xi)\| > b_n\} \to 0$ as $n \to \infty$ holds. □

Such specific properties arising from the study of the CLT in infinite-dimensional spaces are connected with geometrical properties of the spaces. Therefore the following questions are quite natural:

1. *In what spaces does the condition* $E\|\xi\|^2 < \infty$ *imply* $\xi \in \mathrm{CLT}$?
2. *In what spaces does the pre-Gaussian property of r.e.* ξ *imply* $\xi \in \mathrm{CLT}$?

The answers to the above questions are known: they are spaces of type-2 and cotype-2 (see Definition 2.6), respectively. The first is a consequence of a more general result showing the connection between the CLT and operators of type-2.

THEOREM 2.4. *Let* $\mathbf{B}_1$, $\mathbf{B}_2$ *be separable Banach spaces. The following assertions are equivalent:*

(a) *the operator* $u \in \mathbf{U}_2(\mathbf{B}_1, \mathbf{B}_2)$;
(b) *for each r.e.* $\xi \in \mathbf{L}_2^0(\mathbf{B}_1)$, *r.e.* $u(\xi)$ *satisfies the CLT;*
(c) *for each r.e.* $\xi \in \mathbf{L}_2^0(\mathbf{B}_1)$, *r.e.* $u(\xi)$ *is pre-Gaussian.*

Proof. (a) $\Rightarrow$ (b) Let $\xi \in \mathbf{L}_2^0(\mathbf{B}_1)$ and $u \in \mathbf{U}_2(\mathbf{B}_1, \mathbf{B}_2)$. For a given $\varepsilon > 0$ we can construct $\mathbf{B}_1$-r.e. X with zero mean taking only a finite number of values such that

$$E\|\xi - X\|^2 \le \varepsilon^2. \tag{2.5}$$

By u_n and ν_n we denote the distribution of $\mathbf{B}_2$-r.e. $S_n(u(\xi))$ and $S_n(u(X))$, respectively. Applying inequalities (2.5) and (1.2) we obtain

$$\begin{aligned}\rho_{\mathrm{BL}}(\mu_n, \nu_n) &= \sup\{\,|Ef(S_n(u(\xi))) - Ef(S_n(u(X)))|\, : \|f\|_{\mathrm{BL}} \le 1\}\\ &\le E\|S_n(u(\xi)) - S_n(u(X))\|\\ &\le C(E\|\xi - X\|^2)^{1/2} \le C\varepsilon\end{aligned}$$

with some constant C independent of n. Hence, it follows that

$$\begin{aligned}\rho_{\mathrm{BL}}(\mu_n, \mu_m) &\le \rho_{\mathrm{BL}}(\mu_m, \nu_m) + \rho_{\mathrm{BL}}(\nu_n, \nu_m) + \rho_{\mathrm{BL}}(\mu_m, \nu_n)\\ &\le 2C\varepsilon + \rho_{\mathrm{BL}}(\nu_n, \nu_m)\end{aligned} \tag{2.6}$$

for all $n, m \ge 1$. Note, that $\mathbf{B}_2$-r.e. $u(X)$ can be naturally considered as r.e. in a finite-dimensional space. Therefore, the sequence $(\nu_n, n \ge 1)$ is fundamental in the metric space $(\mathcal{P}(\mathbf{B}_2), \rho_{\mathrm{BL}})$. It follows from (2.6) that the sequence $(\mu_n, n \ge 1)$ is also fundamental. This, together with the completeness of the space $(\mathcal{P}(\mathbf{B}_2), \rho_{\mathrm{BL}})$, implies weak convergence of μ_n.

The implication (b) $\Rightarrow$ (c) is obvious, and (c) $\Rightarrow$ (a) is already proved by Theorem 1.4.
□

Assuming $\mathbf{B}_1 = \mathbf{B}_2 = \mathbf{B}$ and $u = I_{\mathbf{B}}$, we obtain the following result.

THEOREM 2.5. *For each* $\mathbf{B}$*-r.e.* ξ *with a zero mean, the condition* $E\|\xi\|^2 < \infty$ *implies* $\xi \in \mathrm{CLT}(\mathbf{B})$ *if and only if* $\mathbf{B}$ *is a space of type-2.*

The spaces of cotype q, $q \ge 2$ are defined as follows.

DEFINITION 2.6. A Banach space $\mathbf{B}$ is a space of cotype q, $q \ge 2$, if there exists a constant $C = C(q, \mathbf{B}) > 0$ such that for any finite set $x_1, \ldots, x_n$ of the elements of the space $\mathbf{B}$, the following inequality is valid:

$$E \left\| \sum_{i=1}^{n} \gamma_i x_i \right\| \geq C \left(\sum_{i=1}^{n} \|x_i\|^q \right)^{1/q}, \tag{2.7}$$

where $\{\gamma_i, i \geq 1\}$ is a sequence of independent identically distributed standard ($E\gamma_1 = 0$, $E\gamma_1^2 = 1$) Gaussian random variables.

It is easy to check that spaces $\mathbf{L}_r$ are spaces of cotype-2 if $1 \leq r \leq 2$, and of cotype r if $r > 2$. Other examples can be found in the supplements. It should be noted that the sequence $\{\gamma_i, i \geq 1\}$ in Definition 2.6 can be substituted by the Rademacher sequence $\{\varepsilon_i, i \geq 1\}$ (see [220]).

One of the specific features of the geometrical properties of cotype-2 spaces is that the pre-Gaussian r.e. have a special structure. This is shown by the following result.

THEOREM 2.7. *If* $\mathbf{B}$ *is a space of cotype-2 and* $\mathbf{B}$*-r.e.* ξ *is pre-Gaussian, then there exists an operator* $u \in \mathbf{L}(\mathbf{l}_2, \mathbf{B})$ *and* $\mathbf{l}_2$*-r.e.* X *such that* $E\|X\|^2 < \infty$ *and* $L(\xi) = L(u(X))$.

The proof can be found in [49] or [89].

It follows from this result, in particular, that in spaces of cotype-2, the pre-Gaussian r.e. ξ satisfies the CLT (the space $\mathbf{l}_2$ is a space of type-2). As in the case of spaces of type-2, the reverse relation also takes place.

THEOREM 2.8. *The following assertions are equivalent:*

(a) *the space* $\mathbf{B}$ *is a space of cotype-2;*

(b) *every pre-Gaussian* $\mathbf{B}$*-r.e.* ξ *satisfies the CLT;*

(c) *for each pre-Gaussian* $\mathbf{B}$*-r.e.* ξ $E\|\xi\| < \infty$.

Before proving the theorem, we shall present the following result.

PROPOSITION 2.9. *If* $\xi \in \mathrm{CLT}(\mathbf{B})$, *then*

$$\sup_{t>0} t^2 P\{\|\xi\| > t\} < \infty. \tag{2.8}$$

Proof. First we assume that ξ is a symmetric $\mathbf{B}$-r.e. Let μ be a Gaussian limit distribution for the sequence $\mu_n = L(S_n(\xi))$, $n \geq 1$. By Theorem 1.3.2, for each $a > 0$ we have

$$\overline{\lim_{n\to\infty}}\, P\{\|S_n(\xi)\| \geq a\} \leq \mu\{x : \|x\| \geq a\}.$$

Hence we conclude that for each $\varepsilon > 0$ there will be the number $a > 0$ such that $P\{\|S_n(\xi)\| \geq a\} \leq \varepsilon$ for each $n \geq 1$. Consequently, by the Levy inequality (Theorem 1.3.9)

$$P\left\{\max_{1\leq i\leq n} \|\xi_i\| \geq 2an^{1/2}\right\} \leq 2\varepsilon$$

or

$$P\{\|\xi\| \geq 2an^{1/2}\} \leq 1 - (1 - 2\varepsilon)^{1/n} \leq C(\varepsilon)n^{-1} \tag{2.9}$$

with some constant $C(\varepsilon) > 0$. From (2.9) there follows (2.8). In the case of non-symmetric r.e. $\xi \in \mathrm{CLT}(\mathbf{B})$, it is sufficient to note that $\tilde{\xi} \in \mathrm{CLT}(\mathbf{B})$, where $\tilde{\xi}$ is symmetrization of r.e. ξ, and to use the inequality

$$P\{\|\xi\| > t + d\} \leq P\{\|\tilde{\xi}\| > t\} P^{-1}\{\|\xi\| \leq d\},$$

which is valid for all $t, d > 0$. The number d can be chosen such that $P\{\|\xi\| \leq d\} \geq 1/2$. □

REMARK 2.10 It is known that for $\xi \in \mathrm{CLT}$ the following condition, which is stronger than (2.8), is necessary (see [3]):

$$\lim_{t\to\infty} t^2 P\{\|\xi\| > t\} = 0.$$

Proof of Theorem 2.8. As mentioned above, the implication (a) $\Rightarrow$ (b) is the corollary of Theorem 2.7. From Proposition 2.9, the implication (b) $\Rightarrow$ (c) is obtained. There remains to show that (c) => (a). Let $\{x_j, j \geq 1\}$ be such a sequence of the elements of space **B** that the series $Y := \Sigma_{i=1}^{\infty}\gamma_i x_i$ converges a.s. where $\{\gamma_i, i \geq 1\}$ is a sequence of i.i.d. standard Gaussian r.v. Consider a sequence $\{\alpha_i, i \geq 1\}$ such that $\Sigma_{i=1}^{\infty}\alpha_i^2 = 1$, and define the $\mathbf{B}$-r.e. ξ as follows:

$$P\{\xi = x_j\alpha_j^{-1}\} = P\{\xi = -x_j\alpha_j^{-1}\} = \frac{\alpha_j^2}{2}.$$

Immediate computations show that $\operatorname{cov}\xi = \operatorname{cov} Y$, i.e. ξ is a pre-Gaussian $\mathbf{B}$-r.e. Therefore, $E\|\xi\| = \Sigma_{j=1}^{\infty}\alpha_j\|x_j\| < \infty$. Hence $\Sigma_{j=1}^{\infty}\|x_j\|^2 < \infty$. Thus, the convergence of the series $\Sigma_{j=1}^{\infty}\gamma_j x_j$ implies the convergence of the series $\Sigma_{j=1}^{\infty}\|x_j\|^2$. Inequality (2.7) is easily obtained by the use of the closed graph theorem for the natural inclusion map of the Banach space

$$\mathbf{G}_2(\mathbf{B}) := \left\{x = (x_j, j \geq 1) : x_j \in \mathbf{B}, \sum_{j=1}^{\infty}\gamma_j x_j < \infty \text{ a.s. } \|x\| := E\|\sum_{j=1}^{\infty}\gamma_j x_j\|\right\}$$

into the Banach space

$$\mathbf{l}_2(\mathbf{B}) := \left\{x = (x_j, j \geq 1) : x_j \in \mathbf{B}, \|x\|^2 := \sum_{j=1}^{\infty}\|x_j\|^2 < \infty\right\}. \square$$

As mentioned above, the pre-Gaussian property is a necessary condition for the CLT to be valid in any Banach space. As far as the necessity of the condition $E\|\xi\|^2 < \infty$, the following theorem holds.

THEOREM 2.11. *A Banach space* **B** *is a space of cotype-2 if and only if for each* $\mathbf{B}$-*r.e.* ξ *the condition* $\xi \in \mathrm{CLT}(\mathbf{B})$ *implies* $E\|\xi\|^2 < \infty$.

Proof. If **B** is a space of cotype-2, then the necessity of the condition $E\|\xi\|^2 < \infty$ follows from Theorem 2.7. For the converse statement it is sufficient to make use of the implication (b) $\Rightarrow$ (a) of Theorem 2.8. □

We shall now consider the CLT in the spaces $\mathbf{c}_0$ and $\mathbf{C}(S)$, where (S, d) is a metric compact. We shall make use of Theorem 2.5 as well as the results (and the notations) of the preceding section. Here it is relevant to stress that, in Theorem 2.5, separability of the space $\mathbf{B}_1$ is an essential requirement. Thus, for instance, if $\lambda \in \Lambda_0$, then it follows from Theorem 1.12 that $u_\lambda \in \mathbf{U}_2(\mathbf{l}_\infty, \mathbf{c}_0)$, where $u_\lambda(x) = (\lambda_1 x_1, \ldots, \lambda_n x_n, \ldots)$. If we assume $\xi = \Sigma_{i=1}^{\infty}\varepsilon_i e_i$, $\lambda_i = (\ln(i+2))^{-1/2}$, $i \geq 1$, then $\xi \in \mathbf{L}_2^0(\mathbf{l}_\infty)$, but $\operatorname{cov} u_\lambda(\xi) \overline{\in} \mathcal{R}_1(\mathbf{c}_0)$. This shows that without an assumption on the separability of the space $\mathbf{B}_1$ Theorem 2.5 is no longer valid.

Recalling that the space $\mathrm{lip}_\rho(S)$ is separable, from Corollary 1.9 and Theorem 2.5 we immediately have the following result.

THEOREM 2.12. *Let* ξ *be* $\mathbf{C}(S)$-*r.e.,* $E\xi = 0$. *If there exists a continuous with respect to* d *semimetric* ρ *satisfying condition (1.7) such that*

$$P\{\xi \in \text{lip}_\rho(S)\} = 1, \tag{2.10}$$

$$E\|\xi\|_\rho^2 < \infty, \tag{2.11}$$

then $\xi \in \text{CLT}(\mathbf{C}(S))$.

Condition (2.10) is relatively difficult to check. Condition (2.11) which states that there exists a r.e. ς such that for all $s, t \in S$ and for all $\omega \in \Omega$.

$$|\xi(s, \omega) - \xi(t, \omega)| \leq \varsigma(\omega)\rho(s, t), \ E\varsigma^2 < \infty, \tag{2.12}$$

only ensures $P\{\xi \in \text{Lip}_\rho(S)\} = 1$. But it is impossible to apply Corollary 1.9 and Theorem 2.4 with the inclusion i_1 because of the non-separability of the space $\text{Lip}_\rho(S)$. The following lemma enables us to relinquish condition (2.10).

LEMMA 2.13. *Let ρ be a semimetric continuous with respect to d such that condition* 1.7 *is satisfied. Then there exists a semimetric $\rho_1 \geq \rho$ continuous with respect to d such that*

$$\rho(s, t)\rho_1^{-1}(s, t) \to 0 \ \text{as} \ \rho(s, t) \to 0 \tag{2.13}$$

and

$$\int_0^a H^{1/2}(S, \rho_1, x)dx < \infty. \tag{2.14}$$

Besides, $(\text{Lip}_\rho(S), \|\cdot\|_{\rho_1})$ *is a separable Banach space.*

Proof. Let $b = \sup_{s,t \in S}\rho(s, t)$. If (1.7) is satisfied, then one can choose a function ψ, which is monotonically decreasing on the interval $(0, b]$, and has the properties $\psi(x) \geq 1$, $\psi(x) \uparrow \infty$ as $x \downarrow 0$, $\int_0^b \psi(x)\,dx < \infty$ and

$$\int_0^a H^{1/2}(S, \rho, x)\psi(x)\,dx < \infty. \tag{2.15}$$

Now assume that $\phi(t) = \int_0^t \psi(u)\,du$ and $\rho_1(s, t) = \phi(\rho(s, t))$. Since $H(S, \rho_1, x) = H(S, \rho, \phi^{-1}(x))$, where ϕ^{-1} is a function, inverse to ϕ, then (2.14) follows from (2.15). It is obvious that ρ_1 is continuous with respect to d, and (2.13) is satisfied since $z^{-1}\phi(z) \to \infty$ as $z \to \infty$. There remains to show that $(\text{Lip}_\rho(S), \|\cdot\|_{\rho_1})$ is a separable space. For this it is sufficient to show that the set $A = \{x : \|x\|_{\rho_1} \leq 1\}$ is compact in the space $\text{Lip}_\rho(S)$. Since the family of functions A is bounded and equicontinuous with respect to the supremum norm $\|\cdot\|_\infty$, then for each $\varepsilon > 0$ we can construct a finite ε-net (with respect to $\|\cdot\|_\infty$). But if we have $x, y \in A$ such that $\|x - y\|_\infty < \varepsilon$, then

$$\|x - y\|_{\rho_1} = \|x - y\|_\infty + \sup_{s \neq t} \frac{|x(s) - y(s) - x(t) + y(t)|}{\rho_1(s, t)}$$

$$\leq \varepsilon + 2\sup\left\{\frac{\rho(s,t)}{\rho_1(s, t)} : \rho(s, t) < \delta\right\} + \frac{2\|x - y\|_\infty}{\delta}$$

$$= \varepsilon + 2\varepsilon\delta^{-1} + 2\sup\left\{\frac{\rho(s, t)}{\rho_1(s, t)} : \rho(s, t) < \delta\right\}.$$

Therefore, at first choosing a sufficiently small δ, then ε, we can make $\|x - y\|_{\rho_1}$ arbitrarily small and this means that we can also construct the ε-net with respect to the norm $\|\cdot\|_{\rho_1}$. □

THEOREM 2.14. *Let ξ be $\mathbf{C}(S)$-r.e., $E\xi = 0$ and $E\|\xi\|_\rho^2 < \infty$, where ρ is a semimetric* *continuous with respect to d satisfying* (1.7). *Then* $\xi \in \mathrm{CLT}(\mathbf{C}(S))$.

Proof. Consider the inclusion operator

$$i_3 : (\mathrm{Lip}_\rho(S), \|\cdot\|_{\rho_1}) \to \mathbf{C}(S),$$

where ρ_1 is the semimetric from Lemma 2.16. Since

$$d_{i_3}(s, t) = \|i_3^*(\delta_s) - i_3^*(\delta_t)\| = \sup_{\|x\|_{\rho_1} \le 1} |x(s) - x(t)|$$

$$= \sup_{\|x\|_{\rho_1} \le 1} \frac{|x(s) - x(t)|}{\rho_1(s, t)} \rho_1(s, t) \le \rho_1(s, t),$$

and it follows from Lemma 2.16 that (2.14) is satisfied, then ρ_1 satisfies condition (1.7). Therefore, from Theorem 1.8 it follows that i_3 is an operator of type-2. Obviously, that from (2.11), $E\|\xi\|_{\rho_1}^2 < \infty$ follows, and we can apply Theorem 2.5, where

$$\mathbf{B}_1 = (\mathrm{Lip}_\rho(S), \|\cdot\|_{\rho_1}), \ \mathbf{B}_2 = \mathbf{C}(S), \ u = i_3. \square$$

Now in the case of the space $\mathbf{c}_0$ we shall formulate theorems analogous to Theorem 2.12 and 2.14.

THEOREM 2.15. *Let* ξ *be* $\mathbf{c}_0$*−r.e.,* $E\xi = 0$ *and* $\lambda \in \Lambda_0$. *If*

$$P\{\xi \in \mathbf{c}_{0,\lambda}\} = 1, \tag{2.16}$$

$$E\|\xi\|_\lambda^2 < \infty, \tag{2.17}$$

then $\xi \in \mathrm{CLT}(\mathbf{c}_0)$.

To prove this theorem we have to apply Corollary 1.13 and Theorem 2.5 where $\mathbf{B}_1 = \mathbf{c}_{0,\lambda}$, $\mathbf{B}_2 = \mathbf{c}_0$. $\square$

Condition (2.16) is complicated to check and condition (2.17) only ensures that $P\{\xi \in \mathbf{l}_{\infty,\lambda}\} = 1$. But the non-separability of the space $\mathbf{l}_{\infty,\lambda}$ does not permit us to apply Theorem 2.5. It is possible to omit condition (2.16) if we strengthen the condition on the sequence λ.

THEOREM 2.16. *Let* ξ *be* $\mathbf{c}_0$*−r.e.,* $E\xi = 0$ *and* $\lambda \in \Lambda_1$. *If* $E\|\xi\|_\lambda^2 < \infty$, *then* $\xi \in \mathrm{CLT}(\mathbf{c}_0)$.

Proof. Making use of Proposition 1.11, we can assert that there exists $a \in \mathbf{c}_0^+$ such that $\tilde{\lambda} = \lambda/a \in \Lambda_1$. Since $\Lambda_1 \subset \Lambda_0$, then it follows from Corollary 1.13 that the embedding $\mathbf{c}_{0,\tilde{\lambda}} \to \mathbf{c}_0$ is an operator of type-2. It is easily checked that $P\{\xi \in \mathbf{c}_{0,\tilde{\lambda}}\} = 1$ and $E\|\xi\|_{\tilde{\lambda}}^2 < \infty$, and we can apply Theorem 2.5. $\square$

It is easily seen that in Theorem 2.16 the condition $\lambda \in \Lambda_1$ cannot be substituted by a weaker $\lambda \in \Lambda_0$. In fact, $\mathbf{c}_0$−r.e. $\xi = (\lambda_1 \varepsilon_1, \ldots, \lambda_n \varepsilon_n, \ldots)$ with $\lambda_n = (\ln(n + 2))^{-1/2}$ satisfies condition (2.17) with $\lambda \in \Lambda_0$, but $\mathrm{cov}\,\xi \,\overline{\in}\, \mathscr{R}_1(\mathbf{c}_0)$. But if $\lambda \,\overline{\in}\, \Lambda_0$, then we can construct the following example.

EXAMPLE 2.17. Let $k \ge 2$, $\ln_k x = \ln\max(e, \ln_{k-1} x)$. There exists a symmetric pre-Gaussian $\mathbf{c}_0$−r.e. ξ with independent coordinates such that $E\|\xi\|_\lambda^2 < \infty$ where $\lambda_n = (\ln n)^{+1/2}(\ln_k n)^{1/2+\varepsilon}$, $n \ge 1$, $\varepsilon > 0$, but $\xi \,\overline{\in}\, \mathrm{CLT}(\mathbf{c}_0)$.

In constructing the example, the scheme used is the same as in Example 2.3, only computations are to be more precise. Now set for sufficiently large n, $a_n = \lambda_n$, $p_n =$

$(\ln_{k-1} n)^{-1}$ (it is obvious that for not large n we can just suppose $a_n = 1$, $p_n = 1/2$, $n = 1, ..., n_0$). Since $\sigma_j^2 : E\xi_j^2 = 2a_j^2 p_j = 2(\ln j \ln_{k-1} j)^{-1}(\ln_k j)^{1+2\varepsilon}$, then $\operatorname{cov} \xi \in \mathcal{R}_1(\mathbf{c}_0)$. Next we consider the same events E_{nk} and A_n, but now we choose $N_n = \exp\{Cn \ln_{k-1} n\}$, $C > 1$. In order to show that

$$P\{A_n\} \to 1 \text{ as } n \to \infty \tag{2.18}$$

we shall make use of the estimate

$$\prod_{j \le N_n} (1 - P\{E_{nj}\}) \le (1 - P\{E_{n,N_n}\})^{N_n}$$

$$= (1 - (\ln_{k-2}(Cn \ln_{k-1} n))^{-n})^{\exp\{Cn \ln_{k-1} n\}}.$$

Taking into account the fact that for any $C > 1$, $(\ln_{k-2}(Cn \ln_{k-1} n))^{-n} \exp\{Cn \ln_{k-1} n\} \to \infty$ as $n \to \infty$, we obtain (2.18). If $\omega \in A_n$, then

$$\|S_n(\xi(\omega))\| \ge n^{-1/2} \sup_{k \le N_n} \left| \sum_{i=1}^{n} a_k \varsigma_{ik}(\omega) \right| > n^{1/2} a_{N_n}$$

$$= (\ln_{k-1}(Cn \ln_{k-1} n))^{\varepsilon+1/2} (C \ln_{k-1} n)^{-1/2} \to \infty \text{ as } n \to \infty.$$

This relation proves that $\xi \,\overline{\in}\, \mathrm{CLT}$. □

It can be noted that, similarly, one can construct an example with $\lambda_n = (\ln n)^{+1/2}(d(n))^{\varepsilon+1/2}$ where $d(n)$ is a monotonically increasing sequence such that $d(n) \to \infty$, and there exists $C > 1$ and the sequence $b(n) \to \infty$ such that for all sufficiently large n

$$Cd(n) - \ln(d(\exp\{Cnd(n)\})) > n^{-1} b(n).$$

3.3. Supplements

3.3.1. Let (S, ν) be a measurable space with σ-finite measure ν, $\Phi : \mathbf{R}^+ \to \mathbf{R}^+$ be a convex continuous non-decreasing function, $\Phi(0) = 0$ and there exists a constant $C > 0$ such that $\Phi(2u) \le C\Phi(u)$ for all $u \in \mathbf{R}^+$.

THEOREM [73]. (a) *If the function $u \to \Phi(u^{1/p})$, $1 < p \le 2$, is convex, then the Orlicz space $\mathbf{L}_\Phi(S, \nu)$ is a space of type p.*

(b) *If the function $u \to \Phi(u^{1/q})$, $2 < q < \infty$, is concave, then $\mathbf{L}_\Phi(S, \nu)$ is a space of cotype $\div q$.*

3.3.2. Let the numbers $p \in (1, \infty)$, $q \in [1, \infty]$. The Lorentz space $\mathbf{L}_{p,q}(S, \nu)$ is a space (of equivalence classes) of measurable functions $f : S \to \mathbf{R}$ for which the following quantity is finite

$$\|f\|_{p,q}^* := \begin{cases} \left(\dfrac{q}{p} \displaystyle\int_0^\infty (t^{1/p} f^*(t))^q t^{-1} \, dt \right)^{1/q}, & 1 < p < \infty, \ 1 \le q < \infty; \\ \sup\limits_{t > 0} t^{1/p} f^*(t), \ 1 < p < \infty, \ q = \infty, & \end{cases}$$

where $f^*(t) = \inf\{y > 0 : \nu\{u \in S : |f(u)| > y\} \le t\}$ is a non-increasing rearrangement of the function f.

The space $\mathbf{L}_{p,q}$ is a Banach space with respect to the norm $\|\cdot\|^*_{p,q}$ in the case $q \leq p$, and in the case $q > p$ with respect to the norm $\|f\|_{p,q} := \|f^{**}\|^*_{p,q}$, where $f^{**}(t) = t^{-1} \int_0^t f^*(s)\,ds$.

THEOREM [50]. *Let the measure ν be atom-free.*

(a) *If $p \neq 2$, $q \neq \infty$, then $\mathbf{L}_{p,q}(S, \nu)$ is a space of type* $\min(2, \min(p, q))$ *and of cotype* $\max(2, \max(p, q))$.

(b) *If $q < 2$, then $\mathbf{L}_{p,q}(S, \nu)$ is a space of type q and of cotype $2 + \varepsilon$ for each $\varepsilon > 0$; if $q > 2$, then $\mathbf{L}_{2,q}(S, \nu)$ is a space of cotype q and of type $2 - \varepsilon$ for each $\varepsilon > 0$.*

(c) *If $1 < p < \infty$, then $\mathbf{L}_{p,\infty}(S, \nu)$ is not a space of type r for any $r > 1$, and is not a space of cotype s for any $s < \infty$.*

3.3.3. THEOREM [109]. *If $\mathbf{B}$ is a space of type-2 and of cotype-2, then $\mathbf{B}$ is isomorphic to a Hilbert space.*

3.3.4 THEOREM (see [112]). *There exists a Banach space $\mathbf{B}$ and a pre-Gaussian $\mathbf{B}$-r.e. ξ such that:*

(a) *$\mathbf{B}$ is a space of cotype $2 + \varepsilon$ and of type $2 - \varepsilon$ for each $\varepsilon > 0$;*

(b) *$\|\xi\| \leq 1$ a.s. and $\xi \overline{\in} \mathrm{CLT}(\mathbf{B})$.*

3.3.5. THEOREM [71]. *Let ξ be a pre-Gaussian $\mathbf{C}(S)$-r.e., $E\xi = 0$, $P\{\|\xi\| \leq 1\} = 1$. If there exists a r.v. ς and a pseudometric ρ on S such that for some $0 < r < 1$ the conditions*

(1) $\lim_{\varepsilon \to 0} \varepsilon H(S, \rho^r, \varepsilon) = 0$; $E\varsigma^r < \infty$ *and*

(2) $|\xi(t, \omega) - \xi(s, \omega)| \leq \varsigma(\omega)\rho(s, t)$ *for all* $s, t \in S, \omega \in \Omega$,

are satisfied, then $\xi \in \mathrm{CLT}(\mathbf{C}(S))$.

3.3.6. THEOREM [72]. *If the norm $\phi(\cdot) = \|\cdot\|$ of the space $\mathbf{B}$ satisfies the conditions:* $\phi \in D^2(\mathbf{B} \setminus \{0\}, \mathbf{R})$,

$$\sup\{\|\phi''(x) - \phi''(y)\| \, \|x - y\|^{-\delta} : \|x\| = \|y\| = 1, x \neq y\} < \infty$$

for some $\delta > 0$, then $\mathbf{B}$-r.e. ξ, $E\xi = 0$ satisfies the CLT if and only if

(1) *ξ is pre-Gaussian;*

(2) $\lim_{t \to \infty} t^2 P\{\|\xi\| > t\} = 0$.

3.3.7. In the space $\mathbf{B}$ a Rosental Λ_2-inequality holds, if there exists a constant C such that for any finite number of symmetric bounded pre-Gaussian $\mathbf{B}$-r.e. $X_1, \ldots, X_n$ the following inequality holds:

$$\sup_{t \geq 0} tP^{1/2}\left\{ \left\| \sum_{i=1}^n X_i \right\| > t \right\} \leq C\left\{ \left(\sup_{t \geq 0} t^2 \sum_{j=1}^n P\{\|X_j\| > t\} \right)^{1/2} + \right.$$
$$\left. + \sup_{t \geq 0} tP^{1/2}\left\{ \left\| \sum_{j=1}^n g(X_j) \right\| > t \right\} \right\},$$

where $g(X_j)$ is a Gaussian $\mathbf{B}$-r.e. with a covariance operator $\operatorname{cov} g(X_j) = \operatorname{cov} X_j$, $j = 1, \ldots, n$.

THEOREM [114]. *The following statements are equivalent:*

(a) *in a space $\mathbf{B}$ a Rosental Λ_2-inequality holds;*

(b) *for each pre-Gaussian $\mathbf{B}$-r.e. ξ such that $\lim_{t \to \infty} t^2 P\{\|\xi\| > t\} = 0$, the CLT holds.*

3.3.8. Let Φ be a class of non-trivial continuous functions $\phi : \mathbf{R}^+ \to \mathbf{R}^+$ satisfying the condition that there exists a constant $C > 0$ such that $\phi(x + y) \leq C\phi(x)\phi(y)$ for all $x, y \in \mathbf{R}^+$.

THEOREM [1]. *Let* $\mathbf{B}$*-r.e.* $\xi \in \mathrm{CLT}(\mathbf{B})$ *and let* μ *be a Gaussian limit distribution of the sequence* $\mathbf{L}(S_n(\xi))$*. If* $\phi \in \Phi$ *and* $E\phi(\xi) < \infty$*, then the following statements are equivalent:*

(a) $\lim_{t\to\infty}\sup_{n\geq 1} nE\phi(\xi/\sqrt{n})\chi\{\|\xi\| > t\sqrt{n}\} = 0$;
(b) $\lim_{t\to\infty} E\phi(S_n(\xi))\chi\{\|S_n(\xi)\| > t\} = 0$;
(c) $\lim_{n\to\infty} E\phi(S_n(\xi)) = \int_{\mathbf{B}} \phi(x)\mu(\mathrm{d}x)$.

In particular, if $E\|\xi\|^p < \infty$, $p \geq 2$*, then*

$$\lim_{n\to\infty} E\|S_n(\xi)\|^p = \int_{\mathbf{B}} \|x\|^p \mu(\mathrm{d}x);$$

if $E\exp\{\alpha\|\xi\|^p\} > \infty$ *for some* $\alpha > 0$ *and* $0 < p \leq 1$*, then*

$$\lim_{n\to\infty} E\exp\{\alpha\|S_n(\xi)\|^p\} = \int_{\mathbf{B}} \exp\{\alpha\|x\|^p\}\mu(\mathrm{d}x).$$

3.3.9. THEOREM [115]. *If the space* $\mathbf{B}$ *is uniformly smoothable, then* $\xi \in \mathrm{CLT}(\mathbf{B})$ *if and only if*

(a) $\lim_{t\to\infty} t^2 P\{\|\xi\| > t\} = 0$;
(b) *for each* $\varepsilon > 0$ $\liminf_{n\to\infty} P\{\|S_n(\xi)\| < \varepsilon\} > 0$.

Recently* in [116] the same result was proved without any assumption on the space $\mathbf{B}$.

3.3.10.* THEOREM [2]. *Let* X *be a* $\mathbf{C}[0, 1]$*-valued centred and weakly square integrable r.e. satisfying the first inequality in* (2.12) *and the following conditions:*

(a) $\sup_{t\geq 0} t^2 P\{\varsigma > t\} < \infty$;
(b) $\lim_{t\to\infty} t^2 P\{\sup_{0\leq u\leq 1} |X(u)| > t\} = 0$;
(c) ρ *from* (2.12) *and the semimetric* τ*, induced by the covariance of* X*, are both pre-Gaussian. Then* $X \in \mathrm{CLT}(\mathbf{C}[0, 1])$.

This theorem in [2] was derived as a corollary from a general result on empirical processes. A straightforward proof of this result was given in [83].

---

* Added during translation.

CHAPTER 4

Gaussian Measure of ε-Strip of Some Sets

4.1. General Remarks

Let μ_n, $n \geq 1$, and μ be probability measures on a separable Banach space **B**. If $\mu_n \Rightarrow \mu$, then $\Delta_n(\mu_n, \mu) := \sup_{A \in A} |\mu_n(A) - \mu(A)| \to 0$ as $n \to \infty$ for each μ-uniformity class $\mathscr{A} \subset \mathscr{B}(\mathbf{B})$. To estimate the rate of this convergence, we are faced with the question of the rate of convergence of $\sup_{A \in \mathscr{A}} \mu((\partial A)_\varepsilon)$ to zero as $\varepsilon \to 0$. Since, in this book, we confine ourselves to limit theorems with a Gaussian limiting law, we shall consider this question only for Gaussian measures. Further, in this chapter η will stand for a Gaussian **B**$-$r.e. with a zero mean and a covariance operator $T = T_\eta$, and μ for the distribution of r.e. η. In the case of a finite-dimensional Euclidean space $\mathbf{R}_k$ and a non-degenerate Gaussian measure μ, it is known (see, e.g., [31]) that the class A of all convex sets $A \subset \mathbf{R}_k$ is μ-uniform and that the estimate

$$\sup_{A \in \mathscr{A}} \mu((\partial A)_\varepsilon) \leq C\varepsilon \tag{1.1}$$

with some constant $C = C(k)$ is valid.

In an infinite-dimensional case the situation becomes rather worse: it is in Hilbert space, which may be considered the best in many respects among infinite-dimensional Banach spaces, that the following result is obtained.

PROPOSITION 1.1. *Let μ be a Gaussian measure in* **H**, *not concentrated in a finite-dimensional space. The class of balls of the space* **H** *is not μ-uniform.*

Proof. Without loss of generality we may consider $\mathbf{H} = \mathbf{l}_2$ and that the covariance matrix corresponding to the measure μ is diagonal with elements σ_i^2, $i \geq 1$, on the diagonal. We shall show that for any $\sigma_i^2 > 0$, satisfying the condition $\Sigma_{i=1}^{\infty}\sigma_i^2 < \infty$, and for any arbitrarily small $\varepsilon > 0$, it is possible to find a ball $V_r(a)$, such that $\mu((\partial V_r(a))_{\varepsilon/2}) > 1/2$. We denote

$$A_{n,\varepsilon,r} = \left\{x = (x_1, x_2, \ldots) \in \mathbf{l}_2 : |x_n| < \frac{\varepsilon}{4}; \sum_{i \neq n} x_i^2 < r^2\right\}$$

and consider the ball $V_r(be_n)$, where, as usual, $\{e_n\}_{n \geq 1}$ is a standard base in $\mathbf{l}_2$. It is easy to make sure that if we choose $r = b - \varepsilon/4$, $r_1^2 = r\varepsilon - 3\varepsilon^2/4$, then

$$\mu(V_{r+\varepsilon}(be_n) \backslash V_r(be_n)) \geq \mu(A_{n,\varepsilon,r_1})$$

$$\geq 1 - \mu\left\{x : |x_n| > \frac{\varepsilon}{4}\right\} - \mu\left\{x : \sum_{i \neq n} x_i^2 \geq r_1^2\right\}$$

$$\geq 1 - 16\varepsilon^{-2}\sigma_n^2 - r_1^{-2} \sum_{i \neq n} \sigma_i^2 \geq 1 - 16\varepsilon^{-2}\sigma_n^2 - (b\varepsilon)^{-1} \sum_{i = n} \sigma_i^2.$$

Hence one can see that n and b may be chosen so that

$$\mu(V_{r+\varepsilon}(be_n)\setminus V_r(be_n)) \geq \frac{1}{2}.\square$$

This example shows that even in Hilbert space the estimate (1.1) is incorrect for the class of all balls.

If the μ-uniform class $\mathscr{A}$ is specified by some function $f:\mathbf{B}\to\mathbf{R}$, i.e. $\mathscr{A} = \mathscr{W}_f = \{W_t(f), t\in\mathbf{R}\}$, then instead of the sets $(\partial W_t(f))_\varepsilon$ it is sufficient to consider the sets $\{t - \varepsilon \leq f(x) \leq t + \varepsilon\}$. (If $f(x) = \|x\|$, then $(\partial W_t(f))_\varepsilon = \{x : t - \varepsilon \leq f(x) \leq t + \varepsilon\}$, but generally the equality is not true.) Then the problem of estimating the rate of convergence

$$\sup_{t\in\mathbf{R}} \mu(x : t - \varepsilon \leq f(x) \leq t + \varepsilon)$$

to zero as $\varepsilon \to 0$ comes to investigating the properties of smoothness of distribution function of r.v. $f(\eta)$. If, for instance, the function $F(r) := \mu(x : f(x) < r) \in \mathrm{Lip}_\alpha(\mathbf{R})$ then the following estimate holds

$$\sup_{t\in\mathbf{R}} \mu(x : t - \varepsilon \leq f(x) \leq t + \varepsilon) \leq C\varepsilon^\alpha. \tag{1.2}$$

In particular, if r.v. $f(\eta)$ has a bounded density, the estimate (1.2) is obtained with $\alpha = 1$.

At present, the smoothness properties of the distribution function of r.v. $f(\eta)$ have been most thoroughly studied for the function $f(x) = \|x - a\|$, $a \in \mathbf{B}$. For other functions, only the questions of existence of the density have been considered to a greater extent. When η is a Wiener process on the interval [0, 1], there are some results that also concern the boundedness of the density of r.v. $f(\eta)$. All these questions are elucidated in detail in a review article [52]. In this chapter we confine ourselves to a consideration of the smoothness properties of the distribution function of r.v. $\|\eta - a\|$, $a \in \mathbf{B}$. It should be noted that boundedness of the density of r.v. $\|\eta - a\|$ yields estimate (1.1) for class $\mathscr{A}$, the balls with the centre at the point $a \in \mathbf{B}$.

Further, by $p_{T,a}(r)$ we denote the density of the distribution

$$P(\|\eta - a\| < r) := F_{T,a}(r) = \mu(x : \|x - a\| < r).$$

Instead of $p_{T,0}$ and $F_{T,0}$ we shall write p_T, F_T respectively. It is convenient to introduce the following notations: $\mathfrak{M}_0(\mathbf{B})$ for the class of all Gaussian measures on $\mathbf{B}$ with a zero mean and $\mathfrak{M}_0^b(\mathbf{B})$ for the subclass of those Gaussian measures whose density p_T is bounded.

Before considering some particular spaces $\mathbf{B}$, we shall present one general result.

THEOREM 1.2. *The distribution function F_T is continuous everywhere, the density p_T exists and is continuous everywhere, except, perhaps, countable number of points at which the function p_T has jumps downwards. There exists constants C, C_1, C_2, depending on the covariance operator T, such that for all $x > C$ the following estimate holds*

$$p_T(x) \leq C_1 \exp\{-C_2 x^2\}.$$

The proof can be found in [207] or in [61].

COROLLARY 1.3. *In a Banach space* $\mathbf{B}$ *the class of balls with the centre at zero is μ-uniform.*

4.2. The Distribution Density of the Norm of a Gaussian Element in $\mathbf{L}_p$

Let (S, Σ, λ) be a measurable space with a measure.

THEOREM 2.1. *For any separable Banach space $\mathbf{L}_p = \mathbf{L}_p(S, \Sigma, \nu)$, $1 < p < \infty$, the equality*

$\mathfrak{M}_0^b(\mathbf{L}_p) = \mathfrak{M}_0(\mathbf{L}_p)$ *is true. In particular,* $\mathfrak{M}_0^b(\mathbf{H}) = \mathfrak{M}_0(\mathbf{H})$ *and* $\mathfrak{M}_0^b(\mathbf{l}_p) = \mathfrak{M}_0(\mathbf{l}_p)$, $1 < p < \infty$.

For the proof see [52].

REMARK 2.2. The conditions guaranteeing separability of the space $\mathbf{L}_p(S, \Sigma, \nu)$ can be found in [98]. In particular, if $S \subset \mathbf{R}_k$ with a Borel σ-algebra Σ, ν is a Lebesgue measure in $\mathbf{R}_k$, then $\mathbf{L}_p(S, \Sigma, \nu)$ is a separable space for all $1 \le p < \infty$.

In proving Theorem 2.1 one makes use of the so-called method of fibring, enabling us to study the properties of distributions of different functionals from random elements in separable metric spaces. It is by this method, considered in detail in a review article [52], that in spaces $\mathbf{L}_p$ for $1 < p < \infty$ one can prove the boundedness of the density $p_{T,a}$ of the random variable $\|\eta - a\|$ and for $a \neq 0$.

To elucidate the dependence of the constant $C(T, a)$ bounding the density $p_{T,a}$ from the covariance operator T and from the centre a, we shall consider the Hilbert space $\mathbf{l}_2$. Before formulating the result, one property inherent in the quantity $C(T, a)$ should be noted. Namely, for all $t > 0$

$$tC(t^2T, ta) = C(T, a).$$

This follows from the evident equality

$$\sup_{r \ge 0} \mu\{r \le \|x - a\| \le r + \varepsilon\} = \sup_{r \ge 0} \mu\{r \le \|tx - ta\| \le r + \varepsilon t\}.$$

THEOREM 2.3. *Let* $\eta = (\eta^j)$ *be a Gaussian* $\mathbf{l}_2$*-r.e.*

$$E\eta = 0,\ E(\eta^i)^2 = \sigma_i^2 > 0,\ \sigma_i \ge \sigma_{i+1},\ E\eta^i\eta^j = 0,\ i \neq j,\ i, j \ge 1.$$

The following estimate is valid,

$$\sup_{x \ge 0} p_{T,a}(x) \le \inf_{m \ge 2} \inf_{I_m} \psi(\sigma, a, I_m), \tag{2.1}$$

where I_m *is a set of integers* $l_1 < \cdots < l_m$,

$$\psi(\sigma, a, I_m) = C(m) \sum_{i \in I_m} \sigma_i^{-1} + C(\sigma_{l_1}\sigma_{l_2})^{-1}\left[\left(\sum_{i \in I_m} \sigma_i^2\right)^{1/2} + \left(\sum_{i \in I_m} a_i^2\right)^{1/2}\right].$$

Finding the double minimum in (2.1) is a rather difficult problem in a general case, but in some simple cases this problem can be solved. Note, that from (2.1) it follows that if the centre of balls runs a finite-dimensional subspace

$$\Pi_k \mathbf{l}_2 = \{x \in \mathbf{l}_2 : x_j = 0, j \ge k + 1\},$$

then,

$$\sup\{p_{T,a}(x) : x \ge 0, a \in \Pi_k \mathbf{l}_2\} \le C(T). \tag{2.2}$$

Evaluating roughly, from (2.1) we obtain the following estimate:

$$\sup_{x \ge 0} p_{T,a}(x) \le C(\sigma_1\sigma_2)^{-1}\left[\left(\sum_{i=1}^{\infty} \sigma_i^2\right)^{1/2} + \|a\|\right]. \tag{2.3}$$

It should be noted that at some assumptions on the covariance operator T, the analogue of estimate (2.1) can be also obtained in spaces $\mathbf{l}_p$, $p \ge 1$ (the exact formulation is given in Supplement 4.4.13). In particular, one can obtain the estimate

$$\sup_{x \ge 0} p_{T,a}(x) \le C(p, T)(C_2(p, T) + \|a\|^{p-1}). \tag{2.4}$$

PROPOSITION 2.4. *In estimates* (2.1) *and* (2.4), *the power indices in the terms depending on a cannot be substituted by smaller ones.*

Proof of Theorem 2.3. To make the writing simpler, we shall estimate $p_{T,a}$ choosing a specific set $I_m = (1, ..., m)$, $m \geq 2$. By ψ_a, $\psi_{m,a}$, $\phi_{j,a}$ we denote the distribution densities of random variables $\|\eta - a\|^2$, $\Sigma_{j=1}^m |\eta^j - a_j|^2$ and $|\eta^j - a_j|^2$, respectively. We have

$$\phi_{j,a}(x) = (2(2\pi)^{1/2}\sigma_j)^{-1}x^{-1/2}\left[\exp\left\{-\frac{(x^{1/2}+a_j)^2}{2\sigma_j^2}\right\} + \right.$$
$$\left. + \exp\left\{-\frac{(-x^{1/2}+a_j)^2}{2\sigma_j^2}\right\}\right], \quad x \geq 0.$$

By a method of mathematical induction, we shall show that for all $x > 0$ and $k \geq 1$ the following estimate is true:

$$\psi_{k,a}(x) \leq C(k)x^{-1/2}\sum_{i=1}^{k}\sigma_i^{-1}, \tag{2.5}$$

where

$$C(k) \leq 2^{(k-2)/2}\pi^{-1/2}. \tag{2.6}$$

Estimate (2.5) is true for $k = 1$ since $\psi_{1,a} = \phi_{1,a}$. For brevity, we denote

$$f_j(x) = \exp\left\{-\frac{(x^{1/2}+a_j)^2}{2\sigma_j^2}\right\} + \exp\left\{-\frac{(-x^{1/2}+a_j^2)}{2\sigma_j^2}\right\}.$$

Let estimate (2.5) be true for all $l \leq k - 1$. Then

$$\begin{aligned}\psi_{k,a}(x) &= \int_0^x \phi_{k,a}(x-y)\psi_{k-1,a}(y)\,dy \\ &\leq (2(2\pi)^{1/2}\sigma_k)^{-1}\int_0^{x/2}(x-y)^{-1/2}f_k(x-y)\psi_{k-1,a}(y)\,dy + \\ &+ \int_{x/2}^{x}\phi_{k,a}(x-y)C(k-1)y^{-1/2}\sum_{i=1}^{k-1}\sigma_i^{-1}\,dy \\ &\leq \pi^{-1/2}\sigma_k^{-1}x^{-1/2} + \sqrt{2}C(k-1)x^{-1/2}\sum_{i=1}^{k-1}\sigma_i^{-1} \\ &\leq \max(\pi^{-1/2}, \sqrt{2}C(k-1))x^{-1/2}\sum_{i=1}^{k}\sigma_i^{-1}.\end{aligned}$$

It is easily seen that $C(k)$ admits the estimate (2.6). Thus inequality (2.5) is proved.

Estimate (2.5) is useless for small values of x, therefore we shall prove that for all $k \geq 2$ and $x \geq 0$

$$\psi_{k,a}(x) \leq C(\sigma_1\sigma_2)^{-1}. \tag{2.7}$$

We have

$$\int_0^{x/2}\phi_{1,a}(x-y)\phi_{2,a}(y)\,dy \leq C(\sigma_1\sigma_2)^{-1}\int_0^{x/2}((x-y)y)^{-1/2}dy \leq C(\sigma_1\sigma_2)^{-1}.$$

The same estimate is also true for the integral over the interval $[x/2, x]$, therefore for all $x \geq 0$ we have $\psi_{2,a}(x) \leq C(\sigma_1\sigma_2)^{-1}$.

Since $\sup_x\psi_{k,a}(x) \leq \sup_x\psi_{k-1,a}(x)$, then (2.7) is proved.

As usual, by Π_k and Q_k we denote the projectors in $\mathbf{l}_2$ defined by the formulas

$$\Pi_k(x) = (x_1, x_2, \dots, x_k, 0, \dots), \; Q_k(x) = x - \Pi_k(x).$$

By G_k we denote the distribution of the random variable $\|Q_k(\eta - a)\|$. Since $p_{T,a}(x) = 2x\psi_a(x^2)$, then

$$\sup_{x\geq 0} p_{T,a}(x) = 2\sup_{x\geq 0} x^{1/2}\int_0^x \psi_{m,a}(x-y)G_m(dy)$$
$$\leq 2\sup_{x\geq 0} x^{1/2}\int_0^{x/2} \psi_{m,a}(x-y)G_m(dy) +$$
$$+ 2\sup_{x\geq 0} x^{1/2}\int_{x/2}^{x} \psi_{m,a}(x-y)G_m(dy).$$

Applying (2.5) for the estimate of the first integral and (2.7) for the estimate of the second integral, we obtain

$$\sup_{x\geq 0} p_{T,a}(x) \leq 2C(m)\sum_{i=1}^{m}\sigma_i^{-1}\sup_{x\geq 0} x^{1/2}\int_0^{x/2}(x-y)^{-1/2}G_m(dy) +$$
$$+ C(\sigma_1\sigma_2)^{-1}\sup_{x\geq 0} x^{1/2}\int_{x/2}^{x} G_m(dy) \leq CC(m)\sum_{i=1}^{m}\sigma_i^{-1} +$$
$$+ C(\sigma_1\sigma_2)^{-1}E\|Q_m(\eta - a)\|.$$

Using estimate (1.2.1) we obtain (2.1). □

Proof of Proposition 2.4. We shall show that for any $\varepsilon > 0$ one can find a ball $V_r(a)$ with $\|a\|^{p-1}\varepsilon = C(T, p)$ and a measure μ such that $\mu((\partial V_r(a))_\varepsilon) > 1/4$ (the fact that $\|Q_m(a)\|$, but not $\|a\|$, is included in estimate (2.1) is of no importance since in the given construction there will be $\|Q_m(a)\| = \|a\|$). First, consider the case $p = 2$, since in this case the proof of Proposition 1.1 can be used. With the same notations as in the proof of Proposition 1.1, and choosing the measure μ with a diagonal covariance matrix with the numbers σ_i^2, $i \geq 1$ on the diagonal, we have

$$\mu(A_{n,\varepsilon,r_1}) \geq \left(\frac{2}{\pi}\right)^{1/2}\int_0^{2\varepsilon\sigma_n^{-1}} \exp\left\{-\frac{u^2}{2}\right\}du\left(1 - r_1^{-2}\sum_{i\neq n}\sigma_i^2\right).$$

We shall choose $a = (r + \varepsilon/4)e_n$ such that there were

$$\left(\frac{2}{\pi}\right)^{1/2}\int_0^{2\varepsilon\sigma_n^{-1}} \exp\left\{-\frac{u^2}{2}\right\}du > \frac{1}{2}, \; r_1^2 = r\varepsilon + \frac{3\varepsilon^2}{4} > 2\sum_{i\neq n}\sigma_i^2$$

(in the case of estimate (2.1), when choosing n we also assume $n \overline{\in} I_m$). Then

$$\mu((\partial V_r(a))_\varepsilon) > \frac{1}{4},$$

and since

$$\|a\| = r + \frac{\varepsilon}{4}$$

and

$$\|a\|\varepsilon = r\varepsilon + \frac{\varepsilon^2}{4} = r_1^2 - \frac{\varepsilon_1^2}{2} \geq C(T),$$

it then follows that in estimates (2.1) and (2.3) one cannot put the quantities $\|Q_m a\|^\beta$ and

$\|a\|^{\beta}$, respectively, where $\beta < 1$.

In the case of $1 < p < \infty, p \neq 2$, the reasonings are quite similar. The measure μ is taken with the diagonal covariance matrix and with the numbers $\sigma_i^2, i \geq 1$, on the diagonal. It can be shown that

$$(\partial V_r(a))_\varepsilon \supset \left\{x \in \mathbf{l}_p : |x_n| < \frac{\varepsilon}{2}, \sum_{i=1, i\neq n}^{\infty} |x_i|^p < r_2\right\}$$

if $a = re_n, r_2^p = pr^{p-1}\varepsilon/2$. We choose n and r such that

$$\left(\frac{2}{\pi}\right)^{1/2} \int_0^{2\varepsilon\sigma_n^{-1}} \exp\left\{-\frac{u^2}{2}\right\} du > \frac{1}{2}, \; r_2^p \geq C(p) \sum_{i\neq n} \sigma_i^{p/2} + \left(\frac{\varepsilon}{2}\right)^p.$$

Then $\mu((\partial V_r(a))_\varepsilon) > 1/4$, and since

$$\|Q_m(a)\|_r^{p-1}\varepsilon = \|a\|^{p-1}\varepsilon = r^{p-1}\varepsilon \geq C(p, T),$$

it follows that the case $p \neq 2$ is also proved. □

4.3. Distribution Density of the Norm of a Gaussian Element in $\mathbf{c}_0$

In this case by $\eta = (\eta_1, \eta_n, \ldots)$ we denote a Gaussian $\mathbf{c}_0$-r.e. with independent coordinates, $E\eta_i = 0, E\eta_i^2 = \sigma_i^2, i \geq 1$. By Theorem 1.2.5, $P\{\eta \in \mathbf{c}_0\} = 1$ if and only if for each $\varepsilon > 0$

$$\sum_{i=1}^{\infty} \exp\{-\varepsilon\sigma_i^{-2}\} < \infty. \tag{3.1}$$

By μ we denote the distribution of r.e. η and by F and p, respectively, we denote the distribution function and the distribution density of the random variable $\|\eta\|$, i.e.

$$F(x) = P\{\sup_{i\geq 1} |\eta_i| < x\} = \prod_{i=1}^{\infty} P\{|\eta_i| < x\} = \prod_{i=1}^{\infty}\left(1 - R\left(\frac{x}{\sigma_i}\right)\right),$$

$$p(x) = F'(x) = \sum_{i=1}^{\infty} \left(\frac{2}{\pi}\right)^{1/2} \sigma_i^{-1} \exp\left\{-\frac{x^2}{2\sigma_i^2}\right\} \prod_{j\neq 1}\left(1 - R\left(\frac{x}{\sigma_j}\right)\right), \tag{3.2}$$

where

$$R(x) = \left(\frac{2}{\pi}\right)^{1/2} \int_x^{\infty} \exp\left\{-\frac{u^2}{2}\right\} du.$$

Later we shall use the following elementary inequalities:

$$\left(\frac{2}{\pi}\right)^{1/2} (1+x)^{-1} \exp\left\{-\frac{x^2}{2}\right\} \leq R(x) \leq \frac{4}{3}\left(\frac{2}{\pi}\right)^{1/2} (1+x)^{-1} \exp\left\{-\frac{x^2}{2}\right\}, \; x \geq 0; \tag{3.3}$$

$$e^{-2} \leq \left(1 - \frac{1}{n}\right)^n \leq e^{-1/2}, \tag{3.4}$$

if n is sufficiently large. First we present the result showing that in 'bad' spaces, such as $\mathbf{C}[0, 1]$ and $\mathbf{c}_0$, there exist Gaussian random elements whose norm has an unbounded distribution density.

THEOREM 3.1. *The inclusion* $\mathfrak{M}_0^b(\mathbf{c}_0) \subset \mathfrak{M}_0(\mathbf{c}_0)$ *is strict.*

COROLLARY 3.2. *If a Banach space* **B** *contains a subspace isometric to the space* $\mathbf{c}_0$, *then the inclusion* $\mathfrak{M}_0^b(\mathbf{B}) \subset \mathfrak{M}_0(\mathbf{B})$ *is strict.*

Theorem 3.1, in its turn, is a consequence of the following result.

THEOREM 3.3. *Let the continuous function* $\psi : (0, \infty) \to [0, \infty)$ *be such that* $\lim_{t\to 0}\psi(t) = 0$. *There exists a Gaussian vector* η *in the space* $\mathbf{c}_0$ *with a zero mean and independent coordinates, such that*

$$\sup_{r>0,\, \varepsilon>0} \frac{(F(r+\varepsilon) - F(r))}{\psi(\varepsilon)} = +\infty.$$

Before proving the theorem, we present an auxiliary statement.

LEMMA 3.4. *Let* $\gamma_1, \ldots, \gamma_k$ *be independent standard Gaussian random variables,* $k \geq 3$,

$$F_k(x) = P\{\max_{1\le i\le k} |\gamma_i| < x\},$$

$$\Delta = (\ln k)^{-1/2}, \; r_0 = (2 \ln k - \ln\ln k)^{1/2}.$$

Then

$$\inf_{r_0 \le x \le r_0+\Delta} F_k'(x) \geq \frac{C}{\Delta}. \tag{3.5}$$

Proof. It can be easily seen that

$$F_k'(x) = k\left(\frac{2}{\pi}\right)^{1/2} \exp\left\{-\frac{x^2}{2}\right\}(1 - R(x))^{k-1}. \tag{3.6}$$

Further, the inequalities $e^{-2x} \leq 1 - x \leq e^{-x}$ hold for $0 \leq x \leq C_1$, where C_1 is some absolute constant. Using these inequalities, as well as (3.3) and (3.6), we obtain

$$F_k'(x) \geq \left(\frac{2}{\pi}\right)^{1/2} k \exp\left\{-\frac{x^2}{2}\right\} \exp\left\{-\frac{C_2(k-1)\exp\{-x^2/2\}}{1+x}\right\}.$$

Since

$$k \exp\left\{\frac{r_0^2}{2}\right\} = (\ln k)^{1/2}, \; r_0\Delta = \left(2 - \frac{(\ln\ln k)}{\ln k}\right)^{1/2},$$

then the elementary estimates lead to the statement of the lemma. □

Proof of Theorem 3.3. Let us choose a Gaussian vector η composed of elongating blocks. In each block the coordinates have an identical distribution, i.e. for the two monotone sequences $\lambda_i \downarrow 0$ and the integers $n_i \uparrow \infty$ we assume

$$\sigma_j = \lambda_i (2 \ln n_i)^{-1/2} \text{ for } j \in I_i, \tag{3.7}$$

where

$$I_i = \left\{j : \sum_{k=1}^{i-1} n_k < j \leq \sum_{k=1}^{i} n_k\right\}, \; i = 1, 2, \cdots$$

Condition (3.1) expressed in the terms λ_i and n_i takes the form: for any $\varepsilon > 0$,

$$\sum_{i=1}^{\infty} n_i^{1-\varepsilon\lambda_i^{-2}} < \infty. \tag{3.8}$$

We set $\lambda_i = 2^{-i}$ and assume that the length of the sth block is $n_s \geq s$. Let the number $s \geq 1$ be fixed. We denote

$$U = \sup\left\{ |\eta_j| : j \in \bigcup_{k=1}^{s-1} I_k \right\},\quad V = \sup_{j \in I_s} |\eta_j|,$$

$$W = \sup\left\{ |\eta_j| : j \in \bigcup_{k=s+1}^{\infty} I_k \right\}.$$

If $s = 1$, then we assume that $U = 0$. We have

$$F(x+\varepsilon) - F(x) \geq P\{U < x\} P\{x < V < x+\varepsilon\} P\{W < x\}. \tag{3.9}$$

Suppose

$$x_0 = (2 \ln n_s - \ln\ln n_s)^{1/2},\quad x = x_0 \sigma_s = 2^{-s}\left(2 - \frac{(\ln\ln s)}{\ln s}\right)^{1/2}.$$

Then we can estimate

$$P\{W > x\} \leq \sum_{k=s+1}^{\infty} n_k R\left(\frac{x}{\sigma_k}\right) \leq \sum_{k=s+1}^{\infty} 2n_k \exp\left\{-\frac{x^2}{2\sigma_k^2}\right\}$$

$$= 2 \sum_{k=s+1}^{\infty} n_k \exp\left\{1 - \frac{x_0^2 4^{k-s}}{2 \ln n_s}\right\}$$

$$\leq 2 \sum_{k=s+1}^{\infty} n_k^{1-4^{k-s}(1-(\ln\ln n_s/\ln n_s))}.$$

Hence we can see that for sufficiently large n_1 for all $s \geq 1$

$$P\{W < x\} \geq \frac{1}{2}. \tag{3.10}$$

To estimate the quantity $P\{x < V < x+\varepsilon\}$, where $\varepsilon = \sigma_s/(\ln n_s)^{1/2}$, Lemma 3.4 will be applied. Then

$$P\{x < V < x+\varepsilon\} = P\left\{x_0 < \frac{\sup_{j \in I_s} |\eta_j|}{\sigma_s} < x_0 + (\ln n_s)^{-1/2}\right\} \geq C. \tag{3.11}$$

From estimates (3.9)−(3.11) for the chosen values of x and ε we obtain

$$\frac{F(x+\varepsilon) - F(x)}{\psi(\varepsilon)} \geq \frac{CP\{U < 2^{-s}\}}{\psi(2^{-s}/\ln n_s)}, \tag{3.12}$$

since $x = x_0\sigma_s \geq 2^{-s}$.

Estimate (3.12) being obtained, it is not difficult to complete the proof by a method of mathematical induction. If the numbers $n_1, \ldots, n_{s-1}$ are already chosen, then we choose n_s so large that $CP\{U < 2^{-s}\}/\psi(2^{-s}/\ln n_s) \geq s$. The latter relation is possible since $P\{U < 2^{+s}\} > 0$ and it depends only on $n_1, \ldots, n_{s-1}$, and $\psi(t) \to 0$ as $t \to 0$. □

The above proof does not enable us to judge how rapidly the length of the blocks n_k of Gaussian measure not belonging to the class $\;_0^b(\mathbf{c}_0)$ has to grow. The information on the behaviour of the sequence $\{n_i\}$ and $\{\lambda_i\}$ is necessary for comparing them with the condition of Theorem 3.7, presented below, which ensures boundedness of the density of the norm of vector η having the same structure of elongating blocks. In the direct proof of Theorem 3.1, as was done in [151], the sequences $\{n_i\}$ and $\{\lambda_i\}$ were so constructed that for sufficiently large i we obtained the inequality

$$p(\lambda_i) \geq a_i, \tag{3.13}$$

where a_i, $i \geq 1$, is some monotonic sequence of infinitely increasing numbers. Without proof we present the conditions on the sequences $\{n_i\}$, $\{\lambda_i\}$ and $\{a_i\}$, under which (3.13) were valid (the computations in detail are available in [151]):

$$n_i \geq \max\{\exp_2 i;\ \exp\{a_i^2\lambda_i^2 \times \exp\{C(i-1)\lambda_{i-1}\lambda_i^{-1}(\ln n_{i-1})^{1/2} n_{i-1}^{1-(\lambda_i/\lambda_1)^2}\}\}\}, \tag{3.14}$$

$$\left(\frac{\lambda_i}{\lambda_j}\right)^2 \geq 1 + j\mathrm{e}^{-j} \ \text{ for all } \ j > i. \tag{3.15}$$

Here and subsequently $\exp_k x = \exp\{\exp_{k-1} x\}$, $k \geq 2$. If we take the concrete sequences $\{\lambda_i\}$ and $\{a_i\}$, then we shall obtain a more visible condition on $\{n_i\}$. Thus, for instance, if $a_i = i^{1/2}$, $\lambda_i = i^{-1/2}$, then

$$n_i \geq \exp_2\{(i(i-1))^{1/2}(\ln n_{i-1})^{-1/2} n_{i-1}^{1-1/i}\}. \tag{3.16}$$

This equality shows that the length of the blocks n_i of Gaussian r.e. with a distribution not belonging to the class $\mathfrak{M}_0^b(\mathbf{c}_0)$, is growing very fast: n_i is comparable with $\exp_{2i}(n_1)$.

The proposition below gives an idea about the qualitative behaviour of density p for the above examples.

PROPOSITION 3.5. *If the density p is unbounded then at $x \downarrow 0$ the function $p(x)$ is oscillating with a rising amplitude an infinite number of times.*

This result follows from a more general fact: the distribution function of the norm of a Gaussian vector in $\mathbf{c}_0$ with independent coordinates can be extended to an analytical function in some cone (the precise formulation is given in Supplement 4.5.6). Besides, one must bear in mind the relation $\liminf_{x \downarrow 0} p(x) = 0$, taking place for every Gaussian vector which has at least two coordinates that are not linearly dependent.

Now, when we already know that in the space $\mathbf{c}_0$ the distribution density of the norm of a Gaussian vector can be unbounded, it is quite natural to ask: 'Under what conditions is this density bounded?' Consider a random vector η with independent coordinates and let $\mu = L(\eta)$. By Λ we denote a class of continuous monotone functions g defined on $(0, K)$, where $0 < K \leq \infty$, and

$$\lim_{x\to 0} g(x) = \infty, \quad \frac{\lim_{x\to 0} \ln g(x)}{g(ax)} = 0$$

for any fixed $a > 1$. First, consider the case when

$$\sigma_1 = b_1, \ \sigma_j = (\ln j)^{-1/2} b_j, \ j \geq 2, \tag{3.17}$$

where $\{b_j\}$ is a monotone sequence tending to zero. Note that condition (3.1) can hold also for the sequence $\{b_j\}$ having, besides one zero limit point, other limit points, for example,

$$b_j = \begin{cases} a_j, & j \neq 2^k, \ k = 1, 2, \cdots \\ 1, & j = 2^k \ k = 1, 2, \cdots. \end{cases}$$

Let $b(x)$ be a monotone continuous function defined on the interval $[1, \infty]$ such that $b(n) = b_n$, and let $\sigma(x) = (\ln x)^{-1/2} \times b(x), x \geq 2$. As usual, b^{-1} denotes a function inverse to b.

THEOREM 3.6. *If the measure* μ *is such that for some* $0 < \alpha < 1/2$, *the function* $(b^{-1})^{\alpha} \in \Lambda$, *then* $\mu \in \mathfrak{M}_0^b(\mathbf{c}_0)$.

Proof. It follows from (3.2) that

$$p(x) \leq C \sum_{j=1}^{\infty} \sigma_j^{-1} \exp\left\{ -\frac{x^2}{2\sigma_j^2} - \left(\frac{2}{\pi}\right)^{1/2} \sum_{k \neq j} \sigma_k (\sigma_k + x)^{-1} \exp\left\{ -\frac{x^2}{2\sigma_k^2} \right\}\right\}.$$

This estimate, together with the known fact that $x \ln x \to 0$ as $x \to 0$, enables us to assert that for any small, but fixed, $x_0 > 0$, the following is valid

$$\sup_{x \geq x_0} p(x) \leq C(x_0, \sigma).$$

It is easily seen that, to prove the theorem, it is sufficient to prove the boundedness of the function $\tilde{p}(x) = \nu(x) \exp\{-g(x)\}$ on the interval $0 < x < x_0$, where

$$\nu(x) = \sum_{j=1}^{\infty} a_j(x), \ a_j(x) = \sigma_j^{-1} \exp\left\{ -\frac{x^2}{2\sigma_j^2} \right\},$$

$$g(x) = \sum_{j=1}^{\infty} \tilde{a}_j(x), \ \tilde{a}_j(x) = \sigma_j(\sigma_j + x)^{-1} \exp\left\{ -\frac{x^2}{2\sigma_j^2} \right\}.$$

Consider the following functions:

$$j(x) := \sigma^{-1}(x), \ k(x) \equiv k_\varepsilon(x) := b^{-1}(x(2(1-\varepsilon))^{-1/2}),$$

$$m(x) := b^{-1}\left(\frac{x}{\sqrt{2}}\right), \ r(x) \equiv r_\delta(x) := b^{-1}(x(2(1+\delta))^{-1/2}),$$

where $0 < \varepsilon < 1/2$ and $\delta > 0$ are some fixed numbers whose choice will be dealt with later.

In addition, we assume that

$$b(x) \geq x^{-\gamma}, \ \gamma > 1. \tag{3.18}$$

Then for each arbitrarily small ε_1,

$$j(x) \geq x^{-(\gamma+\varepsilon_1)^{-1}}.$$

We assume that $j(x) \leq k(x)/2$; otherwise our reasonings would be simpler. To simplify the notation instead of sums like $\Sigma_{j=[j(x)]+1}^{[k(x)]}$ we shall just write $\Sigma_{i=j(x)}^{k(x)}$. We denote

$$\nu_1(x) := \sum_{k=1}^{j(x)} a_k(x), \ \nu_2(x) := \sum_{k=j(x)}^{m(x)} a_k(x), \ \nu_3(x) := \sum_{k=m(x)}^{r(x)} a_k(x),$$

$$\nu_4(x) := \sum_{k=r(x)}^{\infty} a_k(x), \ g_1(x) := \sum_{k=1}^{j(x)} \tilde{a}_k(x), \ g_2(x) := \sum_{k=j(x)}^{k(x)} \tilde{a}_k(x).$$

Taking into account all the assumptions on the sequence $\{b_j\}$, we easily obtain the following estimates:

$$\nu_1(x) \le x^{-1} j(x), \tag{3.19}$$

$$\nu_2(x) \le C(\ln m(x))^{1/2}(b(m(x)))^{-1}(m(x))^{1-(2\ln j(x))^{-1}}, \tag{3.20}$$

$$\nu_3(x) \le C(\ln r(x))^{2/3}(b(r(x)))^{-1}, \tag{3.21}$$

$$\nu_4(x) \le \sum_{j=r(x)}^{\infty} j^{-(1+\delta)+\gamma}(\ln j)^{1/2}, \tag{3.22}$$

$$g_1(x) \ge 2e^{-1/2} j(x), \tag{3.23}$$

$$g_2(x) \ge C\varepsilon^{-1}(k(x))^{\varepsilon}(\ln k(x))^{-1/2}. \tag{3.24}$$

We denote $I_i(x) = \nu_i(x)\exp\{-g(x)\}$, $i = 1, 2, 3, 4$. Then it follows from (3.19) – (3.24)

$$I_1(x) \le x^{-1} j(x)\exp\{-2e^{-1/2} j(x)\} \le C\exp\{-cj(x)\},$$

$$I_2(x) \le C\exp\left\{-cj(x)\left[1 - \frac{\ln(\sqrt{2}x^{-1})}{cj(x)}\right] - k^{\varepsilon-\varepsilon_2}(x)\left[\frac{k^{\varepsilon_2}(x)}{\varepsilon(\ln k(x))^{1/2}} - ck^{\varepsilon_2-\varepsilon}(x)\ln m(x)\right]\right\},$$

$$I_3(x) \le C\exp\left\{-j(x)\left[c - \frac{\ln(2(1+\delta)x^{-2})}{2j(x)}\right] - \right.$$

$$\left. - k^{\varepsilon-\varepsilon_2}(x)\left[\frac{k^{\varepsilon_2}(x)}{\varepsilon(\ln k(x))^{1/2}} - \frac{3\ln\ln r(x)}{2k^{\varepsilon-\varepsilon_2}(x)}\right]\right\},$$

$$I_4(x) \le r^{\gamma+\varepsilon_1-\delta}(x)\exp\{-g(x)\},$$

where $\varepsilon > \varepsilon_2 > 0$. Now by choosing the free parameters δ, ε and ε_2 such that the relations $\delta > \gamma + \varepsilon_1$, $\varepsilon - \varepsilon_2 = \alpha$, take place, we see that $I_i(x) \to 0$ as $x \to 0$, $i = 1, 2, 3, 4$. Since $\tilde{p}(x) = \Sigma_{i=1}^4 I_i(x)$, then the proof of the theorem is completed under condition (3.18). It is clear by intuition that the more rapdily the function σ tends to zero, the nearer we are to the finite-dimensional case; nevertheless, condition (3.18) is essential for the above estimate of the quantity $I_4(x)$. For the estimate of the function $I_4(x)$, in the case $b(x) \le x^{-\gamma}$ we introduce one more notation: $s(x) = b^{-1}(x^2/2)$. Then for all $j > s(x)$, $x^2(2b_j)^{-1} \ge 1$, and for sufficiently small x, we have

$$\nu_5(x) := \sum_{j=r(x)}^{s(x)} a_j(x) \le \sum_{j=r(x)}^{s(x)} b_j^{-1}(\ln j)^{1/2} j^{-1-\delta} \le (b(s(x)))^{-1} = 2x^{-2},$$

$$\nu_6(x) := \sum_{j=s(x)}^{\infty} a_j(x) \le \sum_{j=s(x)}^{\infty} (\ln j)^{1/2}\exp\left\{-\frac{x^2\ln j}{2b_j^2} - \ln b_j\right\}$$

$$\le \sum_{j=s(x)}^{\infty} (\ln j)^{1/2}\exp\left\{-\frac{[(x^2\ln j/2b_j) + b_j\ln b_j]}{b_j}\right\}$$

$$\le \sum_{j=s(x)}^{\infty} (\ln j)^{1/2}\exp\left\{-\frac{j^{\gamma}}{2}\right\}.$$

To compensate for the growth of the function $\nu_5(x)$ as $x \to 0$, it is useless to consider the function $\exp\{-g(x)\}$. It is preferable to return to formula (3.2) and to single out from the infinite product a cofactor of three terms

$$\prod_{i=1}^{3}\left(1 - R\left(\frac{x}{\sigma_i}\right)\right) \le Cx^3. \square$$

Next we consider the case of elongating blocks used in the proof of Theorem 3.3. We shall show that the density p is bounded if the lengths of the blocks are increasing not too rapidly. Let the sequence $\{\sigma\}$ be defined by formula (3.7) with monotone sequences $\lambda_i \downarrow 0$ and $n_i \uparrow \infty$, satisfying (3.8). By $n(x)$ and $\lambda(x)$ we denote continuous monotone functions with the properties $\lambda(k) = \lambda_k$ and $n(k) = n_k$, $k \geq 1$.

THEOREM 3.7. *If the measure* $\mu \in \mathfrak{M}_0(\mathbf{c}_0)$ *is such that for some* $0 < \alpha < 1/2$ *the function* $(n(\lambda^{-1}(\cdot)))^{\alpha} \in \Lambda$, *then* $\mu \in \mathfrak{M}_0^b(\mathbf{c}_0)$.

Proof. Now it is convenient to make use of the following estimate

$$p(x) \leq C(\sigma)\hat{p}(x), \tag{3.25}$$

where $\hat{p}(x) = x^{m/2} w(x) \exp\{-cq(x)\}$ and m is either an integer or 0,

$$w(x) := \sum_{j=1}^{\infty} d_j(x), \;\; d_j(x) := \lambda_j^{-1}(2 \ln n_j)^{1/2} n_j^{1-x^2\lambda_j^{-2}},$$

$$q(x) := \sum_{j=1}^{\infty} \tilde{d}_j(x), \;\; \tilde{d}_j(x) := \lambda_j(\lambda_j + (2\ln n_j)^{1/2}x)^{-1} n_j^{1-x^2\lambda_j^{-2}}.$$

The cofactor $x^{m/2}$ in (3.25) was obtained in the same way as at the end of the proof of Theorem 3.6, i.e. by separating from each product in the sum (3.2) the cofactor

$$\prod_{j=1}^{m} \left(1 - R\left(\frac{x}{\sigma_i}\right)\right)^{1/2} \leq C(\sigma)x^{m/2}.$$

This operation will have influence only on the constant at the first two summands $\tilde{d}_1(x)$ and $\tilde{d}_2(x)$. As in the proof of the preceding theorem, it is sufficient to show that the function $\hat{p}(x)$ is bounded for $0 < x < x_0$, where x_0 is some small number. We assume, in addition, that

$$n(x) \geq x^{\gamma_1}, \;\; \gamma_1 > 0. \tag{3.26}$$

For positive ε, δ_1 and δ_2 we denote:

$$K(x) \equiv K_{\varepsilon}(x) := \lambda^{-1}(x(1-\varepsilon)^{1/2}), \;\; M(x) := \lambda^{-1}(x),$$

$$S(x) := \lambda^{-1}(x(1+\delta_1)^{-1/2}), \;\; T(x) := \lambda^{-1}(x^2(1+\delta_2)^{-1}).$$

From the definition of these functions we have

$$w_1(x) := \sum_{j=1}^{K(x)} d_j(x) \leq Cx^{-1}(\ln n(K(x)))^{1/2} K(x) n(K(x)), \tag{3.27}$$

$$w_2(x) := \sum_{j=k(x)}^{M(x)} d_j(x) \leq Cx^{-1}(\ln n(M(x)))^{1/2} Z(x), \tag{3.28}$$

$$w_3(x) := \sum_{j=M(x)}^{S(x)} d_j(x) \leq Cx^{-1}(\ln n(S(x)))^{1/2} S(x), \tag{3.29}$$

$$w_4(x) := \sum_{j=S(x)}^{T(x)} d_j(x) \leq (1+\delta_2)x^{-2} \sum_{j=S(x)}^{T(x)} (2\ln n_j)^{1/2} n_j^{-\delta_1}$$

$$\leq Cx^{-2}(S(x))^{-\gamma_1\delta_1+1+\varepsilon_1}, \tag{3.30}$$

$$q_1(x) := \sum_{j=1}^{K(x)} \tilde{d}_j(x) \geq C(1 + (\ln n(K(x)))^{1/2})^{-1} n^{\varepsilon}(K(x)), \tag{3.31}$$

$$q_2(x) := \sum_{j=K(x)}^{M(x)} \widetilde{d}(x) \geq C(\ln n(m(x)))^{-1/2} Z(x), \tag{3.32}$$

where

$$Z(x) = \sum_{j=K(x)}^{M(x)} n_j^{1-x^2\lambda_j^{-2}}.$$

Next, setting $m = 4$ in (3.25), from (3.27)-(3.32) we obtain

$$J_1(x) := x^2 w_1(x) \exp\{-q(x)\}$$
$$\leq CxK(x)n(K(x))(\ln n(K(x)))^{1/2} \exp\{-q(x)\}$$
$$\leq Cx_0 \exp\{-c(n(K(x)))^{\varepsilon}(\ln n(K(x)))^{-1/2}[1 - c(\ln n(K(x)))^{3/2} \times$$
$$\times (n(K(x)))^{-\varepsilon} - c \ln K(x)(\ln n(K(x)))^{1/2}(n(K(x)))^{-\varepsilon}]\}, \tag{3.33}$$

$$J_2(x) := x^2 w(x) \exp\{-q(x)\} \leq CxZ(x)(\ln n(M(x)))^{1/2} \exp\{-q(x)\}$$
$$\leq Cx_0 \exp\{-cZ(x)(\ln n(M(x)))^{-1/2} [1 - (Z(x))^{-1} \ln Z(x)(\ln n(M(x)))^{1/2}] -$$
$$- c(n(K(x)))^{\varepsilon}(\ln n(K(x)))^{-1/2} \times$$
$$\times [1 - c(\ln n(M(x)) \ln n(K(x)))^{1/2}(n(K(x)))^{-\varepsilon}]\}, \tag{3.34}$$

$$J_3(x) := x^2 w_3(x) \exp\{-q(x)\}$$
$$\leq Cx_0 \exp\{-c(n(K(x)))^{\varepsilon}(\ln n(K(x)))^{1/2}[1 - c(n(K(x)))^{-\varepsilon} \times$$
$$\times [(\ln n(S(x)) \ln n(K(x)))^{1/2} + \ln S(x)(\ln n(K(x)))^{1/2}]]\}, \tag{3.35}$$

$$J_4(x) := x^2 w_4(x) \leq C(S(x))^{-\gamma_1\delta_1+1+\varepsilon_1}. \tag{3.36}$$

We shall now deal with the choice of free parameters: set $\delta_1 \geq \gamma_1^{-1}(1 + \varepsilon_1)$, $\varepsilon = \alpha + \varepsilon_1$, where α is a number taking part in the formulation of the theorem, and $0 < \varepsilon_1 < 1/4$. Then, using the condition of the theorem $n^{\alpha}(\lambda^{-1}(\cdot)) \in \Lambda$, one can show that the quantities in square brackets in (3.33) – (3.36) tend to zero as $x \to 0$, for instance

$$\frac{\ln S(x)(\ln n(K(x)))^{1/2}}{(n(K(x)))^{\varepsilon}} \leq \frac{\ln n(S(x))}{\gamma_1(n(K(x)))^{\varepsilon-\varepsilon_1}} \frac{(\ln n(K(x)))^{1/2}}{(n(K(x)))^{\varepsilon_1}}$$
$$= \frac{1}{\gamma_1\alpha} \frac{\ln(n(\lambda^{-1}(x(1+\delta_2)^{-1/2})))^{\alpha}}{(n(\lambda^{-1}(x(1-(\alpha+\varepsilon_1))^{-1/2})))^{\alpha}} \frac{(\ln n(K(x)))^{1/2}}{(n(K(x)))^{\varepsilon_1}} \to 0.$$

Therefore, $\Sigma_{i=1}^4 J_i(x) \to 0$ as $x \to 0$. It remains to estimate the quantity $w_5(x) := \Sigma_{j=T(x)}^{\infty} d_j(x)$. From the definition of the function $T(x)$ we have

$$w_5(x) \leq \sum_{j=T(x)}^{\infty} \lambda_j^{-1} n_j^{1+\varepsilon_1-x^2\lambda_j^{-2}} \leq \sum_{j=T(x)}^{\infty} \exp\{(1 + \varepsilon_1 - x^2\lambda_j^{-2})\ln n_j - \ln \lambda_j\}$$
$$\leq \sum_{j=T(x)}^{\infty} \exp\left\{\left[1 + \varepsilon_1 - (1+\delta_2)\lambda_j^{-1}\left(1 - \frac{\lambda_j}{1+\delta_2} \ln \frac{1}{\lambda_j}\right)\right] \ln n_j\right\}$$
$$\leq \sum_{j=T(x)}^{\infty} \exp\left\{\left[-\frac{1+\delta_2}{2\lambda_j} + 1 + \varepsilon_1\right] \ln n_j\right\} \to 0$$

as $x \to 0$, since, for $j \geq T(x)$,

$$\frac{1+\delta_2}{2\lambda_j} \geq \frac{(1+\delta_2)^2}{2x^2} \geq \frac{(1+\delta_2)^2}{2x_0^2},$$

then one can choose δ_2 such that the following inequality is valid:

$$\delta_2 > \max(((1 + \varepsilon_1 + \gamma_1^{-1})2)^{1/2}x_0 - 1, 0).$$

Thus, the theorem is proved if condition (3.26) is satisfied. In the case when $n(x)$ increases more slowly than the power function, $\lambda(x)$ cannot decrease as slowly as possible since condition (3.8) is to be satisfied. For instance, if $(\ln k)^{\gamma_1} \leq n_k \leq (\ln k)^{\gamma_2}$, where γ_1 and γ_2 are two positive numbers, then

$$\lambda_k^2 = \frac{\ln \ln k}{b_k \ln k}, \quad b_k \to \infty \text{ as } k \to \infty.$$

It is clear that the slower n_k grows, the nearer we are to the case considered in the preceding theorem, and since no essentially new estimates are required and the computations are rather cumbersome we omit the proof for the case $n_k \leq k^{\gamma}$.

4.4. Supplements

4.4.1. Let γ_n, $n \geq 1$ be a Gaussian sequence for which $E\gamma_n \geq 0$, $\max_{n\leq 1} E\{\gamma_n - E\gamma_n)^2 > 0$, $\varsigma := \sup_{n\geq 1}\gamma_n < \infty$ a.s.

THEOREM. (a) *The distribution of r.v.* ς *is concentrated on a half-line* $[b_0, \infty)$, *where*

$$b_0 = \inf\{b : P\{\varsigma \leq b\} > 0\},$$

and on the interval (b_0, ∞) *it is absolutely continuous.*

(b) *The distribution density of r.v.* ς *is continuous on* (b_0, ∞) *with the exception, perhaps, of a countable number of points where it has jumps downwards. On* $[b_0 + \varepsilon, \infty)$, $\varepsilon > 0$, *the density has a bounded variation.*

This result from [207] is more general as compared to Theorem 1.1, since not only the norm but also the large class of continuous convex functionals on the Banach space can be expressed as

$$f(x) = \sup_{k\geq 1} g_k(x),$$

where g_k, $k \geq 1$, are linear continuous functionals.

4.4.2. THEOREM [179]. *If the modulus of uniform convexity* $\delta_{\mathbf{B}}$ *of a Banach space* **B**, *defined by the equality*

$$\delta_{\mathbf{B}}(\varepsilon) = \inf\left\{1 - \frac{\|x+y\|}{2} : \|x\| = \|y\| = 1, \|x-y\| = \varepsilon\right\},$$

satisfies the condition $\delta_{\mathbf{B}}(\varepsilon) \geq C\varepsilon^p$ *for some* $p > 0$, *then* $\mathfrak{M}_0^b(\mathbf{B}) = \mathfrak{M}_0(\mathbf{B})$.

4.4.3. THEOREM [221]. *Let* η *be a Gaussian* $\mathbf{C}(S)$*−r.e. where* S *is a metric compact. If* $E\eta(t) = 0$, $0 < \sigma_0^2 < E\eta^2(t) < \sigma_1^2$ *for all* $t \in S$, *then the distribution of r.v.* $\|\eta\|$ *has a bounded density and for each* $k \geq 0$ *there exists a constant* $C(k, T_\eta)$ *such that for any* $\varepsilon > 0$

$$\sup_{r\geq 0} r^k P\{r \leq \|\eta\| \leq r + \varepsilon\} \leq C(k, T_\eta)\varepsilon.$$

4.4.4. The condition of separation from zero of the quantity $\inf_{t\in S} E\eta^2(t)$ in the above result is rather strong (it is enough to note that the standard Wiener process does not satisfy this condition). In [55], [119], [120] further studies of the distribution of the quantities $\sup_{s\in S} \eta(s)$ and $\sup_{s\in S} |\eta(s)|$, where S is some parametric set, are carried out. We shall formulate one result from [120].

Let $\eta(s)$, $s \in S$, be a Gaussian process, $E\eta(t) = 0$, $E\eta^2(t) = 1$ for all $t \in S$,

$$\tau(s, t) = (E(\eta(s) - \eta(t))^2)^{1/2}, \quad \psi(\varepsilon) = \int_0^\varepsilon H^{1/2}(S, \tau, x)\, dx.$$

By $\nu(\varepsilon)$, $\varepsilon > 0$, we denote the largest length of the chain of elements $t_1, \ldots, t_k \in S$ such that $E\eta^2(t_1) \geq \varepsilon^2$, and for all $2 \leq i \leq k$,

$$E(\eta^2(t_i) \mid \eta(t_1), \ldots, \eta(t_{i-1})) - (E(\eta(t_i) \mid \eta(t_1), \ldots, \eta(t_{i-1}))^2 \geq \varepsilon^2.$$

THEOREM [120]. *If for some $\delta > 0$ and any $\gamma > 0$,*

$$\psi(\varepsilon) = O(\varepsilon^\delta), \quad \ln \nu(\varepsilon) = O(\nu(\varepsilon^\gamma)),$$

as $\varepsilon \to 0$, then the density of distribution of the random variable $\sup_{s\in S} |\eta(s)|$ is bounded.

If the results of [120] are applied in the situation considered in Theorem 3.7, then we obtain the following result (which was kindly provided by M.A. Lifshits). Let $\xi = (\xi_1, \ldots, \xi_n, \ldots)$ be $\mathbf{c}_0$-r.e. with a distribution μ and independent coordinates $E\xi_i = 0$, $i \geq 1$ and let $\sigma_i^2 = E\xi_i^2$ be defined by (3.7). If, for some $\Theta > 0$, $\ln n_i \leq c\lambda_i^{-\Theta}$, then $\mu \in \mathfrak{M}_0^b(\mathbf{c}_0)$.

Applying the functions $n(\cdot)$ and $\lambda(\cdot)$ introduced on page 72, the above condition can be rewritten as

$$\ln n(\lambda^{-1}(y)) \leq cy^{-\Theta}, \quad \Theta > 0.$$

It is obvious that this condition, generally speaking is incomparable with the condition $n(\lambda^{-1}(\cdot))^\alpha \in \Lambda$ from Theorem 2.7.

4.4.5. Let $f : \mathbf{C}[0, 1] \to \mathbf{R}$ be an integral functional defined by the formula

$$f(x) = \int_0^1 q(x(t), t)\, dt,$$

where $q : \mathbf{R} \times [0, 1] \to \mathbf{R}$ satisfies some requirements. The works [34], [51], [118], [119], [184] are devoted to the studies of distribution of r.v. $f(\eta)$ for different continuous Gaussian processes η. As an example, we shall present the following result from [118] for the case when q depends only on the first argument.

THEOREM. *Let $q: \mathbf{R} \to \mathbf{R}$ be a measurable function, monotone on the intervals $[-a, 0)$ and $(0, a]$ for some $a > 0$. If for some $A > 0$, $C < \pi^2/8$*

$$|q'(x)| \geq A \exp\{-C |x|^{-2}\}$$

for almost all $x \in [-a, a]$, then the distribution of r.v. $f(w) = \int_0^1 q(w(t))\, dt$ has a bounded density. Here $w(\cdot)$ is a standard Wiener process.

4.4.6. In the case of $\mathbf{C}[0, 1]$ there is an explicit expression of the distribution function of the norm of a standard Wiener process (see [64]):

$$P\{\sup_{0\le t\le 1} |w(t)| < x\} = \frac{4}{\pi}\sum_{n=0}^{\infty}\frac{(-1)^n}{2n+1}\exp\left\{-\frac{\pi^2(2n+1)^2}{8x^2}\right\}.$$

The density of the distribution of r.v. $\sup_{0\le t\le 1}|w(t)|$ and all its derivatives are bounded functions.

4.4.7. THEOREM [52]. *Let* $(\mathbf{B}, \|\cdot\|)$ *be a finite-dimensional Banach space, and let* η *be a Gaussian* $\mathbf{B}$*−r.e. If* $E\eta = 0$ *or if the linear support of the distribution* $L(\eta)$ *(defined as the least closed affine manifold* K *such that* $L(\eta)(K) = 1$*) coincides with* $\mathbf{B}$*, then the distribution of r.v.* $\|\eta\|$ *has a bounded density.*

4.4.8. THEOREM [178]. *In a Hilbert space* $\mathbf{H}$*, for any* $\varepsilon > 0$ *one can construct a norm* $N(\cdot)$ *and a Gaussian* $\mathbf{H}$*−r.e.* η, $E\eta = 0$*, with the following properties:*

1. $(1+\varepsilon)^{-1}\|x\| \le N(x) \le (1+\varepsilon)|x|$ *for all* $x \in \mathbf{H}$;
2. $N(x)$ *is an infinitely differentiable function at all points* $x \ne 0$ *and its derivatives are bounded on a unit sphere;*
3. *the density of distribution of r.v.* $N(\eta)$ *is unbounded.*

THEOREM [176]. *Let* $\mathbf{B}$ *be a separable Banach space. For each* $\varepsilon > 0$ *there exists a norm* $N(\cdot)$ *in* $\mathbf{B}$ *and a Gaussian* $\mathbf{B}$*−r.e.* η, $E\eta = 0$*, such that*

1. $(1+\varepsilon)^{-1}\|x\| \le N(x) \le (1+\varepsilon)\|x\|$ *for all* $x \in \mathbf{B}$;
2. *the density of distribution of r.v.* $N(\eta)$ *is unbounded.*

4.4.9. Let $p_{T,a}$ be a density of distribution of r.v. $\|\eta - a\|$, where η is a Gaussian $\mathbf{l}_2$−r.e. with a zero mean and with a diagonal covariance operator T. It follows from Theorem 2.3 (see formula (2.3)) that

$$\sup_{x\ge 0} p_{T,a}(x) \le C(\sigma_1\sigma_2)^{-1}\left(\left(\sum_{i=1}^{\sigma}\sigma_i^2\right)^{1/2} + \|a\|\right).$$

It turns out (see [46]) that it has already proved impossible to obtain the estimate with one dispersion σ_1^{-1}. On the other hand, this estimate is not good in all cases. Among others, the following estimate is given in [211]:

$$p_{T,a}(x) \le C\{1 + \min(x, \|a\|)\}\psi,$$

where $\psi = \psi(T) = (\psi_1\psi_2\psi_3)^{-2/3}$, $\psi_i = (\Sigma_{j\ge i}\sigma_j^4)^{1/4}$. It is easily seen that if $\sigma_i^2 = n^{-1}$, $i = 1, \ldots, n$, $n \ge 3$, then $\psi\sigma_1\sigma_2 \to 0$ as $n \to \infty$.

4.4.10. THEOREM [139]. *There exists a Gaussian* $\mathbf{c}_0$*−r.e.* $\eta = (\eta_1, \eta_2, \ldots)$ *with independent coordinates such that the density* p *of the distribution r.v.* $\|\eta\|$ *is bounded but the derivative* p' *is unbounded.*

4.4.11. THEOREM [138]. *Let* $\eta = (\eta_1, \ldots, \eta_n, \ldots)$ *be a Gaussian vector with independent coordinates* $E\eta_i = 0$, $E\eta_i^2 = \sigma_i^2$, $i \ge 1$*, and* $(\sigma_i, i \ge 1) \in \Lambda_0$*. Then the function*

$$F(x) = P\{\sup_{i\ge 1}|\eta_i| < x\}$$

can be analytically prolonged to the region of complex plain $\{z = x + iy : x > 0, |y| < 12^{1/4}x/(1+\sqrt{3})\}$.

4.4.12. Let η be a Gaussian $\mathbf{l}_2$-r.e. with a zero mean and a covariance operator T whose eigenvalues are $\lambda_1 \geq \lambda_2 \geq \cdots$. Then

$$E\exp\{it\|\eta + a\|^2\} = \prod_{j=1}^{\infty}(1 - 2it\lambda_j)^{-1/2}\exp\left\{\frac{ita_j}{1 - 2it\lambda_j}\right\},$$

$a = (a_1, a_2, ...) \in \mathbf{l}_2$.

4.4.13. Let $1 \leq p < \infty$. We denote

$$m_0(p) = \begin{cases} p, & \text{if } p \text{ is an integer,} \\ 1 + [p], & \text{if } p \text{ is not integer.}\end{cases}$$

THEOREM. *Let $\eta = (\eta_1, \eta_2, ...)$ be a Gaussian $\mathbf{l}_p$-r.e., $1 \leq p < \infty$, $p \neq 2$, $E\eta = 0$, $E\eta_i^2 = \sigma_i^2$, $i \geq 1$. Let the coordinates η_i, $i \in I_k := \{i_1, ..., j_k\}$, $k \geq m_0(p)$, be independent of one another and of the rest of the coordinates. Then for any set of indices $I_m = \{l_1, ..., l_m\} \subset I_k$ the following estimate is valid*

$$\sup_{x \geq 0} P_{T,a}(x) \leq C(p, m)\sum_{i \in I_m}\sigma_i^{-1} +$$

$$+ C(p)\left(\prod_{i=1}^{m_0(p)}\sigma_{l_i}^{-1}\right)^{p/m_0(p)}\left\{\left(\sum_{i \in I_m}\sigma_i^p\right)^{(p-1)/p} + \left(\sum_{i \in I_m}|a_i|^p\right)^{(p-1)/p}\right\}.$$

The proof of the theorem follows the scheme of the proof of Theorem 2.3. In [145] a rather rough estimate (2.4), presented in the main text, is proved. Note that the requirement of the independence of at least some coordinates of the vector η is a restriction for all $p \neq 2$, since only in $\mathbf{l}_2$ one can reduce, by means of unitary transformation, a covariance matrix to a diagonal form. In spaces $\mathbf{l}_p$, $p \neq 2$, unitary transforms correspond to matrices (a_{ij}) having the following structure: either $a_{ij} = 0$ or $|a_{ij}| = 1$, besides, only one element differs from zero in each column and each line.

4.4.14. Let η be a Gaussian $\mathbf{B}$-r.e. with a zero mean, $F(x) = P\{\|\eta\| < x\}$. The behaviour of the function F as $x \to \infty$ is fully described in Theorem 1.2, and the behaviour of F as $x \to 0$ (the so-called 'Problem of small balls') is a rather complicated problem. There is a number of works [60], [88], [204], [205] on this problem. We present one result from [88].

THEOREM. *Let η be a Gaussian $\mathbf{l}_2$-r.e. with a diagonal covariance matrix with numbers $\sigma_1^2 \geq \sigma_2^2 \geq \cdots$ on the diagonal. Let $\psi(x, a)$ denote the distribution density of r.v. $\|\eta - a\|^2$. If $\sum_{j=1}^{\infty} a_j^2\sigma_j^{-2} < \infty$, then as $x \uparrow 0$*

$$\psi(x, a) \sim (4\pi)^{-1/2}\left(\sum_{j=1}^{\infty}\sigma_j^4(1 + 2r\sigma_j^2)^{-2}\right)^{-1/2}\prod_{j=1}^{\infty}(1 + 2r\sigma_j^2)^{-1/2} \times$$

$$\times\exp\left\{-\frac{1}{2}\sum_{j=1}^{\infty}a_j^2\sigma_j^2 + \sum_{j=1}^{\infty}r\sigma_j^2(1 + 2r\sigma_j^2)^{-1}\right\},$$

where $r = r(x, a)$ is defined from the relation

$$x = \sum_{j=1}^{\infty}\sigma_j^2(1 + 2r\sigma_j^2)^{-1} + \sum_{j=1}^{\infty}a_j^2(1 + 2r\sigma_j^2)^{-2}.$$

Hence, applying the relation (see [88])

$$P\{\|\eta - a\| < x\} \sim \psi(x, a)(r(x, a))^{-1}$$

as $x \uparrow 0$ which is true if $\Sigma_{j=1}^{\infty} a_j^2 \sigma_j^{-2} < \infty$, we obtain asymptotics of Gaussian measure of small balls.

4.4.15. Let μ_T be a Gaussian measure on **H** with a zero mean and a covariance operator T. In this chapter the properties of $\mu_T(V_r(a))$ as a function of r have been studied. When dealing with the sums of a random number of random elements, the CLT for the mixtures or approximation of distributions of sums of independent random elements by infinitely divisible laws (see [9], [21], [29], [140]) we have to consider $\mu_T(V_r(a))$ as a function of T. We shall present some qualitative results concerning the function $f: S_1 \to \mathbf{R}$, $f(T) = \mu_T(V_r)$, where $r > 0$ is a fixed number and S_1 is a set of Gaussian covariance operators, and is also the convex cone of a Banach space of nuclear operators with a nuclear norm $\|\cdot\|_1$.

THEOREM [137]. *The function f is continuous but not uniformly continuous. It is Fréchet differentiable, but $\|f'(T)\|$ is not bounded on the sphere $\|T\|_1 = r^2$.*

Let μ_{T_1} and μ_{T_2} be two Gaussian measures on **H** with zero means and with covariance operators T_1 and T_2, respectively. It is known that $\mu_{T_1} \sim \mu_{T_2}$ if and only if $T - I_{\mathbf{H}}$ is a Hilbert–Schmidt operator, where T is such that $T_1 = T_2^{1/2} T T_2^{1/2}$.

THEOREM [10]. *If $\mu_{T_1} \sim \mu_{T_2}$, then there exist two absolute constants C_1 and C_2 such that*

$$C_1 \|T - I_{\mathbf{H}}\|_2 \le \|\mu_{T_1} - \mu_{T_2}\| \le C_2 \|T - I_{\mathbf{H}}\|_2,$$

where $\|\cdot\|_2$ is a Hilbert–Schmidt norm.

Let $U = T_1 - T_2$, $S = (U^* U)^{1/2}$, $\lambda_{ij} \ge \lambda_{2j} \ge \lambda_{3j} \ge \cdots$ be the eigenvalues of the operator T_j, $j = 1, 2$.

THEOREM [10]. *The following estimates take place:*

(a)

$$\Delta(T_1, T_2; a) := \sup_{r>0} | \mu_{T_1}(V_r(a)) - \mu_{T_2}(V_r(a)) |$$
$$\le C_3 \operatorname{Tr} S + C_4 \|U\| \, \|a\|^2 (1 + (T_1 a, a) + T_2 a, a))^{-1},$$

where

$$C_3 = C_3(\lambda_{11}, \lambda_{21}, \lambda_{31}, \lambda_{12}, \lambda_{22}, \lambda_{32}), \quad C_4 = C_4(\lambda_{11}, \ldots, \lambda_{51}, \lambda_{12}, \ldots, \lambda_{52}).$$

(b) *If T_1 and T_2 are commutative than*

$$\Delta(T_1, T_2; a) \le C_3 \operatorname{Tr} S + C_4 (Sa, a)(1 + (T_1 a, a) + (T_2 a, a))^{-1}.$$

(c) *If there exists a bounded operator $T_1^{-1/2} T_2 T_1^{-1/2}$, then*

$$\sup_{a \in \mathbf{H}} \Delta(T_1, T_2; a) \le C_5 \|T_1^{-1/2} T_2 T_1^{-1/2} - I_{\mathbf{H}}\|,$$

where $C_5 = C_5(\lambda_{ij}, i = 1, \ldots, 5; j = 1, 2)$.

CHAPTER 5

Estimates of Rate of Convergence in the Central Limit Theorem

In this chapter we shall study the rate of convergence in the CLT in infinite-dimensional separable Banach spaces. The distance on the sets, the Prokhorov metric and the bounded Lipschitz metric are considered as measures of closeness of the distribution of the normed sum of i.i.d. random elements to a Gaussian measure.

5.1. Estimates of the Rate of Convergence on Sets

Throughout this section we shall denote by η a Gaussian $\mathbf{B}$-r.e. with a zero mean and a covariance operator $T = T_\eta$, and we shall denote by $\mathcal{D}(\eta)$ a domain of normal attraction of r.e. η, i.e. a class of all $\mathbf{B}$-r.e. ξ for which the relation $L(S_n(\xi)) \Rightarrow L(\eta)$ as $n \to \infty$ holds. We also denote

$$\Delta_n(\xi, \eta; \mathcal{A}) = \sup_{A \in \mathcal{A}} |P\{S_n(\xi) \in A\} - P\{\eta \in A\}|,$$

where $\mathcal{A} \subset \mathcal{B}(\mathbf{B})$.

In this section two principal questions will be considered:

1. For which $L(\eta)$-uniform classes $\mathcal{A} \subset \mathcal{B}(\mathbf{B})$ is the rate of convergence $\Delta_n(\xi, \eta; \mathcal{A})$ to zero as $n \to \infty$ of power order if $\mathbf{B}$-r.e. $\xi \in \mathcal{D}(\eta)$ satisfies some moment conditions, in particular, if for some $s > 2$ the absolute $E\|\xi\|^s$ or pseudomoment

$$\nu_s(\xi, \eta) := \int_{\mathbf{B}} \|x\|^s \, |L(\xi) - L(\eta)| \, (dx)$$

is finite?

2. If $\mathcal{A}$ is a fixed $L(\eta)$-uniform class in a particular Banach space $\mathbf{B}$, e.g. a class of balls with its centre in zero, then what properties of $\mathbf{B}$-r.e. η and $\xi \in \mathcal{D}(\eta)$ ensure some order of the rate of convergence of $\Delta_n(\xi, \eta; \mathcal{A})$ to zero as $n \to \infty$?

In general, we shall confine ourselves to r.e. $\xi \in \mathcal{D}(\eta)$ with a finite pseudomoment $\nu_3(\xi, \eta)$. For such r.e. in a finite-dimensional space $\Delta_n(\xi, \eta; \mathcal{A}) = O(n^{1/2})$ where $\mathcal{A}$ is a class of all convex sets (see, e.g., [31], [188]). The situation is quite different in infinite-dimensional spaces. Even in the case of the class $\mathbf{W}_0(\mathbf{B})$ (of balls with its centre in zero) it is possible, for different spaces $\mathbf{B}$, to have different rates of convergence $\Delta_n(\xi, \eta; \mathbf{W}_0(\mathbf{B}))$ to zero as $n \to \infty$. If, for instance, in a Hilbert space $\mathbf{H}$ there can be $\Delta_n(\xi, \eta; \mathbf{W}_0(\mathbf{H})) = o(n^{-1/2})$ (see Supplement 5.4.3), then in the space $\mathbf{l}_1$ the order $O(n^{-1/6})$ is optimal (see Theorem 1.24).

One elementary estimate. We shall present the estimate of distance $\Delta_n(\xi, \eta; A)$, where A is some Borel set. The derivation of this estimate, on the one hand, is rather elementary, and, on the other hand, well demonstrates the essence of the method of convolutions successfully applied in infinite-dimensional spaces.

The following elementary inequality of smoothing underlies this method.

LEMMA 1.1. *Let A_1, A_2, A be Borel sets of the space* **B** *and $A_1 \subset A \subset A_2$. For an arbitrary pair of measures $\mu_1, \mu_2 \in \mathcal{P}(\mathbf{B})$ the following inequality takes place:*

$$|\mu_1(A) - \mu_2(A)| \le \max_{i=1,2} \left| \int_{\mathbf{B}} g_i(x)(\mu_1 - \mu_2)(dx) \right| + \min_{i=1,2} \mu_i(A_2 \backslash A_1), \tag{1.1}$$

where the function $g_1 \in \mathcal{F}(A_1, A)$, $g_2 \in \mathcal{F}(A, A_2)$.

Proof. By the definition of the class $\mathcal{F}(A_1, A)$, we have

$$\mu_2(A) - \mu_1(A) \le \int_{\mathbf{B}} g_2(x)(\mu_2 - \mu_1)(dx) + \int_{\mathbf{B}} g_2(x)\mu_1(dx) - \mu_1(A)$$
$$\le \left| \int_{\mathbf{B}} g_2(x)(\mu_2 - \mu_1)(dx) \right| + \mu_1(A_2 \backslash A).$$

Similarly,

$$\mu_2(A) - \mu_1(A) \ge - \left| \int_{\mathbf{B}} g_1(x)(\mu_2 - \mu_1)(dx) \right| - \mu_1(A \backslash A_1).$$

The two inequalities above prove inequality (1.1). □

In this chapter we shall often make use of the following identical decomposition: if μ_i, $\nu_i \subset \mathcal{P}(\mathbf{B})$, $i = 1, ..., n$, then

$$\mu_1 * \cdots * \mu_n - \nu_1 * \cdots * \nu_n = \sum_{i=1}^{n} \mu_1 * \cdots * \mu_{i-1} * (\mu_i - \nu_i) * \nu_{i+1} * \cdots * \nu_n. \tag{1.2}$$

If A is an $L(\eta)$-continuous set, then the quantity

$$\Delta_n(\xi, \eta; A) = |P\{S_n(\xi) \in A\} - P\{\eta \in A\}|$$

tends to zero as $n \to \infty$. We shall estimate this convergence assuming that $E\|\xi\|^3 < \infty$. To this aim, we take in inequality (1.1) $A_1 = A_\varepsilon$, $A_2 = A_{-\varepsilon} := \mathbf{B} \backslash (\mathbf{B} \backslash A)_\varepsilon$, where $\varepsilon > 0$ is some number. We have

$$\Delta_n(\xi, \eta; A) \le I + P\{\eta \in (\partial A)_\varepsilon\}, \tag{1.3}$$

where

$$I = \max_{i=1,2} \left| \int_{\mathbf{B}} g_i(x)(L(S_n(\xi)) - L(\eta))(dx) \right|,$$

$$g_1 \in \mathcal{F}(A, A_\varepsilon), \quad g_2 \in \mathcal{F}(A_{-\varepsilon}, A).$$

Let $\eta_1, ..., \eta_n$ be independent copies of $\mathbf{B}$-r.e. η. Since

$$L(\eta) = L\left(n^{-1/2} \sum_{i=1}^{n} \eta_i\right),$$

we can apply the identical decomposition (1.2) and write

$$L(S_n(\xi)) - L(\eta) = \sum_{k=1}^{n} L(W_{k,n}) * (L(\xi n^{-1/2}) - L(\eta n^{-1/2})), \tag{1.4}$$

where $W_{k,n} := n^{-1/2}(\xi_1 + \cdots + \xi_{k-1} + \eta_{k+1} + \cdots + \eta_n)$. Applying (1.4) we obtain

$$I \le \sum_{k=1}^{n} \max_{i=1,2} \left| \int_{\mathbf{B}} \int_{\mathbf{B}} g_i(n^{-1/2}x + y)(L(\xi) - L(\eta))(dx) L(W_{k,n})(dy) \right|. \tag{1.5}$$

In order to estimate each summand from (1.5) we assume that the functions g_1 and g_2 are not only continuous but also have bounded derivatives up to the third order. To be more exact, we assume that the functions g_1 and g_2 belong to the class $(\mathbf{Q}^3(\mathbf{B}, \mathbf{R}), C_0\mathbf{K}_{1,3})$.

Now we consider the internal integral

$$\int_{\mathbf{B}} g_i(n^{-1/2}x + y)(L(\xi) - L(\eta))(dx).$$

For a fixed $y \in \mathbf{B}$ we write

$$\int_{\mathbf{B}} g_i(n^{-1/2}x + y)(L(\xi) - L(\eta))(dx)$$

$$= \int_{\mathbf{B}} [g(n^{-1/2}x + y) - g_i(y) - g_i'(y)(xn^{-1/2}) - g_i''(y)(xn^{-1/2})^2](L(\xi) - L(\eta))(dx). \quad (1.6)$$

Here the following equalities are used:

$$\int_{\mathbf{B}} g_i'(y)(x)(L(\xi) - L(\eta))(dx) = 0,$$

$$\int_{\mathbf{B}} g_i''(y)(x)^2(L(\xi) - L(\eta))(dx) = 0.$$

The first is a consequence of equality to zero of the mean of r.e. ξ and η, and the second will be proved.

LEMMA 1.2. *If $\xi \in \mathscr{D}(\eta)$ and $E\|\xi\|^2 < \infty$, then, for each $R \in \mathbf{L}^2(\mathbf{B}, \mathbf{R})$, $ER(\xi)^2 = ER(\eta)^2$ is valid.*

Proof. Consider the continuous functions $g_1(x) = \max(0, R(x)^2)$ and $f \in \mathscr{F}(V_r, V_{r+\varepsilon})$, where $r > 0$, $\varepsilon > 0$. By $\tilde{g}_1$ we denote the product $f \cdot g_1$. Taking the same route as in the proof of the equality of smoothing, we easily obtain the following inequality:

$$|Eg_1(S_n(\xi)) - Eg_1(\eta)| \le |E\tilde{g}_1(S_n(\xi)) - E\tilde{g}_1(\eta)| + \sup_{n\ge1}\int_{\|x\|\ge r} g_1(x)L(S_n(\xi))(dx) +$$

$$+ \int_{\|x\|\ge r} \tilde{g}_1(x)L(\eta)(dx) \le |E\tilde{g}_1(S_n(\xi)) - E\tilde{g}_1(\eta)| +$$

$$+ \|R\| \sup_{n\ge1}\int_{\|x\|>r} \|x\|^2 L(S_n(\xi))(dx) +$$

$$+ \|R\| \int_{\|x\|\ge r} \|x\|^2 L(\eta)(dx).$$

Since $\tilde{g}_1 \in \mathbf{C}_0^b(\mathbf{B}, \mathbf{R})$ then the first summand on the right side tends to zero as $n \to \infty$. The second summand tends to zero as $r \to \infty$ due to the conditions $E\|\xi\|^2 < \infty$, $\xi \in \mathscr{D}(\eta)$ (see Supplement 3.3.9). Besides,

$$\lim_{r\to\infty}\int_{\|x\|\ge r} \|x\|^2 L(\eta)(dx) = 0.$$

Hence

$$\lim_{n\to\infty} |Eg_1(S_n(\xi)) - Eg_1(\eta)| = 0.$$

Similarly

$$\lim_{n\to\infty} |Eg_2(S_n(\xi)) - Eg_2(\eta)| = 0,$$

where the function $g_2(x) = \max(0, -R(x)^2)$.

Thus,

$$ER(\xi)^2 = \lim_{n\to\infty} ER(S_n(\xi))^2 = ER(\eta)^2. \square$$

We now return to equality (1.6). Applying the Taylor formula (Theorem 2.1.13) from (1.4) and (1.5) we obtain:

$$\left| \int_{\mathbf{B}} \int_{\mathbf{B}} g_i(xn^{-1/2} + y)(L(\xi) - L(\eta))(dx)L(W_{k,n})(dy) \right|$$
$$\leq C\varepsilon^{-3}n^{-3/2} \int_{\mathbf{B}} \|x\|^3 \,|L(\xi) - L(\eta)|\,(dx) = C\varepsilon^{-3}n^{-3/2}\nu_3(\xi, \eta),$$

for each $k = 1, ..., n$ and for each $i = 1, 2$. Hence we arrive at the estimate

$$I \leq C\varepsilon^{-3}n^{-1/2}\nu_3(\xi, \eta). \tag{1.7}$$

Taking into account the results of Chapter 4, it is natural to assume that the Gaussian $\mathbf{B}-$r.e. η and the set A are such that for each $\varepsilon > 0$

$$P\{\eta \in (\partial A)_\varepsilon\} \leq C(T)\varepsilon$$

with some constant $C(T) > 0$. Hence, and from relations (1.3), (1.5), (1.7), it follows that

$$\Delta_n(\xi, \eta; A) \leq C\varepsilon^{-3}n^{-3/2}\nu_2(\xi, \eta) + C(T)\varepsilon.$$

By minimizing the right side with respect to $\varepsilon > 0$ we arrive at the following result.

THEOREM 1.3. *Let the Gaussian* $\mathbf{B}-$*r.e.* η *and the Borel set* $A \subset \mathbf{B}$ *be such that the following two conditions are satisfied:*

(a) *there exists a constant* $C(T) > 0$ *such that for each* $\varepsilon > 0$ $P\{\eta \in (\partial A)_\varepsilon\} \leq C(T)\varepsilon$;

(b) *for each* $\varepsilon > 0$ *there exists the functions* $g_1, g_2 \in (\mathbf{Q}^3(\mathbf{B}, \mathbf{R}); C\mathbf{K}_1(\varepsilon))$ *such that* $g_1 \in \mathscr{F}(A, A_\varepsilon)$, $g_2 \in \mathscr{F}(A_{-\varepsilon}, A)$.

Then for each $\mathbf{B}-$*r.e.* $\xi \in D(\eta)$ *and for each* $n \geq 1$, *the following estimate takes place:*

$$\Delta_n(\xi, \eta; A) \leq 2C^{1/4}C^{3/4}(T)\nu_3^{1/4}(\xi, \eta)n^{-1/8}. \tag{1.8}$$

Some notes should be given concerning conditions (a) and (b).

If, for instance, $\mathbf{B} = \mathbf{C}[0, 1]$ and A is a bounded set, then by the results of Chapter 2, condition (b) is not satisfied. In this case Theorem 1.3 becomes meaningless. To avoid that, it would be natural to substitute the requirement for Fréchet differentiability of the functions g_1, g_2 by differentiability with respect to some operator $u \in \mathbf{L}(\mathbf{F}, \mathbf{B})$. The corresponding substitution is possible if $L(\xi) = L(u(X))$, $L(\eta) = L(u(Y))$, where X, Y are $\mathbf{F}-$r.e. As we can judge by the results of Chapter 3, such a situation often arises (in spaces of cotype 2 or in a space $\mathbf{C}(S)$ if ξ is concentrated in some $\mathrm{Lip}_\rho(S)$). It gives us the idea to consider the distance $\Delta_{n,u}(X, Y; \mathscr{A}) := \Delta_n(u(X), u(Y) : \mathscr{A})$ instead of the distance $\Delta_n(\xi, \eta; \mathscr{A})$. It is to this statement of the problem that our attention will be given in the items below.

As far as condition (a) is concerned, no additional information on the structure of the constant $C(T)$ is required for estimate (1.8). As we shall see (condition 2 of Theorem 1.7), the 'good' behaviour of the function $s \to C(sT)$ in many respects contributes to making estimate (1.8) more precise for some sets $A \in \mathscr{B}(\mathbf{B})$.

Estimate on uniform classes of sets. Let $\mathbf{F}$ be a Banach space (not necessarily separable), the operator $u \in \mathbf{L}(\mathbf{F}, \mathbf{B})$. By Y we shall denote a Gaussian $\mathbf{F}-$r.e. with a zero mean, $T = T_Y := \operatorname{cov} Y$, and let $\mathscr{D}^*(Y)$ be a class of all $\mathbf{F}-$r.e. X such that $EX = 0$, $ES(X)^2 = ES(Y)^2$ for each symmetric operator $S \in \mathbf{L}^2(\mathbf{F}, \mathbf{R})$.

Consider the class $\mathscr{W} = \{W_t, t \in \mathbf{R}\}$, indexed by a one-dimensional parameter and we assume that the relation $W_t \subseteq W_{t'}$, if $t < t'$, holds.

DEFINITION 1.4. The class $\mathscr{W} = \{W_t, t \in \mathbf{R}\}$ will be called absorbing if the following relations hold:

(1) For all $x \in \mathbf{B}$, $t \in \mathbf{R}$, $W_{t-\|x\|} \subset W_t + x \subset W_{t+\|x\|}$;

(2) For all $t \in \mathbf{R}$, $\delta > 0$, $\overline{W}_t \subset W_{t+\delta}$, $\dot{W}_t \supset W_{t-\delta}$.

EXAMPLE 1.5. If the function $f: \mathbf{B} \to \mathbf{R}$ satisfies the Lipschitz condition $|f(x) - f(y)| \leq \|x - y\|$, then the class $\mathscr{W}_f = \{W_t(f), t \in \mathbf{R}\}$ is absorbing.

EXAMPLE 1.6. Let A be a non-empty measurable set of the space $\mathbf{B}$. We shall define the class $\mathscr{W} = \mathscr{W}_A = \{W_{A,t}, t \in \mathbf{R}\}$ assuming that

$$W_{A,t} = \begin{cases} \varnothing & \text{if } t < 0, \\ A_t & \text{if } t \geq 0, \end{cases} \quad (A_t = \widetilde{A}, \text{ if } t = 0).$$

The class $\mathscr{W}_A$ is absorbing. The check of properties (1) and (2) for $\mathscr{W}_f$ and $\mathscr{W}_A$ offers no difficulty. Note that in the case $A = \{a\}$, $a \in \mathbf{B}$, the class (which will hereafter be denoted by $\mathscr{W}_a$ or $\mathscr{W}_a(\mathbf{B})$ if we wish to point out a space) consists of balls (considering an empty set as a ball with a negative radius) with the centre at the point a. But if $A = V_r(a)$, then $\mathscr{W}_A$ consists of an empty set and balls with a centre a and radius $t \geq r$.

THEOREM 1.7. *If the class* $\mathscr{W} = \{W_t, t \in \mathbf{R}\}$, *the operator* $u \in \mathbf{L}(\mathbf{F}, \mathbf{B})$, $\|u\| \leq 1$, *and the Gaussian* $\mathbf{F}$*-r.e. satisfy the conditions:*

(1) *the class* $\mathscr{W}$ *is absorbing and* $(\mathbf{Q}_u^3, \mathbf{CK}_{1,3})$*-uniform;*

(2) *for all* $s \in (0, 1]$ *and* $\varepsilon > 0$.

$$\sup_{t \in \mathbf{R}} P\{s^{1/2} u(Y) \in W_{t+\varepsilon} \backslash W_{t-\varepsilon}\} \leq \phi(sT)\varepsilon, \tag{1.9}$$

where the function $\phi: s \to \phi(sT)$, $s \in (0, 1]$, *satisfies the condition*

$$\tau(T) \equiv \tau := \sup\left\{\frac{\phi(s'T)}{\phi(sT)} : \frac{s}{2} < s' \leq s \leq 1\right\} < \infty,$$

then for each $\mathbf{F}$*-r.e.* $X \in \mathscr{D}^*(Y)$ *and all* $n \geq 1$ *the following estimate takes place:*

$$\Delta_{n,u}(X, Y; \mathscr{W}) \leq 2\left(\frac{31\tau}{6}\right)^{1/3} \max\left\{\frac{\delta_u(\mathscr{W}, n)n^{1/2}}{\phi(n^{-1}T)}; C^{1/3}\nu_3^{1/3}(X, Y)\right\}\phi(T)n^{-1/6}, \tag{1.10}$$

where

$$\delta_u(\mathscr{W}, n) := \sup\{\sup_{t \in \mathbf{R}} |P\{u(Z) \in n^{1/2} W_t\} -$$
$$- P\{u(Y) \in n^{1/2} W_t\}| : Z \in \mathscr{D}^*(Y) : \nu_3(Z, Y) \leq \nu_3(X, Y)\}.$$

Before proving the theorem we shall present one auxiliary statement which will be of use subsequently.

For an integer $n \geq 1$ and a positive number $L > 0$, we define the class $U_{n,L}(Y)$, consisting of all possible vectors $(Z_1, \ldots, Z_n)$, where $\mathbf{F}$-r.e. $Z_i \in \mathscr{D}^*(Y)$ are independent and such that $\nu_3(Z_i, Y) \leq L$, $i = 1, \ldots, n$. We shall assume that the number $n \geq 1$ is fixed. We also introduce the following notations:

$$\kappa_k := \kappa_k(X, Y; \mathscr{W}, n, L) := \sup\left\{\sup_{t\in\mathbf{R}} \left| P\left\{\sum_{i=1}^{k} u(Z_i)n^{-1/2} \in W_t\right\} - \right.\right.$$

$$\left.\left. - P\left\{\sum_{i=1}^{k} u(Y_i)n^{-1/2} \in W_t\right\} \right| : (Z_1, ..., Z_k) \in U_{k,L}(Y)\right\},$$

$k = 1, ..., n$; $Y_1, ..., Y_n$ are independent copies of r.e. Y

$$\psi(\varepsilon, sT) := \sup_{t\in\mathbf{R}} P\{s^{1/2}u(Y) \in W_{t+\varepsilon}\backslash W_{t-\varepsilon}\},$$

$$\psi_1(\varepsilon, sT) := \sup_{r>\varepsilon/2} r^{-1}\psi(r, sT).$$

LEMMA 1.8. *Let $n \geq 2$. If the class $\mathscr{W}$ is absorbing and $(\mathbf{Q}_u^3, \mathbf{K})$-uniform with the function $\mathbf{K}(\varepsilon) = (K_1(\varepsilon), K_2(\varepsilon), K_3(\varepsilon))$, then for each $k = 2, ..., n$ and for each $\varepsilon > 0$ the following recurrent inequality takes place:*

$$\kappa_k \leq 2\max(K_2(\varepsilon)\varepsilon^{-1}; K_3(\varepsilon))n^{-3/2}\kappa_{k-1}Lk +$$

$$+ \frac{7}{6}\max\left\{K_2(\varepsilon)\psi_1\left(\varepsilon, \frac{k-1}{n}T\right); K_3(\varepsilon)\psi\left(\frac{\varepsilon}{2}, \frac{k-1}{n}T\right)\right\}n^{-3/2}Lk +$$

$$+ \psi\left(\varepsilon, \frac{k}{n}T\right)\Bigg\}. \tag{1.11}$$

Proof. Let $(Z_1, ..., Z_k) \in U_{k,L}(Y)$. We shall make use of the inequality of smoothing (1.1) with $A = W_t, A_1 = W_{t-\varepsilon}, A_2 = W_{t+\varepsilon}, t \in \mathbf{R}, \varepsilon > 0$. We have

$$\left| P\left\{\sum_{i=1}^{k} u(Z_i)n^{-1/2} \in W_t\right\} - P\left\{\sum_{i=1}^{k} u(Y_i)n^{-1/2} \in W_t\right\} \right| \leq I + \psi\left(\varepsilon, \frac{k}{n}T\right).$$

where

$$I := \max_{j=1,2} \left| \int_{\mathbf{B}} g_j(x)\left\{L\left(\sum_{i=1}^{k} u(Z_i)n^{-1/2}\right) - L\left(\sum_{i=1}^{k} u(Y_i)n^{-1/2}\right)\right\}(dx) \right|,$$

$$g_1 \in (\mathbf{Q}_u^3, \mathbf{K}(\varepsilon)) \cap \mathscr{F}(W_t, W_{t+\varepsilon}),\ g_2 \in (\mathbf{Q}_u^3, \mathbf{K}(\varepsilon)) \cap \mathscr{F}(W_{t-\varepsilon}, W_t).$$

The existence of the functions g_1, g_2 is ensured by the condition $(\mathbf{Q}_u^3, \mathbf{K})$-uniformity of the class $\mathscr{W}$. By g we shall denote the function equal either to g_1 or to g_2. Note that from the definition of the class $\mathscr{F}(W_t, W_{t+\varepsilon})$ we have

$$|g_u^{(l)}(y)(x)^l| \leq K_l(\varepsilon)\chi_{\overline{W_{t+\varepsilon}\backslash W_{t-\varepsilon}}}(y + u(x)),\ l = 1, 2, 3 \tag{1.12}$$

for all $y \in \mathbf{B}$ and $x \in \mathbf{F}$.

Then let

$$\rho_{l,k} := L\left(\sum_{i=1}^{l-1} u(Z_i)n^{-1/2}\right) * L\left(\sum_{i=1}^{k-l} u(Y_i)n^{-1/2}\right),$$

where the number $l = 1, ..., k$. Applying the identical expansion (1.2), we obtain

$$I \leq \sum_{l=1}^{k} \left| \int_{\mathbf{F}}\int_{\mathbf{B}} g(n^{-1/2}u(x) + y)\rho_{l,k}(dy)(L(Z_l) - L(Y))(dx) \right|$$

$$= \sum_{l=1}^{k} \left| \int_{\mathbf{B}}\int_{\mathbf{B}} R(x, y)\rho_{l,k}(dy)(L(Z_l) - L(Y))(dx) \right|,$$

where

$$R(x, y) = g(y + n^{-1/2}u(x)) - g(y) - g_u'(y)(n^{-1/2}x) - \frac{1}{2}g_u''(y)(n^{-1/2}x)^2.$$

Denoting

$$I_{1,l} = \left| \int_{\|x\| \geq \varepsilon n^{1/2}} \int_{\mathbf{B}} R(x, y)\rho_{l,k}(dy)(L(Z_l) - L(Y))(dx) \right|,$$

$$I_{2,l} = \left| \int_{\|x\| < \varepsilon n^{1/2}} \int_{\mathbf{B}} R(x, y)\rho_{l,k}(dy)(L(Z_l) - L(Y))(dx) \right|,$$

we have

$$I \leq \sum_{l=1}^{k} (I_{1,l} + I_{2,l})$$

and the integrals $I_{1,l}$, $i = 1, 2$; $l = 1, \dots, k$ must be estimated. Therefore, we shall apply the Taylor formula (2.1.14) and relation (1.12). We obtain

$$|R(x, y)| \leq \frac{1}{2} \, |g_u''(y)(n^{-1/2}x)^2| + \left| \int_0^1 (1 - v)g_u''(y + n^{-1/2}vu(x))(n^{-1/2}x)^2 \, dv \right|$$

$$\leq 2^{-1}K_2(\varepsilon)n^{-1}\|x\|^2 \chi_{\overline{W_{t+\varepsilon}\backslash W_{t-\varepsilon}}}(y) +$$

$$+ \int_0^1 (1 - v)K_2(\varepsilon)\|x\|^2 n^{-1}\chi_{\overline{W_{t+\varepsilon}\backslash W_{t-\varepsilon}}}(y + n^{-1/2}vu(x)) \, dv,$$

$$I_{1,l} \leq 2^{-1}K_2(\varepsilon)n^{-1} \int_{\|x\| > \varepsilon n^{1/2}} \|x\|^2 \rho_{l,k}(\overline{W_{t+\varepsilon}\backslash W_{t-\varepsilon}}) \, |L(Z_l) - L(Y)| \, (dx) +$$

$$+ K_2(\varepsilon)n^{-1} \int_0^1 \int_{\|x\| > \varepsilon n^{1/2}} \|x\|^2 \rho_{l,k}(\overline{W_{t+\varepsilon}\backslash W_{t-\varepsilon}} - n^{-1.2}vu(x)) \times$$

$$\times \, |L(Z_l) - L(Y)| \, (dx) \, dv. \tag{1.13}$$

Note that the vector $(Z_1, \dots, Z_{l-1}, Y_1, \dots, Y_{k-l}) \in U_{k-1,L}(Y)$. According to the properties of the class $\mathcal{W}$, for each $x \in \mathbf{B}$ and for each $\delta > 0$

$$\rho_{l,k}(\overline{W_{t+\varepsilon}\backslash W_{t-\varepsilon}} + x) \leq \rho_{l,k}(W_{t+\varepsilon+\delta+\|x\|}) - \rho_{l,k}(W_{t-\varepsilon-\delta-\|x\|})$$

$$\leq 2\kappa_{k-1} + \psi\left(\varepsilon + \delta + \|x\|, \frac{k-1}{n}T\right).$$

Taking into account the arbitrariness of the number $\delta > 0$ and continuity of the function ψ, we obtain the estimate

$$\rho_{l,k}(\overline{W_{t+\varepsilon}\backslash W_{t-\varepsilon}} + x) \leq 2\kappa_{k-1} + \psi\left(\varepsilon + \|x\|, \frac{k-1}{n}T\right). \tag{1.14}$$

By substituting this estimate in (1.13), we obtain the following inequalities:

$$I_{1,l} \leq 2^{-1}K_2(\varepsilon)n^{-1} \int_{\|x\| > \varepsilon n^{1/2}} \|x\|^2 \left(2\kappa_{k-1} + \psi\left(\varepsilon, \frac{k-1}{n}T\right)\right) \times$$

$$\times \, |L(Z_l) - L(Y)| \, (dx) + \int_0^1 \int_{\|x\| > \varepsilon n^{1/2}} K_2(\varepsilon)n^{-1}\|x\|^2(1 - v) \times$$

$$\times \left(2\kappa_{k-1} + \psi\left(\varepsilon + n^{-1/2}v\|x\|, \frac{k-1}{n}T\right)\right) \, |L(Z_l) - L(Y)| \, (dx) \, dv$$

$$\leq \left\{2\varkappa_{k-1}K_2(\varepsilon)\varepsilon^{-1}n^{-3/2} + \frac{7}{6}K_2(\varepsilon)\psi_1\left(\varepsilon, \frac{k-1}{n}T\right)n^{-3/2}\right\} \times$$

$$\times \int_{\|x\| > \varepsilon n^{1/2}} \|x\|^3 \ |L(Z_l) - L(Y)|(dx). \tag{1.15}$$

Similarly we also estimate the integral $I_{2,l}$, but now we shall use the estimate

$$|R(x, y)| \leq \left| \int_0^1 \frac{(1-v)^2}{2} g_u'''(y + vu(x))(n^{-1/2}x)^3 \, dv \right|$$

$$\leq K_3(\varepsilon)\|x\|^3 n^{-3/2} \int_0^1 2^{-1}(1-v)^2 \chi_{\overline{W_{t+\varepsilon} \ W_{t-\varepsilon}}}(y + n^{-1/2}v \cdot u(x)) \, dv.$$

Taking into account inequality (1.14), we obtain

$$I_{2,l} \leq \left\{3^{-1}K_3(\varepsilon)n^{-3/2}\delta_{k-1} + 6^{-1}K_3(\varepsilon)\psi\left(\frac{\varepsilon}{2}, \frac{k-1}{n}T\right)n^{-3/2}\right\} \times$$

$$\times \int_{\|x\| < \varepsilon n^{1/2}} \|x\|^3 \ |L(Z_l) - L(Y)|(dx). \tag{1.16}$$

From estimates (1.15) and (1.16) we easily obtain (1.11). □

Proof of Theorem 1.7. For $n = 1$, estimate (1.10) is obvious. In the sequel, let the number $n \geq 2$ be fixed. We shall prove that for each $k = 1, \ldots, n$ the following estimate is valid

$$\varkappa_k \leq 2\left(\frac{31}{6}\tau\right)^{1/3} V_{k,n} \tag{1.17}$$

where

$$V_{k,n} = \max\left\{\frac{\delta_u n^{1/2}}{\phi(n^{-1}T)}, K^{1/3}\nu_3^{1/3}(X, Y)\phi\left(\frac{k}{n}T\right)k^{1/3}n^{-1/2}\right\}.$$

The theorem will be proved by induction on k. For $k = 1$, estimate (1.17) is obvious. Assume that

$$\varkappa_{k-1} \leq 2\left(\frac{31}{6}\tau\right)^{1/3} V_{k-1,n}, \ 2 \leq k \leq n.$$

From condition (2) we get $V_{k-1,n} \leq \tau V_{k,n}$. Applying this relation and the recurrent inequality (1.11), we easily obtain the inequality

$$\varkappa_k \leq 2Kn^{-3/2}\varepsilon^{-3}\tau C_\tau V_{k,n} k\nu_3(X, Y) +$$

$$+ \frac{7}{6}Kn^{-3/2}\tau\phi\left(\frac{k}{n}T\right)k\nu_3(X, Y)\varepsilon^{-1} + \phi\left(\frac{k}{n}T\right)\varepsilon, \tag{1.18}$$

where $C_\tau := 2(31\tau/6)^{1/3}$.

Assuming in (1.18) $\varepsilon = 2^{-1}C_\tau(K\nu_3(X, Y)kn^{-3/2})^{1/3}$, we have $\varkappa_k \leq C_\tau V_{k,n}$. Thus, (1.17) is proved. There remains to note that $\Delta_{n,u}(X, Y, \mathscr{W}) \leq \varkappa_n$ and to use (1.17) with $k = n$. □

Two remarks concerning estimate (1.10) are necessary.

REMARK 1.9. If in the conditions of Theorem 1.7 instead of (1.9) the inequality

$$\sup_{t\in\mathbf{R}} P\{s^{1/2}u(Y)\in W_{t+\varepsilon}\setminus W_{t-\varepsilon}\}\le\phi(sT)\psi(\varepsilon)$$

holds, where $\psi:\mathbf{R}\to\mathbf{R}$ is a monotone functions, $\psi(0)=0$, $\lim_{t\to\infty}\psi(t)t^{-1}=0$, $\lim_{t\to 0}\psi(t)t^{-1}=\infty$, then by making in the proof of Theorem 1.7 some necessary changes of a technical character, we obtain the estimate

$$\Delta_{n,u}(X,Y;\mathcal{W})\le 2\left(\frac{31r}{6}\right)^{1/3}\max\left\{\frac{\delta_u(\mathcal{W},n)n^{1/2}}{\phi(n^{-1}T)n^{1/6}},\ \psi(K_{1/3}\nu_3^{1/3}(X,Y)n^{-1/6})\right\}\phi(T).$$

REMARK 1.10. The condition that the class $\mathcal{W}$ is absorbing can be weakened by substituting condition (2) in Definition 1.4 for the following relation: for some $\alpha\in(0,1]$ and for all $t\in\mathbf{R}$, $x\in\mathbf{B}$

$$W_{t-\|x\|^\alpha}\subset W_t+x\subset W_{t+\|x\|^\alpha}.$$

The class $\mathcal{W}_f$ can be an example where the function $f:\mathbf{B}\to\mathbf{R}$ satisfies the Hölder condition with the index α:

$$|f(x)-f(y)|\le\|x-y\|^\alpha.$$

The estimate of the distance $\Delta_{n,u}(X,Y;\mathcal{W})$ for such a class $\mathcal{W}$ will differ from (1.10).

Some consequences. Theorem 1.7 is rather general. One can obtain from it a number of estimates, both uniform and non-uniform. It is appropriate here to explain what is meant under the terms 'uniform' or 'non-uniform' estimates, and the easiest way to do this is by means of an example. If we consider the quantity

$$\sup_{r>0}|p\{S_n(\xi)\in V_r(a)\}-P\{\eta\in V_r(a)\}|,$$

then the estimate of such a quantity as a function of n will be called uniform (though here one should bear in mind that there remains non-uniformity, usually in the bad sense of the word, depending on the centre of the balls a), and if we estimate

$$|P\{S_n(\xi)\in V_r(a)\}-P\{\eta\in V_r(a)\}|\le\phi(n)f(r),$$

where $\phi(n)\to 0$ and $f(r)\to 0$ as $n\to\infty$, $r\to\infty$, then such an estimate is called non-uniform. We shall show that in spite of the fact that estimate (1.10) is closer to uniform estimates, by choosing the class $\mathcal{W}$ in a special way, one can also obtain non-uniform estimates.

At first we shall present an inequality useful at estimating the quantity $\delta_u(\mathcal{W},n)$, and also being of independent interest.

THEOREM 1.11. *If* $\mathbf{F}$*−r.e.* Y *is infinite-dimensional, then for each* $\varepsilon>0$ *there exists a constant* $C(\varepsilon,T_Y)$ *such that for any* $\mathbf{F}$*−r.e.* Z *the following inequality takes place:*

$$|L(Z)-L(Y)|(\mathbf{F})\le C(\varepsilon,T_Y)\nu_3^{1-\varepsilon}(Z,Y). \tag{1.19}$$

Proof. For each $N\ge 1$ there exist elements $f_1,\dots,f_N\in\mathbf{F}^*$ such that $\gamma_N:=(f_1(Y),\dots,f_N(Y))$ is a non-degenerate Gaussian $\mathbf{R}_N$−r.e. Besides, one can assume $\Sigma_{i=1}^N\|f_i\|^2=1$. We shall define the semi-norm in the space $\mathbf{F}$ assuming for $x\in\mathbf{F}$, $q(x)=(\Sigma_{i=1}^N f_i^2(x))^{1/2}$. Note that $q(x)\le\|x\|$.

Suppose, first, that $0<\nu:=|L(Z)-L(Y)|(\mathbf{F})<2$. Let $\mathbf{F}=\mathbf{F}^+\cup\mathbf{F}^-$ be a Hahn decomposition of the space $\mathbf{F}$ with respect to $L(Z)-L(Y)$. We shall define the measure μ assuming

$$\mu(A) = L(Z)(A \cap \mathbf{F}^+) + L(Y)(A \cap \mathbf{F}^-),\ A \in \mathscr{B}(\mathbf{F}).$$

Then

$$\int_{\mathbf{F}} \|x\|^3\ |L(Z) - L(Y)|(dx) \geq \int_{\mathbf{F}} q^3(x)(L(Y) - \mu)(dx). \tag{1.20}$$

Choosing $a > 0$ such that $2^{-1}\nu = P(q(Y) \leq a)$ and following the proof of Lemma 1 from [188, p. 12], we obtain

$$\int_{\mathbf{F}} q^3(x)(L(Y) - \mu)(dx) \geq \int_{q(x) \leq a} q^3(x)L(Y)(dx). \tag{1.21}$$

From relations (1.20), (1.21) we have

$$|L(Z) - L(Y)|(\mathbf{F})$$
$$\leq 2P\{q(Y) \leq a\}\left(\int_{q(x) \leq a} q^3(x)L(Y)(dx)\right)^{-N/N+1} \nu_3^{N/N+1}(Z, Y). \tag{1.22}$$

It is easy to check that

$$\sup_{a>0} 2P\{q(Y) \leq a\}\left(\int_{q(x) \leq a} q^3(x)L(Y)(dx)\right)^{-N/N+1} < \infty.$$

Thus, the statement of the theorem in the case $0 < \nu < 2$ follows from (1.22). If $\nu = 0$, then (1.19) is obvious. But if $\nu = 2$, then $L(Z)(\mathbf{F}^+) = 1$, $L(Z)(\mathbf{F}^-) = 0$ and

$$\nu_3^{1-\varepsilon}(Z, Y) \geq \left(\int_{\mathbf{F}} \|x\|^3 L(Y)(dx)\right)^{1-\varepsilon}$$
$$= 2^{-1}\,|L(Z) - L(Y)|(\mathbf{F})\left(\int_{\mathbf{F}} \|x\|^3 L(Y)(dx)\right)^{1-\varepsilon}. \square$$

REMARK 1.12. It can easily be seen that $C(\varepsilon, T_Y) \to \infty$ as $\varepsilon \to 0$. Now, applying estimate (1.19) instead of $\delta_u(\mathscr{W}, n)$ in (1.10), we can substitute the quantity $C(\varepsilon, T_Y)\nu_3^{1-\varepsilon}(X, Y)$. But in order to obtain a uniform estimate (in the sense that the dependence on n in the estimate is separated from other parameters) of the quantity $\Delta_{n,u}(X, Y; \mathscr{W})$, it is necessary to have an explicit dependence on s in condition (1.9).

THEOREM 1.13. *If condition* (1) *of Theorem* 1.7 *is satisfied and for all* $\varepsilon > 0$ *and* $s \in (0, 1]$

$$\sup_{t \in \mathbf{R}} P\{\sqrt{s}\, u(Y) \in W_{t+\varepsilon} \setminus W_{t-\varepsilon}\} \leq \phi(T)s^{-1/2}\varepsilon,$$

then for all $n \geq 1$ *and for all* $X \in \mathscr{D}^*(Y)$ *the following estimate takes place:*

$$\Delta_{n,u}(X, Y; \mathscr{W}) \leq C_0 \max\{C(\varepsilon, T)\nu_3^{1-\varepsilon}(X, Y),\ C^{1/3}\phi(T)\nu_3^{1/3}(X, Y)\}n^{-1/6}, \tag{1.23}$$

where $C_0 = 2(31\sqrt{2}/6)^{1/3}$. *In particular,*

$$\Delta_{n,u}(X, Y; \mathscr{W}) \leq C(T)\nu_3^{1/3}(X, Y)n^{-1/6}.$$

Proof. Estimate (1.23) directly follows from (1.10) and from Theorem 1.11. □

COROLLARY 1.14. *Let the function* $f \in \mathrm{Lip}_1(\mathbf{B})$ *be positively homogeneous, i.e.* $f(tx) = tf(x)$ *for all* $x \in \mathbf{B}$ *and* $t > 0$. *If the class* $\mathscr{W}_f$ *is* $(\mathbf{Q}_u^3, \mathbf{CK}_{1,3})$*-uniform and r.v.* $f(u(Y))$ *has a bounded density, then for all* $n \geq 1$ *and for all* $X \in \mathscr{D}^*(Y)$ *the following estimate takes place:*

$$\Delta_{n,u}(X, Y; \mathscr{W}_f) \leq C(T, C)\nu_3^{1/3}(X, Y)n^{-1/6}.$$

Proof. Since $f \in \mathrm{Lip}_1(\mathbf{B})$, the class $\mathscr{W}_f$ is absorbing (see Example 1.5). Due to the homogeneity of f and the boundedness of the density of r.v. $f(u(Y))$, (1.9) holds, where

$\phi(sT) = s^{-1/2}\phi(T)$. Consequently, it is sufficient to choose $\varepsilon = 2/3$ in estimate (1.23). □

In the particular case $f = \|\cdot\|$ we obtain the following corollaries.

COROLLARY 1.15. *If the class* $\mathbf{W}_0(\mathbf{B})$ *is* $(\mathbf{Q}_u^3, C_0\mathbf{K}_1)$*-uniform and r.v.* $\|u(Y)\|$ *has a bounded density, then there exists a constant* $C(T)$ *such that for all* $X \in \mathscr{D}^*(Y)$ *and for all* $n \geq 1$ *the following estimate takes place:*

$$\Delta_{n,u}(X, Y; \mathbf{W}_0(\mathbf{B})) \leq C(T)\nu_3^{1/3}(X, Y)n^{-1/6}.$$

COROLLARY 1.16. *Let the norm* $\phi(\cdot) = \|\cdot\|$ *of the space* $\mathbf{B}$ *satisfy the conditions* $\phi \in \mathbf{D}^3(\mathbf{B}\ \{0\}, \mathbf{R})$ *and* $\sup_{\|x\|=1}\|\phi^{(i)}(x)\| \leq C$, $i = 1, 2, 3$.

If the density of r.v. $\|\eta\|$ *is bounded, then there exists a constant such that for each r.e.* $\xi \in \mathscr{D}(\eta)$ *the following estimate takes place:*

$$\Delta_n(\xi, n; \mathbf{W}_0(\mathbf{B})) \leq C(T_\eta)\nu_3^{1/3}(\xi, \eta)n^{-1/6}. \tag{1.24}$$

Proof. By Theorem 2.2.22 the class $\mathbf{W}_0(\mathbf{B})$ is $(\mathbf{Q}^3, C\mathbf{K}_{1,3})$-uniform. Thus, estimate (1.24) is a particular case of Corollary 1.15. □

Consider the class $\mathbf{W}_a(\mathbf{B})$ of balls with the centre at the point $a \in \mathbf{B}$. According to the results of Chapter 4, in a general case the density $p_{T,a}$ of r.v. $\|u(Y) + a\|$ depends on the vector a. If this dependence is of a 'power character', then the estimate of the quantity $\Delta_{n,u}(X, Y; \mathbf{W}_a(\mathbf{B}))$ can be obtained as a consequence of Theorem 1.7.

COROLLARY 1.17. *Let the class* $\mathbf{W}_0(\mathbf{B})$ *be* $(\mathbf{Q}_u^3, C\mathbf{K}_{1,3})$*-uniform and let there exists a constant* $C(T, a)$ *such that for each* $\varepsilon > 0$

$$\sup_{t>0} P\{t - \varepsilon \leq \|u(Y) + a\| \leq t + \varepsilon\} \leq C(T, a)\varepsilon.$$

If the function ϕ: $\mathbf{R}^+ \to \mathbf{R}^+$, $\phi(s) = C(T, as)$ *is non-decreasing and satisfies the conditions that there exists a constant* C_1 *such that* $\phi(2s) \leq C_1\phi(s)$ *for all* $s \in \mathbf{R}^+$, *then for all* $X \in \mathscr{D}^*(Y)$ *and for all* $n \geq 1$ *the following estimate takes place:*

$$\Delta_{n,u}(X, Y; \mathbf{W}_a(\mathbf{B})) \leq$$
$$\leq C_2 \max\{C(\varepsilon, T)\nu_3^{1-\varepsilon}(X, Y); C^{1/3}C(T, a)\nu_3^{1/3}(X, Y)\}n^{-1/6}, \tag{1.25}$$

where $C_2 = 2(31\sqrt{2}\,C_1/6)^{1/3}$. *In particular,*

$$\Delta_{n,u}(X, Y; \mathbf{W}_a(\mathbf{B})) \leq C_2\max\left\{C\left(\frac{2}{3}; T\right); C^{1/3}C(T, a)\right\}\nu_3^{1/3}(X, Y)n^{-1/6}.$$

Proof. The $(\mathbf{Q}_u^3, C_0\mathbf{K}_{1,3})$-uniformity of the class $\mathbf{W}_a(\mathbf{B})$ follows from the $(\mathbf{Q}_u^3, C_0\mathbf{K}_{1,3})$-uniformity of the class $\mathbf{W}_o(\mathbf{B})$. It is easy to check that condition (2) of Theorem 1.7, where $r \leq c_1\sqrt{2}$, is satisfied. Now estimate (1.25) follows from (1.10) and Theorem 1.11. □

In order to obtain a non-uniform estimate from (1.10), we assume that the class under consideration, $\mathscr{W}$, has the following property: there exists some number $t_0 \geq 0$ such that $W_t = \emptyset$, if $t < t_0$ and $W_{t_0} \neq \emptyset$. Besides, we assume that $d(\mathscr{W}) := d(\partial W_{t_0}, 0) \leq d(\partial W_t, 0)$ for all $t > t_0$ where, as usual,

$$d(A, x) = \inf\{\|x - y\| : y \in A\}.$$

THEOREM 1.18. *If the class* $\mathscr{W}$ *has the property formulated above, condition* (1) *of Theorem* 1.7 *is satisfied and for all* $\varepsilon > 0$ *and* $s \in (0, 1]$

$$\sup_{t\in\mathbf{R}} P\{s^{1/2}u(Y)\in W_{t+\varepsilon}\setminus W_{t-\varepsilon}\}\le C(T)s(d(\mathcal{W}))^{-3}\varepsilon,$$

then, for all $n\ge 1$ and for all $X\in\mathcal{D}^*(Y)$, the following estimate takes place:

$$\Delta_{n,u}(X, Y;\mathcal{W})\le C_0\max(\nu_3(X, Y), C(T)\nu_3^{1/3}(X, Y)C^{1/3})d^{-3}(\mathcal{W})n^{-1/6}, \tag{1.26}$$

where $C_0 = 2(31/6)^{1/3}$.

Proof. We denote

$$\widetilde{W}_t = \begin{cases} W_t, & \text{if } 0\,\overline{\in}\, W_t,\\ W_t^c, & \text{if } 0\in W_t.\end{cases}$$

Then for each $Z\in\mathbf{L}_0(\mathbf{F})$

$$\begin{aligned} &n^{3/2}d^3(\mathcal{W})\ |P\{Z\in n^{1/2}W_t\} - P\{Y\in n^{1/2}W_t\}|\\ &\quad\le n^{3/2}d^3(0,\partial W_t)\ |P\{Z\in n^{1/2}\widetilde{W}_t\} - P\{Y\in n^{1/2}\widetilde{W}_t\}|\\ &\quad= d^3(0,\partial(n^{1/2}W_t))\ |P\{Z\in n^{1/2}\widetilde{W}_t\} - P\{Y\in n^{1/2}\widetilde{W}_t\}|\le\nu_3(Z, Y).\end{aligned}$$

Hence we obtain the estimate for $\delta_u(\mathcal{W}, n)$:

$$\delta_u(\mathcal{W}, n)\le n^{-3/2}d^{-3}(\mathcal{W})\nu_3(X, Y).$$

Now, taking into account that $\phi(n^{-1}T) = C(T)n^{-1}d^{-3}(\mathcal{W})$, estimate (1.26) follows from (1.10). □

COROLLARY 1.19. *Let the class $\mathbf{W}_0(\mathbf{B})$ be $(\mathbf{Q}_u^3, C\mathbf{K}_{1,3})$-uniform. There exist constants $C_1 = C_1(T)$ and $C_2 = C_2(T)$ such that for all $X\in\mathcal{D}^*(Y)$, $n\ge 1$ and $r\ge C_1$, the following estimate takes place:*

$$\begin{aligned} &|P\{\|S_n(u(X))\| < r\} - P\{\|u(Y)\| < r\}|\\ &\quad\le C_2\max\{\nu_3(X, Y);\nu_3^{1/3}(X, Y)\}r^{-3}n^{-1/6}.\end{aligned} \tag{1.27}$$

Proof. By Theorem 4.12, there exists constants $C_1 = C_1(T)$ and $\widetilde{C}_2 = \widetilde{C}_2(T)$ such that

$$\sup_{t\ge r} P\{(t-\varepsilon)s^{-1/2}\le\|u(Y)\|\le(t+\varepsilon)s^{-1/2}\}\le\widetilde{C}_2(T)s\varepsilon r^{-3}$$

for all $r > C_1$. Also taking into account that $d(\mathcal{W}_{V_r}) = r$, estimate (1.27) is obtained from (1.26). □

Method of approximation. In accordance with the results of Chapter 2 (see Corollary 2.2.13 and Theorem 2.2.15), for every Banach space $\mathbf{B}$ there exists another space $\mathbf{F}_1$ (in fact, it is a Hilbert space) and an operator $v\in\mathbf{L}(\mathbf{F}_1,\mathbf{B})$ such that the class $\mathcal{W}_f$ is $\mathbf{Q}_v^3$-uniform for a rather wide class of functions $f:\mathbf{B}\to\mathbf{R}$ (for $f\in\mathrm{BL}(\mathbf{B})$, for instance). It turns out that the operator v can be substituted for the other operator $u\in\mathbf{L}(\mathbf{F},\mathbf{B})$, if the latter, in a sense, can be approximated by the operator v. An exact formulation of this approximation will be presented in Theorem 1.21 below. First, we shall present one general inequality which is the basis of the method of approximation.

LEMMA 1.20. *Let the class $\mathcal{W} = \{W_t, t\in\mathbf{R}\}\subset\mathbf{B}$ be such that $(W_t)_\varepsilon\subset W_{t+\varepsilon}$ for all $t\in\mathbf{R}$, $\varepsilon > 0$. For the operators $u_1, u_2\in\mathbf{L}(\mathbf{F},\mathbf{B})$, the measures $\mu_1,\mu_2\in\mathcal{P}(\mathbf{F})$ and the arbitrary number $\varepsilon > 0$, the following inequality is true:*

$$\sup_{t\in\mathbf{R}} |\mu_1\cdot u_1^{-1}(W_t) - \mu_2\cdot u_1^{-1}(W_t)| \le \sup_{t\in\mathbf{R}} |\mu_1\cdot u_2^{-1}(W_t) - \mu_2\cdot u_2^{-1}(W_t)| +$$
$$+ \min\{\min_{i=1,2}\sup_{t\in\mathbf{R}} \mu_i\cdot u_1^{-1}(W_{t+2\varepsilon}\backslash W_{t-2\varepsilon});\ \min_{i=1,2}\sup_{t\in\mathbf{R}}\mu_i\cdot u_2^{-1}(W_{t+\varepsilon}\backslash W_{t-\varepsilon})\} +$$
$$+ (\mu_1+\mu_2)\{x\in\mathbf{F}: \|(u_1-u_2)(x)\| > \varepsilon\}. \tag{1.28}$$

Proof. For the measure $\mu\in\mathscr{P}(\mathbf{F})$ we have

$$\mu\cdot u_1^{-1}(W_t) \le \mu\{(x: u_1(x)\in W_t)\cap(x: \|(u_1-u_2)(x)\|\le\varepsilon)\} +$$
$$+ \mu\{x: \|(u_1-u_2)(x)\| > \varepsilon\} \le \mu\cdot u_2^{-1}(W_{t+\varepsilon}) + \mu\{x: \|(u_1-u_2)(x)\| > \varepsilon\}. \tag{1.29}$$

By reversing the operators u_1 and u_2 and by assuming $t-\varepsilon$ instead of t, we obtain

$$\mu\cdot u_1^{-1}(W_t) \ge \mu\cdot u_2^{-1}(W_{t-\varepsilon}) - \mu\{x: \|(u_1-u_2)(x)\| > \varepsilon\}. \tag{1.30}$$

By applying relations (1.29) and (1.30) in a proper order, we shall easily obtain (1.28). □

THEOREM 1.21. *Let the positively homogeneous function $f\in\mathrm{Lip}_1(\mathbf{B})$ and the operator $v\in\mathbf{L}(\mathbf{F}_1,\mathbf{B})$ be such that the class $\mathscr{W}_f$ is $(\mathbf{Q}_v^3, C_0\mathbf{K}_{1,3})$-uniform. Let $u\in\mathbf{L}(\mathbf{F},\mathbf{B})$. We assume that for each $\varepsilon>0$ there exists an operator $u_\varepsilon\in\mathbf{L}(\mathbf{F},\mathbf{F}_1)$ such that*

$$\|u_\varepsilon\|\le C_1\max(1,\varepsilon^{-\gamma}),\ \gamma\ge 0,\ \|u-v\cdot u_\varepsilon\|_2\le\varepsilon.$$

If for each $\varepsilon>0$

$$\sup_{t\in\mathbf{R}} P\{t\le f(u(Y))\le t+\varepsilon\}\le\phi(T)\cdot\varepsilon,$$

then for all $n\ge 1$, $0<\delta<1$, and for all $X\in\mathscr{D}^*(Y)$ the following estimate takes place:

$$\Delta_{n,u}(X,Y:\mathscr{W}_f)\le C\max\{C(\delta,T)\nu_3^{1-\delta}(X,Y)n^{-1/6};$$
$$\phi(T)\nu_3(X,Y)A^{-2}n^{-1/2};\ [\phi(T)^{6+2\gamma}A^{2\gamma}\nu_3^2(X,Y)n^{-1}]^{1/6+3\gamma}\},$$

where $A=(E\|X\|^2+E\|Y\|^2)^{1/2}$, $C(\delta,T_Y)$ is the constant from the inequality (1.19).

Proof. Let the integer $n\ge 2$ be fixed, $\varepsilon>0$ and u_ε be an operator from the conditions of the theorem. We denote by $U_k(Y,A)$, $k=1,\dots,n$, the class of all possible collections of $(Z_1,\dots,Z_k)$, of independent $\mathbf{F}$-r.e. $Z_i\in\mathscr{D}^*(Y)$, satisfying the inequalities $E\|Z_i\|^2\le A^2$ and $\nu_3(Z_i,Y)\le\nu_3(X,Y)$, $i=1,\dots,k$. We shall introduce the following notations: if the operator $w\in\mathbf{L}(\mathbf{F},\mathbf{B})$, then

$$\delta_{k,w} := \sup\left\{\sup_{t\in\mathbf{R}}\left|P\left\{f\left(\sum_{i=1}^{k} w(Z_i)\right)\le t\right\} - P\left\{f\left(\sum_{i=1}^{k} w(Y_i)\right)\le t\right\}\right| : (Z_1,\dots,Z_k)\in U_k(Y,A)\right\},$$

$$\delta_{k,\varepsilon} := \delta_{k,v}\cdot u_\varepsilon,\ s_k = s_k(\varepsilon,A,\phi) := \varepsilon^{2/3}A^{2/3}\phi^{-1/3}(T_Y)\left(\frac{k}{n}\right)^{1/2}.$$

By making a slight change in the proof of Lemma 1.8, we obtain the following recurrent inequality:

$$\delta_{k,\varepsilon}\le 2C_0 s_k^{-3}kn^{-3}\|u_\varepsilon\|^3\nu_3(X,Y)\delta_{k-1,\varepsilon} +$$
$$+\frac{7}{6}C_0\max\left\{s_k^{-2}\tilde{\psi}_1\left(s_k,\frac{k-1}{n}T\right), s_k^{-3}\tilde{\psi}\left(\frac{1}{2}s_k,\frac{k-1}{n}T\right)\right\}\times$$

$$\times n^{-3/2}\|u_\varepsilon\|^3 k\nu_3(X, Y) + \tilde{\psi}\left(s_k, \frac{k}{n}T\right), \tag{1.31}$$

where

$$\tilde{\psi}\left(\tau, \frac{k}{n}T\right) := \sup_{t\in\mathbf{R}} P\left\{\left(\frac{k}{n}\right)^{1/2} v\cdot u_\varepsilon(Y) \in W_{t+\tau}(f)\setminus W_{t-\tau}(f)\right\},$$

$$\tilde{\psi}_1\left(\tau, \frac{k-1}{n}T\right) := \sup_{r\geq \tau/2} r^{-1}\tilde{\psi}\left(r, \frac{k-1}{n}T\right), \; k\geq 2.$$

By the use of inequalities (1.29), (1.30), for all $t\in\mathbf{R}$, $\tau > 0$ we obtain

$$P\left\{\left(\frac{k}{n}\right)^{1/2} v\cdot u_\varepsilon(Y)\in W_{t+\tau}\setminus W_{t-\tau}\right\} \leq P\left\{\left(\frac{k}{n}\right)^{1/2} u(Y)\in W_{t+\tau+s_k}\setminus W_{t-\tau-s_k}\right\} +$$

$$+ P\left\{\|(u - v\cdot u_\varepsilon)(Y)\|\left(\frac{k}{n}\right)^{1/2} \geq s_k\right\} \leq \phi(T)\left(\frac{n}{k}\right)^{1/2}(\tau + s_k) +$$

$$+ \frac{n}{ks_k^2}E\|(u - v\cdot u_\varepsilon)(Y)\|^2 \leq \phi(T)\left(\frac{n}{k}\right)^{1/2}(\tau + s_k) + \frac{n}{ks_k^2}\varepsilon^2 A^2.$$

Here the property of the operator of type 2 has been used. By the definition of the number s_k, we can write

$$\tilde{\psi}\left(\tau, \frac{k}{n}T\right) \leq 2\phi(T)\left(\frac{n}{k}\right)^{1/2}(s_k + \tau),$$

whence

$$\tilde{\psi}\left(s_k, \frac{k}{n}T\right) \leq 4\phi(T)\left(\frac{n}{k}\right)^{1/2} s_k,$$

$$\tilde{\psi}_1\left(s_k, \frac{k-1}{n}T\right) \leq 12\phi(T)\left(\frac{n}{k}\right)^{1/2}.$$

By substituting these estimates in (1.31), we obtain

$$\delta_{k,\varepsilon} \leq 2C_0 s_k^{-3} kn^{-3/2}\|u_\varepsilon\|^3\nu_3(X, Y)\delta_{k-1,\varepsilon} +$$
$$+ 14C_0 s_k^{-2}\phi(T)\left(\frac{n}{k}\right)^{1/2} n^{-3/2}\|u_\varepsilon\|^3 k\nu_3(X, Y) + 4\phi(T)\left(\frac{n}{k}\right)^{1/2} s_k. \tag{1.32}$$

By means of the following inequalities the transition from the distance $\delta_{k,\varepsilon}$ to the distance $\delta_{k,u}$ is realized:

$$\delta_{k,u} \leq \delta_{k,\varepsilon} + 6\phi(T)\left(\frac{n}{k}\right)^{1/2} s_k, \tag{1.33}$$

$$\delta_{k-1,\varepsilon} \leq \delta_{k-1,u} + 4s_k\left(\frac{n}{k}\right)^{1/2}\phi(T). \tag{1.34}$$

We shall prove inequality (1.33), and inequality (1.34) is proved similarly.

Let $(Z_1, \ldots, Z_k)\in U_k(Y, A)$. Applying Lemma 1.20 we obtain

$$\left| P\left\{ f\left(\sum_{i=1}^{k} u(Z_i) n^{-1/2} \right) \le t \right\} - P\left\{ f\left(\sum_{i=1}^{k} u(Y_i) n^{-1/2} \right) \le t \right\} \right|$$
$$\le \delta_{k,\varepsilon} + P\left\{ \left\| \sum_{i=1}^{k} (u - v \cdot u_\varepsilon)(Z_i) \right\| > n^{1/2} s_k \right\} +$$
$$+ P\left\{ \left\| \sum_{i=1}^{k} (u - v \cdot u_\varepsilon)(Y_i) \right\| > n^{1/2} s_k \right\} + 4\phi(T) \left(\frac{n}{k} \right)^{1/2} s_k$$
$$\le \delta_{k,\varepsilon} + 2 s_k^{-2} \varepsilon^2 A^2 k n^{-1} + 4\phi(T) \left(\frac{n}{k} \right)^{1/2} s_k = \delta_{k,\varepsilon} + 6\phi(T) \left(\frac{n}{k} \right)^{1/2} s_k.$$

Here we have again used the definition of the operator of type 2 and the estimate $\|u - v \cdot u_\varepsilon\|_2 \le \varepsilon$. From (1.32)–(1.34) we obtain

$$\delta_{k,u} \le 2 C_0 s_k^{-3} k n^{-3/2} \|u_\varepsilon\|^3 \nu_3(X, Y) \delta_{k-1,u} +$$
$$+ 22 C_0 s_k^{-2} \phi(T) \left(\frac{n}{k} \right)^{1/2} n^{-3/2} \|u_\varepsilon\|^3 k \nu_3(X, Y) + 10\phi(T) \left(\frac{n}{k} \right)^{1/2} s_k$$
$$= 2 C_0 \varepsilon^{-2} A^2 \phi(T) k^{-1/2} \|u_\varepsilon\|^3 \nu_3(X, Y) \delta_{k-1,u} +$$
$$+ 22 C_0 \varepsilon^{-4/3} A^{-4/3} \phi^{-5/3}(T) k^{-1/2} \|u_\varepsilon\|^3 \nu_3(X, Y) + 10 \phi^{2/3}(T) \varepsilon^{2/3} A^{2/3}.$$

By use of the recurrent inequality obtained as well as by Theorem 1.11, the completion of the proof is similar to the proof of Theorem 1.7. □

REMARK 1.22. If in Theorem 1.21 the operator u_ε admits the estimate $\|u_\varepsilon\| \le C \max(1, (\log 1/\varepsilon)^\gamma)$, then for all r.e. $X \in \mathcal{D}(Y)$ such that $\nu_3(X, Y) < \infty$, we have

$$\Delta_{n,u}(X, Y; \mathcal{W}_f) = O(n^{-1/6} (\ln n)^{3\gamma}).$$

Two Examples. As far as we could ascertain, in all the above formulated theorems on the rate of convergence to zero of the quantity $\Delta_{n,u}(X, Y; \mathcal{W})$ two requirements (besides others) were present: the smoothness of the function $F(t) = P\{u(Y) \in W_t\}$ and some uniformity of the class $\mathcal{W}$ (in the case $\mathcal{W} = \mathcal{W}_f$, requirement for smooth approximation of the function f). We shall show that these requirements are essential and that no reinforcements of the other conditions (moments, for instance) will enable us to refuse either of them. In the examples below (formulated in the form of theorems since the constructions of examples and proofs are rather complicated), it will be shown that non-fulfilment of at least one of the above two requirements results in the following: the quantity $\Delta_{n,u}(X, Y; \mathcal{W})$ may tend to zero arbitrarily slowly even for bounded $\mathbf{F}$–r.e. $X \in \mathcal{D}^*(Y)$. Moreover, by the second example (Theorem 1.24) it is proved that in a general case the order $O(n^{-1/6})$ of the rate of convergence of $\Delta_{n,u}(X, Y; \mathcal{W})$ to zero as $n \to \infty$ is optimal (the exponent 1/6 cannot be substituted by a larger one). This gives evidence of an essential difference between the estimates of the rate of convergence on the sets in finite-dimensional and infinite-dimensional Banach spaces. As a first example we shall consider the spaces $\mathbf{B} = \mathbf{c}_0$, $\mathbf{F} = l_2$, the operator $u = j$ being an identical inclusion $l_2 \to \mathbf{c}_0$ and the function $g(x) = \sup_{i \ge 1} x_i$, $x \in \mathbf{c}_0$.

THEOREM 1.23. (1) *The class* $\mathcal{W}_g$ *is* $(\mathbf{Q}_j^k, \mathbf{CK}_{1,k})$*-uniform for each* $k \ge 1$;

(2) *For any monotonically converging to zero sequence* $b_n > 0$, $n \geq 1$, *there exist* l_2*-r.e.* Y *and* l_2*-r.e.* $X \in \mathcal{D}^*(Y)$, $P(\|X\| \leq 1) = 1$, *such that*

(a) $\Delta_{n,j}(X, Y, \mathcal{W}_g) \to 0$ *as* $n \to \infty$;
(b) $\lim_{n \to \infty} \sup b_n^{-1} \Delta_{n,j}(X, Y, \mathcal{W}_g) = \infty$.

Proof. Statement (1) is the consequence of Theorem 2.2.29. Let $X = (X_1, ..., X_n, ...)$, $Y = (Y_1, ..., Y_n, ...)$, where $X_i = \lambda_i \varepsilon_i$, $Y_i = \lambda_i \gamma_i$. We shall choose a sequence $\{\lambda_i\}$ such that $\Sigma_{i=1}^{\infty} \lambda_i^2 < \infty$ and statement (b) are satisfied. Statement (a) follows from the uniform continuity of the function $F(t) = P\{\sup \lambda_i \gamma_i < t\}$, which is easy to check.

We shall split the set of natural numbers into non-intersecting blocks $I_1, I_2, \cdots$ of lengths $n_1 < n_2 < \cdots$ setting

$$I_s = \left\{ k \in N : \sum_{i=1}^{s-1} n_i < k \leq \sum_{i=1}^{s} n_i \right\}, \quad s \geq 1.$$

Let $(m_i)_{i \in N}$ be a strictly increasing sequence of natural numbers and let $(\sigma_i)_{i \in N}$ be a sequence of positive numbers that is monotonically decreasing to zero. We assume $n_i = 2^{m_i^2}$, $i \geq 1$, $\lambda_i = \sigma_s m_s^{-1}$, if $i \in I_s$. We denote:

$$\delta_n(t) = P\{\sup_{i \geq 1} S_n(X_i) = t\},$$

$$\delta_n = \sup_t \delta_n(t), \quad Z_n = n^{-1/2} \sum_{i=1}^{n} \varepsilon_i.$$

Taking into account the independence of the coordinates X_i one can write the inequalities:

$$\delta_n \geq \sum_{k \in I_s} \prod_{i \leq k} P\{Z_n < t\lambda_i^{-1}\} P\{Z_n = t\lambda_i^{-1}\} \prod_{j > k} P\{Z_n \leq t\lambda_j^{-1}\}$$

$$\geq \prod_{i=1}^{s-1} (P\{Z_n < tm_i\sigma_i^{-1}\})^{n_i} \prod_{i=s+1}^{\infty} (P\{Z_n \leq tm_i\sigma_i^{-1}\})^{n_i} \times$$

$$\times \sum_{l=1}^{n_s} (P\{Z_n < tm_s\sigma_s^{-1}\})^{l-1} P\{Z_n = tm_s\sigma_s^{-1}\} (P\{Z_n \leq tm_s\sigma_s^{-1}\})^{n_s - l}.$$

Assuming $t = \sigma_s$, $n = m_s^2$ and noting that

$$P\{Z_n \leq 0\} \geq \frac{1}{2}, \quad P\{Z_n < tm_i\sigma_i^{-1}\} = 1$$

for $i > s$ we obtain

$$\delta_n \geq 2^{-(n_2 + ... + n_{s-1})} P\{Z_n = n^{1/2}\} \sum_{l=1}^{n_s} (P\{Z_n < n^{1/2}\})^l$$

$$\geq 2^{-(n_1 + ... + n_{s-1})} (1 - e^{-1}) \geq 2^{-(1 + 2^{m_2^2} + ... + 2^{m_s^2 - 1})}. \tag{1.35}$$

Let the sequence $\{b_n, n \geq 1\}$ be fixed. We assume $m_1 = 1$. If the numbers $m_1 < \cdots < m_{s-1}$ are already chosen, then $m_s > m_{s-1}$ will be chosen such as to hold

$$2^{-(1 + 2^{m_1^2} + ... + 2^{m_s^2 - 1})} > b_{m_s^2}^{1/2}.$$

This is possible since $\lim_{n \to \infty} b_n = 0$. Then from (1.35) we obtain $\overline{\lim}\, \delta_n b_n^{-1} = \infty$.

Since the function $F(t)$ is continuous, $\Delta_{n,u}(X, Y, \mathcal{W}_g) \geq \delta_n / 2$, which completes the proof of statement (b). There remains to note that the numbers σ_i can be chosen such that the following inequality holds:

$$\sum_{i=1}^{\infty} \lambda_i^2 = \sum_{s=1}^{\infty} \sum_{k \in I_s} \lambda_k^2 = \sum_{s=1}^{\infty} \sigma_s^2 m_s^{-2} 2^{m_s^2} \leq 1.\square$$

In the second example we choose $\mathbf{B} = \mathbf{l}_1$, $\mathbf{F} = \mathbf{l}_2$. We recall that $\mathbf{W}_0(\mathbf{l}_1)$ is the class of balls of the space $\mathbf{l}_1$ with its centre in zero.

THEOREM 1.24. *Let the sequence $b_i \downarrow 0$, $\limsup_{i \to \infty} b_i \sqrt{i} = \infty$, and $m \geq 2$ be a natural number. There exists the operator $u \in \mathbf{L}(\mathbf{l}_2, \mathbf{l}_1)$, a Gaussian $\mathbf{l}_2$-r.e. Y and a symmetric $\mathbf{l}_2$-r.e. $X \in \mathscr{D}^*(Y)$ such that*

(a) *the class $\mathbf{W}_0(\mathbf{l}_1)$ is $(\mathbf{Q}_u^m, C\mathbf{K}_{1,m})$-uniform;*

(b) $P\{\|X\| \leq 1\} = 1$;

(c) *the function $F(t) = P\{\|u(Y(\| < t\}$ has a bounded density;*

(d) *for infinitely many n we get the estimate*

$$\Delta_{n,u}(X, Y; \mathbf{W}_0(\mathbf{l}_1)) \geq b_n n^{1/m - 1/2}.$$

Proof. As in the previous example, let I_s, $s \geq 1$, be blocks of the set of natural numbers of length k_s, and let the sequence $(\lambda_i, i \geq 1)$ satisfy the condition

$$\sum_{i=1}^{\infty} k_i \lambda_i^{m/m-1} < \infty. \tag{1.36}$$

We shall define the diagonal operator $u \in \mathbf{L}(\mathbf{l}_2, \mathbf{l}_1)$, $u(x) = y = (y_1, \ldots, y_n, \ldots)$, $y_i = \lambda_s x_i$, if $i \in I_s$. Statement (a) follows from Theorem 2.2.31 (see also Remark 2.2.32) and from (1.36). Hereafter, let $\{e_i, i \geq 1\}$ be a basis of unit vectors of the space $\mathbf{l}_2$. We shall define the independent r.e. $X^{(s)}$, $s \geq 1$, assuming that

$$P\{X^{(s)} = a_s \lambda_s^{-1} e_i\} = P\{X^{(s)} = -a_s \lambda_s^{-1} e_i\} = (2k_s)^{-1}, \ i \in I_s,$$

and that $X = (X^{(1)}, X^{(s)}, \ldots)$. It is easy to make sure that

$$P\left\{ \|X\|_{\mathbf{l}_2} = \left(\sum_{s=1}^{\infty} a_s^2 \lambda_s^{-2} \right)^{1/2} \right\} = 1$$

and the covariance matrix $\mathbf{l}_2$-r.e. X is diagonal:

$$E(X, e_i)(X, e_j) = \begin{cases} 0, & \text{if } i \neq j, \\ \sigma_s^2, & \text{if } i = j \in I_s, \end{cases}$$

where $\sigma_s = a_s \lambda_s^{-1} k_s^{-1/2}$.

Therefore the condition

$$\sum_{s=1}^{\infty} a_s^2 \lambda_s^{-2} \leq 1 \tag{1.37}$$

gives us (b) and the pre-Gaussian property of $\mathbf{l}_2$-r.e. X. The corresponding Gaussian $\mathbf{l}_2$-r.e. can be taken as $Y = \sum_{i=1}^{\infty} \tilde{\sigma}_i \gamma_i e_i$, where $\tilde{\sigma}_i = \sigma_s$ if $i \in I_s$. Note that $X \in \mathscr{D}^*(Y)$ follows from the pre-Gaussian property.

The proof of (c) follows from the independence of the coordinates of r.e. Y (see Theorem 4.4.13), therefore there remains to prove (d).

Assume

$$\lambda_s = 2^{-(s(m-1)/m)} k_s^{-(m-1)/m},$$
$$a_s = 2^{-(s(2m-1)/m)} k_s^{-(m-1)/m}.$$

Then conditions (1.36) and (1.37) are satisfied.

To prove (d) we shall introduce the following notations:

$$\xi = u(X),\ \xi^{(i)} = u(X^{(i)}),\ \eta = u(Y),$$
$$\eta^{(i)} = \sum_{j \in I_i} \sigma_i \lambda_i \gamma_j e_j = a_i k_i^{-1/2} \sum_{j \in I_i} \gamma_j e_j,$$
$$U_s = \sum_{i=1}^{s-1} \|S_n(\xi^{(i)})\|,\ V_s = \sum_{i=1}^{s-1} \|\eta^{(i)}\|,\ a_s = \sqrt{n} \sum_{i \geq} s\, a_i.$$

Since

$$\|S_n(\xi^{(s)})\| \leq a_s n^{1/2},\ s \geq 1,$$

then

$$\Delta_n(r) := P\{\|S_n(u(X))\| < r\} - P\{\|u(Y)\| < r\}$$
$$\geq P\{U_s < r - a_s\} - P\{V_s + \|\eta^{(s)}\| < r\}. \tag{1.38}$$

We shall use the estimate of the rate of convergence in the CLT in a finite-dimensional space, and by a lemma we shall present the following result.

LEMMA 1.25 [193], [188]. *Let Z_i, $i \geq 1$, be i.i.d. $\mathbf{R}_k$-r.e. with $EZ_1 = 0$ and with a unit covariance matrix. Then*

$$\sup_{A \in \mathfrak{N}} |P\{S_n(Z_1) \in A\} - \Phi(A)| \leq Ck\beta_{3,k} n^{-1/2}, \tag{1.39}$$

where C is an absolute constant, and $\mathfrak{N}$ is the class of all convex sets in $\mathbf{R}_k$,

$$\beta_{3,k} = E\left(\sum_{i=1}^{k} (Z_1^{(i)})^2\right)^{3/2}.$$

In our case we will use estimate (1.39) assuming $k = k_1 + \cdots + k_{s-1}$.

$$Z_1 = \left(\frac{k_1^{1/2}}{a_1}\xi^{(1)}, \ldots, \frac{k_{s-1}^{1/2}}{a_{s-1}}\xi^{(s-1)}\right).$$

It is easy to see that $\beta_{3,k} \leq (k_1 + \cdots + k_{s-1})^{3/2}$, therefore from (1.39) we easily obtain

$$P\{U_s < r - \alpha_s\} \geq P\{V_s < r - \alpha_s\} - C(k_1 + \cdots + k_{s-1})^{5/2} n^{-1/2}. \tag{1.40}$$

From relations (1.38), (1.40) it follows

$$\Delta_n(r) \geq P\{V_s < r - \alpha\} - P\{V_s + \|\eta^{(s)}\| < r\} - C(k_1 + \cdots + k_{s-1})^{5/2} n^{-1/2}. \tag{1.41}$$

Since

$$P\{V_s + \|\eta^{(s)}\| < r\} \leq P\{\|\eta^{(s)}\| < 2\alpha_s\} + P\{(V_s + \|\eta^{(s)}\| < r) \cap (\|\eta^{(s)}\| \geq 2\alpha_s)\}$$
$$\leq P\{\|\eta^{(s)}\| < 2\alpha_s\} + P\{V_s < r - 2\alpha_s\},$$

then from (1.41) we have

$$\Delta_{n,u}(X, Y, \mathbf{W}_0(\mathbf{l}_1)) \geq \sup_{r \geq 0} P\{r \leq V_s \leq r + \alpha_s\} - P\{\|\eta^{(s)}\| < 2\alpha_s\} -$$

$$- Cn^{-1/2}(k_1 + \cdots + k_{s-1})^{5/2}. \quad (1.42)$$

To estimate the first summand in (1.42) we shall use the following inequality (see [158, p. 56]): for all $a > 0$ and $\varepsilon > 0$,

$$\sup_{r \geq 0} P\{r \leq V_s \leq r + \varepsilon\} \geq C\varepsilon(1 + a\varepsilon)^{-1} \int_{-a}^{a} \phi_s(t)\, dt, \quad (1.43)$$

where $\phi_s(t)$ is the characteristic function of the symmetrized r.v. V_s. For all values of t

$$\phi_{s(t)} \geq 1 - t^2 EV_s^2 \geq 1 - t^2 E\left(\sum_{i=1}^{s-1} \|\eta^{(i)}\|\right)^2 \geq 1 - t^2 \left(\sum_{i=1}^{s-1} a_i\right)^2.$$

Assuming, in (1.43), $a = (\Sigma_{i=1}^{s-1} a_i)^{-1}$, we obtain

$$\sup_{r \geq 0} P\{r \leq V_s \leq r + \alpha_s\} \geq C\alpha_s \left(\sum_{i=1}^{s-1} a_i + \alpha_s\right)^{-1}. \quad (1.44)$$

To estimate the second summand in (1.42) we assume that

$$\alpha_s \leq \frac{1}{4}\left(\frac{2}{\pi}\right)^{1/2} a_s k_s^{1/2}. \quad (1.45)$$

Since

$$\|\eta^{(s)}\| = \sum_{i=1}^{k_s} |\gamma_i|\, a_s k_s^{-1/2},$$

then

$$E\|\eta^{(s)}\| = \left(\frac{2}{\pi}\right)^{1/2} a_s k_s^{1/2};$$

$$E\,|\,\|\eta^{(s)}\| - E\|\eta^{(s)}\|\,|^2 = \left(1 - \frac{2}{\pi}\right) a_s^2.$$

Therefore from (1.45) and from Chebyshev's inequality we have

$$P\{\|\eta^{(s)}\| < 2\alpha_s\} \leq P\{|\,\|\eta^{(s)}\| - E\|\eta^{(s)}\|\,| > 2^{-1}E\|\eta^{(s)}\|\}$$
$$\leq 4(E\|\eta^{(s)}\|)^{-2} E\,|\,\|\eta^{(s)}\| - E\|\eta^{(s)}\|\,|^2 \leq 2\pi k_s^{-1}. \quad (1.46)$$

From relations (1.42)–(1.44) and (1.46) we obtain

$$\Delta_{n,u}(X, Y, \mathbf{W}_0(\mathbf{l}_1)) \geq C\alpha_s \left(\sum_{i=1}^{s-1} a_i + \alpha_s\right)^{-1} - 2\pi k_s^{-1} -$$

$$- Cn^{-1/2}(k_1 + \cdots + k_{s-1})^{5/2}, \quad (1.47)$$

if (1.45) is satisfied.

Defining the subsequence $n_s = C_0 k_s$, $s \geq 1$, where C_0 is chosen such that (1.45) takes place, we obtain

$$\alpha_s \geq n_s^{1/2} \sum_{i \geq} s\, a_i \geq n_s^{1/2} a_s = n_s^{1/m - 1/2} 2^{-(s(2m-1)/m)} C_0^{(m-1)/m}. \quad (1.48)$$

Note that the subsequence n_s can be chosen such that $b_{n_s} \leq 2C_1 2^{-(s(2m-1)/m)}$, therefore statement (d) follows from (1.47) and (1.48). □

Estimates of the Rate of Convergence for the Mixtures of Gaussian Measures. It was mentioned above, that when constructing the estimates of the rate of convergence in the CLT on the sets, two conditions are essential: the existence of the smooth approximation of indicator functions of the sets under consideration and the behaviour of the Gaussian measure of the ε-strip of the set. We shall show that having summands of special structure, it is possible to refuse the first condition and that a sufficiently good behaviour of the Gaussian measure of the ε-strip of sets ensures an optimal, with respect to n, estimate of the rate of convergence.

As summands we shall consider $\mathbf{B}$-r.e. whose distributions are mixtures of Gaussian measures. Of course, it is a very strong limitation, but we have decided to include these results for the following reasons: firstly, for such special summands one succeeds in obtaining an optimal non-uniform estimate, even on such non-smooth sets as the balls in the space $\mathbf{C}[0, 1]$; secondly, such an assumption on the particular form of distribution of summands enables us to solve the problem in more general spaces than Banach ones (see [21]).

Let η be a Gaussian $\mathbf{B}$-r.e. with a zero mean, $\mu = L(\eta)$, and let m be a probability measure on $[0, \infty)$. A mixture of Gaussian distributions with a mixing measure (or mixer, for short) m will be called a measure on $\mathbf{B}$, defined by formula

$$\mu_m(A) = \int_0^\infty \mu(t^{-1/2}A)m(\mathrm{d}t). \tag{1.49}$$

If $g: \mathbf{B} \to \mathbf{R}$ is a measurable function, then it is easy to verify the validity of the equality

$$\int_{\mathbf{B}} g(x)\mu_m(\mathrm{d}x) = \int_0^\infty \int_{\mathbf{B}} g(x(t)^{1/2})\mu(\mathrm{d}x)m(\mathrm{d}t).$$

In particular, for any $f \in \mathbf{B}^*$ and $\alpha > 0$

$$E\,|\langle f, \xi\rangle|^\alpha = E\,|\langle f, \eta\rangle|^\alpha \int_0^\infty t^{\alpha/2}m(\mathrm{d}t), \tag{1.50}$$

where $L(\xi) = \mu_m$. It follows from (1.49), (1.50) that if ξ is a Gaussian mixture with a mixer m, then $\mathbf{B}$-r.e. $a\xi$, $a \in \mathbf{R}$, will be a mixture with a mixer $m_{(a)}$ defined by the equality $m_{(a)}(A) = m(a^{-2}A)$. If ξ_1 and ξ_2 are independent $\mathbf{B}$-r.e., being Gaussian mixtures with mixers m_1 and m_2, respectively, then the sum $\xi_1 + \xi_2$ will be a mixture with a mixer $m = m_1 * m_2$.

Let $\xi, \xi_1, \ldots, \xi_n, \cdots$ be i.i.d. $\mathbf{B}$-r.e. with the distribution μ_m. Since the summands are of specific structure, so the CLT, in fact is reduced to the law of large numbers on $\mathbf{R}$.

PROPOSITION 1.26. *Let there exist $f \in \mathbf{B}^*$, $f \neq 0$, such that $E\langle f, \xi\rangle^2 < \infty$. Then the distribution $L(S_n(\xi))$ weakly converges to $L(a\eta)$, where $a = (\int_0^\infty tm(\mathrm{d}t))^{1/2}$.*

Proof. It follows from the condition $E\langle f, \xi\rangle^2 < \infty$ and (1.50) that the quantity a is finite. If by ς_i, $i \geq 1$, we denote i.i.d. non-negative r.v. with a distribution m, then $L(S_n(\xi))$ is a Gaussian mixture with mixer $M_n = L(Z_n)$, where $Z_n = n^{-1}\Sigma_{i=1}^n \varsigma_i$. Since $E\varsigma_i = a^2 < \infty$, a strong law of large numbers can be applied for Z_n; therefore for any function $g \in C_0^b(\mathbf{B})$ we have

$$Eg(S_n(\xi)) = \int_{\mathbf{B}} g(x)\mu_{M_n}(\mathrm{d}x) = \int_{\mathbf{B}}\int_0^\infty g(t^{1/2}x)M_n(\mathrm{d}t)\mu(\mathrm{d}x) \to$$

$$\to \int_{\mathbf{B}} g(ax)\mu(\mathrm{d}x) = \int_{\mathbf{B}} g(x)\mu(a^{-1}\mathrm{d}x). \square$$

Consequently, the question of the rate of convergence in the CLT for mixtures reduces to an estimate of the rate of convergence in the strong law of large numbers. Therefore, first, we shall present two auxiliary results of the strong law of large numbers.

Let ς_i, $i \geq 1$, be i.i.d. random variables,

$$E\,|\varsigma_1|^{1+\alpha} < \infty,\ 0 < \alpha \leq 1,\ E\varsigma_1 = 0,\ Z_n := n^{-1}\sum_{i=1}^{n}\varsigma_i.$$

PROPOSITION 1.27. *Let the function* $f : \mathbf{R} \to \mathbf{R}$ *be bounded, differentiable and*

$$\|f\|_{1,\alpha} := \sup_{x \neq 0} |x|^{-1}\,|f'(x) - f'(0)| < \infty.$$

Then

$$\delta_n(f) := |Ef(Z_n) - f(0)| \leq C(\alpha)\|f\|_{1,\alpha} n^{-\alpha} E\,|\varsigma_1|^{1+\alpha}. \tag{1.51}$$

Proof. By the Taylor formula we have

$$|f(x) - f(0) - f'(0)x| \leq C(\alpha)\|f\|_{1,\alpha}\|x\|^{1+\alpha}.$$

Hence

$$\delta_n(f) = \left|\int_{-\infty}^{\infty}(f(x) - f(0) - f'(0)x)L(Z_n)\mathrm{d}x)\right|$$

$$\leq C(\alpha)\int_{-\infty}^{\infty}|x|^{1+\alpha}L(Z_n)(\mathrm{d}x) \leq C(\alpha)\|f\|_{1,\alpha}E\,|\varsigma_1|^{1+\alpha}n^{-\alpha}.\square$$

PROPOSITION 1.28. *Let the function* $f : \mathbf{R} \to \mathbf{R}$ *be differentiable at the interval* $[-\varepsilon, \varepsilon]$ *and satisfy the conditions:*

$$\|f\|_\infty := \sup_x |f(x)| < \infty,$$

$$\|f\|_{1,\alpha,[-\varepsilon,\varepsilon]} := \sup_{|x|<\varepsilon} |x|^{-\alpha}\,|f'(x) - f'(0)| < \infty.$$

Then

$$\delta_n(f) \leq$$

$$\leq C(\alpha)n^{-\alpha}E\,|\varsigma_1|^{1+\alpha}\{\|f\|_\infty\varepsilon^{-1-\alpha} + \|f\|_{1,\alpha,[-\varepsilon,\varepsilon]} + \varepsilon^{-\alpha}\,|f'(0)|\}. \tag{1.52}$$

Proof. Let $\psi: \mathbf{R} \to [0, 1]$ be an infinitely differentiable function such that $\psi(x) = 1$ if $|x| < 1/2$ and $\psi(x) = 0$ if $|x| > 1$. We set $\psi_\varepsilon(x) = \psi(\varepsilon^{-1}x)$, $\varepsilon > 0$. Then

$$\delta_n(f) \leq \delta_n((1 - \psi_\varepsilon)f) + \delta_n(\psi_\varepsilon f).$$

Since $1 - \psi_\varepsilon(x) = 0$ if $|x| < \varepsilon 2^{-1}$, then

$$\delta_n((1 - \psi_\varepsilon)f) \leq \left|\int_{|x|>\varepsilon/2} f(x)(1 - \psi_\varepsilon(x))L(Z_n)(\mathrm{d}x)\right|$$

$$\leq C(\alpha)\varepsilon^{-1-\alpha}\|f\|_\infty E\,|\varsigma_1|^{1+\alpha}n^{-\alpha}. \tag{1.53}$$

We denote $f_1 = \psi_\varepsilon \cdot f$. On the interval $[-\varepsilon, \varepsilon]$ we have

$$\begin{aligned}|f_1'(x) - f_1'(0)| &= |\psi_\varepsilon'(x)f(x) + \psi_\varepsilon(x)f'(x) - \psi_\varepsilon'(0)f(0) - \psi_\varepsilon(0)f'(0)|\\ &\leq |\psi_\varepsilon'(x) - \psi_\varepsilon'(0)|\ |f(x)| + |\psi_\varepsilon(0)|\ |f(x) - f(0)| +\\ &\quad + |f'(x) - f'(0)|\ |\psi_\varepsilon(x)| + |f'(0)|\ |\psi_\varepsilon(x) - \psi_\varepsilon(0)|\\ &\leq C\varepsilon^{-2}|x|\ \|f\|_\infty + \|f\|_{1,\alpha,[-\varepsilon,\varepsilon]}|x|^\alpha + C\,|f'(0)|\,\varepsilon^{-1}|x|\\ &\leq C\,|x|^\alpha\{\varepsilon^{-1-\alpha}\|f\|_\alpha + \|f\|_{1,\alpha,[-\varepsilon,\varepsilon]} + |f'(0)|\,\varepsilon^{-\alpha}\}.\end{aligned}$$

By applying Proposition 1.27, where $f = f_1$, we obtain

$$\delta_n(\psi_\varepsilon f) \leq C(\alpha)n^{-\alpha}E\,|\varsigma_1|^{1+\alpha}\{\|f\|_\infty \varepsilon^{-1-\alpha} + \|f\|_{1,\alpha,[-\varepsilon,\varepsilon]} + \varepsilon^{-\alpha}|f'(0)|\}. \quad (1.54)$$

(1.52) follows from (1.53) and (1.54). □

Having Propositions 1.26, 1.27 proved, we can formulate the main results of this section. Note that we consider i.i.d. **B**−r.e. ξ_i, $i \geq 1$, with a distribution μ_m. We denote $\beta_r = \int_0^\infty t^r m(dt)$ and, without loss of generality, we shall assume that $\beta_1 = 1$. Let

$$\Delta_n(A) := |L(S_n(\xi))(A) - \mu(A)|,$$

$$a^+ := \begin{cases} 0, & a < 0, \\ a, & a \geq 0, \end{cases} \quad \phi_A(t) = \mu(t^{-1/2}A).$$

In the next theorem A will be considered as a fixed set.

THEOREM 1.29. *Let* $\beta_\alpha < \infty$, *where* $1 < \alpha \leq 2$. *If* $\phi_A(t)$ *is differentiable on the interval* $[1-\varepsilon, 1+\varepsilon]$ *and*

$$\|\phi_A\|_{1,\alpha-1,[(1-\varepsilon)^+,1+\varepsilon]} := \sup_{t\in[(1-\varepsilon)^+,1+\varepsilon]} |t-1|^{1-\alpha}|\phi_A'(t) - \phi_A'(1)| < \infty,$$

then

$$\Delta_n(A) \leq C(\alpha)n^{1-\alpha}\beta_\alpha\{\varepsilon^{-\alpha} + \|\phi_A\|_{1,\alpha-1,[(1-\varepsilon)^+,1+\varepsilon]} + |\phi_A'(1)|\,\varepsilon^{1-\alpha}\}. \quad (1.55)$$

If ϕ_A *satisfies the condition*

$$\|\phi_A\|_{\gamma,[1-\varepsilon_0,1+\varepsilon_0]} := \sup_{t\in[1-\varepsilon_0,1+\varepsilon_0]} |\phi_A(t) - \phi_A(1)|\;|t-1|^{-\gamma} < \infty,$$

where $0 < \gamma \leq 1$, $0 < \varepsilon_0 < 1$, *then for n satisfying inequality* (1.58) *the following estimate takes place:*

$$\Delta_n(A) \leq C(\alpha,\gamma)\beta_\alpha^{\gamma/\alpha+\gamma}\|\phi_A\|_{\gamma,[1-\varepsilon_0,1+\varepsilon_0]}^{\alpha/\alpha+\gamma} n^{-\gamma(\alpha-1)/\alpha+\gamma}. \quad (1.56)$$

Proof. Since the measure $L(S_n(\xi))$ is a Gaussian mixture with mixer $M_n(L(Z_n)$, then by formula (1.49) we have

$$\Delta_n(A) = \left|\int_0^\infty \mu(t^{-1/2}A)M_n(dt) - \mu(A)\right|$$
$$= \left|\int_0^\infty \phi_A(t)(M_n(dt) - \delta_1(dt))\right|$$

and we can (after an obvious substitution of the variable $t = u + 1$) apply Proposition 1.28 and obtain estimate (1.55). To prove estimate (1.56) we shall continue the above equality

$$\Delta_n(A) = \left|\int_0^\infty (\phi_A(t) - \phi_A(1))M_n(dt)\right|$$
$$\leq \left|\int_{|t-1|>\varepsilon} (\phi_A(t) - \phi_A(1))M_n(dt)\right| +$$
$$+ \left|\int_{|t-1|\leq\varepsilon} (\phi_A(t) - \phi_A(1))M_n(dt)\right|$$
$$\leq 2P\{|Z_n - 1| > \varepsilon\} + \|\phi_A\|_{\gamma,[1-\varepsilon_0,1+\varepsilon_0]}\varepsilon^\gamma$$
$$\leq 2\varepsilon^{-\alpha}n^{1-\alpha}\beta_\alpha + \|\phi_A\|_{\gamma,[1-\varepsilon_0,1+\varepsilon_0]}\varepsilon^\gamma, \quad (1.57)$$

where $0 < \varepsilon < \varepsilon_0$. Minimizing the latter expression with respect to ε, we obtain

$$\varepsilon_{\min} = (2\alpha\gamma^{-1}\beta_\alpha n^{1-\alpha}\|\phi_A\|^{-1}_{\gamma,[1-\varepsilon_0,1+\varepsilon_0]})^{1/\alpha+\gamma},$$

and to make this value less than ε_0, the inequality must be satisfied:

$$n > (\varepsilon_0^{-|\alpha+\gamma|}2\alpha\gamma^{-1}\|\phi_A\|^{-1}_{\gamma,[1-\varepsilon_0,1+\varepsilon_0]})^{1/\alpha-1}. \tag{1.58}$$

By substituting $\varepsilon_{\min}$ in (1.57) we obtain (1.56). □

It is easy to see that the value $\alpha = 3/2$ corresponds to the case when the summands ξ_i have the third moment. Then for the fixed set A, inequality (1.55) – under the condition that the quantity $\|\phi\|_{1,1/2,[1/2,3/2]}$ is finite – yields the order $n^{-1/2}$, and estimate (1.56) for $\gamma = 1$ gives $n^{-1/5}$.

In the next theorem not a single set but a uniparametric family of sets is considered.

THEOREM 1.30. *Let the set A be such that the function $F(r) = \mu(rA)$ on the interval* $[0, \infty)$ *has the following properties:*

(1) *the function F is twice-differentiable (we denote $\psi(r) = F'(r)$);*

(2) *the functions $|\psi(r)|$, $|\psi'(r)|$, $|r^6\psi'(r)|$, $|r^5\psi(r)|$ are bounded.*

If $\beta_\alpha < \infty$, $1 < \alpha \le 2$, then

$$\Delta_n(rA) \le C(\alpha, F)\beta_\alpha n^{1-\alpha}(1 + r^{2\alpha})^{-1}. \tag{1.59}$$

In particular, for $\alpha = 3/2$

$$\Delta_n(rA) \le C\left(\frac{3}{2}, F\right)\beta_{3/2}n^{-1/2}(1 + r^3)^{-1},$$

where the constant $C(\alpha, F)$ depends on F and α.

Proof. Denote $\phi_{rA}(T) = F(rt^{-1/2})$. We have

$$\phi'_{rA}(T) = -2^{-1}rt^{-3/2}\psi(t^{-1/2}r),$$

$$\phi''_{rA}(t) = 4^{-1}r^2t^{-3}\psi'(t^{-1/2}r) + \frac{3}{4}rt^{-5/2}\psi(t^{-1/2}r).$$

If $0 < r \le 1$, $|t - 1| \le 1/2$, then from these expressions we have

$$|\phi''_{rA}(t)| \le C(\sup_x |\psi'(x)| + \sup_x |\psi(x)|).$$

Whence for $|t - 1| \le 1/2$ and $1 < \alpha \le 2$

$$|\phi'_{rA}(t) - \phi'_{rA}(1)| \le C(\|\psi\|_\infty + \|\psi'\|_\infty)|t - 1|^{\alpha-1}. \tag{1.60}$$

Having (1.60) we can apply estimate (1.55) of Theorem 1.29 (where $\varepsilon = 1/2$). Then for all $0 < r \le 1$ we obtain

$$\Delta_n(rA) \le C(1 + \|\psi\|_\infty + \|\psi'\|_\infty)\beta_\alpha n^{1-\alpha}. \tag{1.61}$$

In the case $r > 1$

$$\sup_{0<t<1+r^2} |\phi''_{rA}(t)| = \sup_{0<t<1+r^2} \Big|(4r^2)^{-1}(r\cdot t^{-1/2})^6\psi(t^{-1/2}r) +$$

$$+\left(\frac{3}{4}r^4\right)^{-1}(t^{-1/2}r)^5\psi(t^{-1/2}r)\Bigg|$$

$$\leq Cr^{-4}(|u^6\psi'(u)| + |u^5\psi(u)|).$$

Applying estimate (1.55) again, but now with $\varepsilon = r^2$, we obtain

$$\Delta_n(rA) \leq C(\alpha)n^{1-\alpha}\beta_\alpha[r^{-2\alpha} + r^{-4}\sup_u(|u^6\psi'(u)| + |u^5\psi(u)|) + |r\psi(r)|].$$

From (1.61) and (1.62), remembering that $|r\psi(r)| \leq r^{-4}\sup_r|r^5\psi(r)|$ and $2\alpha < 4$, we obtain (1.59). □

We shall present some results, following from Theorem 1.30, in particular spaces. The most important consequence is the following result.

PROPOSITION 1.31. *Let* ξ_i, $i \geq 1$, *be i.i.d.* **H**−*r.e. with the distribution* μ_m, $E\xi_1 = 0$, $\operatorname{cov}\xi_1 = T$, *and let* μ *be a centred Gaussian measure on* **H** *with a covariance operator* T. *If* $E|\xi_1|^{2\alpha} < \infty$, $1 < \alpha \leq 2$, *then*

$$\Delta_n(V_r) \leq C(T)\beta_\alpha n^{1-\alpha}(1 + r^{2\alpha})^{-1}.$$

In particular, if $\alpha = 3/2$, *then*

$$\Delta_n(V_r) \leq C(T)\beta_{3/2}n^{-1/2}(1 + r^3)^{-1}.$$

Proof. In Theorem 1.30 we take $A = V_1$. Let $g(x)$ and $\hat{g}(t)$ be the distribution density and characteristic function of r.v. $\|\eta\|^2$, where $L(\eta) = \mu$. Since $\psi(u) = 2ug(u^2)$, then to check the second condition of Theorem 1.30 we must show the boundedness of the quantity $\sup_{u\geq 0}(|u^3g(u)| + |u^4g'(u)|)$. But the functions $|u^kg(u)|$ and $|u^mg^{(r)}(u)|$ are bounded if $|(d^k/dt^k)\hat{g}(t)|$ and $|(d^m/dt^m)(t^r\hat{g}(t))|$ are integrable functions. Verification of this fact causes no difficulty due to the explicit expression of the function $\hat{g}(t)$ (see Supplement 4.4.12). □

Another case in which one can check all the conditions of Theorem 1.30 is the space **C**[0, 1] with a Wiener measure. Assuming $A = V_1$ and having an explicit expression of the function $F(r)$ (see Supplement 4.4.6), we obtain the following result.

PROPOSITION 1.32. *Let* μ *be a Wiener measure in the space* **C**[0, 1], *and let* ξ_i, $i \geq 1$, *be i.i.d.* **C**[0, 1]−*r.e. with a distribution* μ_m. *If* $\beta_1 = 1$ *and* $\beta_\alpha < \infty$, $1 < \alpha ,= 2$, *then*

$$\Delta_n(V_r) \leq C(\alpha)\beta_\alpha n^{1-\alpha}(1 + r^{2\alpha})^{-1}.$$

In addition, we shall present one more general result, not for all the balls but only for those balls with a sufficiently large radius.

PROPOSITION 1.33. *Let* μ *be a centred Gaussian measure on* **B**. *For sufficiently large r there exists two constants* $C_1 = C_1(\mu)$, $C_2 = C_2(\mu)$ *such that for any* $0 < \varepsilon < 1/2$ *and* $1 < \alpha \leq 2$ *the following estimate takes place:*

$$\Delta_n(V_r) \leq C_1n^{1-\alpha}\beta_\alpha\varepsilon^{-\alpha} + C_1e^{-C_2r^2}\varepsilon. \tag{1.63}$$

Proof. Applying Theorem 4.1.2 we can assert that for a sufficiently large r the derivative of the function $F(r) = \mu(V_r)$ satisfies the inequality

$$\psi(r) \leq C_3 e^{-C_4 r^2} \tag{1.64}$$

with constants $C_3 > 0$ and $C_4 > 0$ depending on μ. Since $\phi_{V_r}(t) = F(t^{-1/2}r)$, then

$$\sup_{1/2<t<3/2,\, r>r_0} |\phi'_{V_r}(t)| = \sup_{1/2<t<3/2,\, r>r_0} |rt^{-3/2}\psi(t^{-1/2}r)|$$
$$\leq 2 \sup_{u>2/3r_0} |u\psi(u)|.$$

Therefore, if r is sufficiently large, i.e. if (1.64) takes place, then from (1.56), where $r = 1$, we obtain (1.63). □

REMARK 1.34. The number ε in (1.63) remains a free parameter sine n and r are independent parameters, and in the interval $0 < \varepsilon < 1/2$ one cannot always successfully minimize the expression standing on the right side of inequality (1.63).

Estimate in a Hilbert Space. Let η be a $\mathbf{H}$−r.e. with a zero mean and a covariance operator T, and let $\lambda_1 \geq \lambda_2 \geq \cdots$ be a collection of eigenvalues of the operator T. The estimate of the rate of convergence in the CLT in the Hilbert space $\mathbf{H}$ will be obtained in the terms of so-called 'truncated moments'. For $\mathbf{H}$−r.e. ξ and the number $p \geq 0$ we define

$$\Lambda_p := n^{-(p-2)/2} E\,|\xi|^p \chi\{|\xi| > n^{1/2}\},$$
$$\tilde{\Lambda}_p := n^{-(p-2)/2} E\,|\xi|^p \chi\{|\xi| \leq n^{1/2}\}.$$

THEOREM 1.35. *Let* $\mathbf{H}$*−r.e.* $\xi \in \mathcal{D}(\eta)$, $E\,|\xi|^2 = 1$. *There exists a constant* $C(T)$ *such that for all* $n \geq 1$ *and for all* $a \in \mathbf{H}$ *the following estimate takes place:*

$$\Delta_n(a) := \Delta_n(\xi, \eta; \mathbf{W}_a(\mathbf{H})) \leq C(T)(1 + |a|^3)(\Lambda_2 + \tilde{\Lambda}_3). \tag{1.65}$$

Note that in a Hilbert space the condition $\lambda_1 \geq \lambda_2 \geq \cdots$ does not limit generality. By renorming the summands in the sum $S_n(\xi)$ it is not difficult to reject the limitation $E\,|\xi|^2 = 1$. If there exists a moment $\beta_p := E\,|\xi|^p$, $2 \leq p \leq 3$, then by applying the Chebyshev inequality one can estimate $\Lambda_2 + \tilde{\Lambda}_3 \leq n^{-(p-2)/2}\beta_p$. In the case $\beta_p < \infty$, $2 < p < 3$, a stronger result $\Lambda_2 + \tilde{\Lambda}_3 = o(n^{-(p-2)/2})$ takes place. One can make sure of this by applying Lebesgue's dominated convergence theorem.

The method of proof of the theorem differs from the other methods of this section and is based on the application of characteristic functions.

When proving Theorem 1.35, the lemmas below will be applied.

By $\xi^{(t)} = \xi\chi\{|\xi| < n^{1/2}\}$, $\xi_{(t)} = \xi - \xi^{(t)}$, we shall denote the sections of r.e. ξ on the level $t = n^{1/2}$. The following lemma is obvious.

LEMMA 1.36. *Let* $S_n^{(t)} := S_n(\xi^{(t)})$. *Then*

$$\Delta_n(a) \leq \Delta_n^{(t)}(a) + 2nP\{|\xi| > n^{1/2}\} \leq \Delta_n^{(t)}(a) + \Lambda_2,$$

where

$$\Delta_n^{(t)}(a) := \sup_{r \geq 0} |P\{|S_n^{(t)} + a| < r\} - P\{|Y + a| < r\}|.$$

Applying the well-known Esseen inequality for characteristic functions (see [158]) we obtain the following lemma.

LEMMA 1.37. *For any* $T_0 > 0$

$$\Delta_n^{(t)}(a) \leq \frac{C\widetilde{K}}{T_0} + \int_{-T_0}^{T_0} \frac{|f(s) - g(s)|}{|s|} \, ds,$$

where C is the absolute constant,

$$\widetilde{K} := \sup_{a \in \mathbf{H}} \sup_{r \geq 0} \frac{d}{dr} P\{|Y + a|^2 < r\},$$

$$f(s) := E \exp\{is \, |S_n^{(t)} + a|^2\}, \; g(s) := E \exp\{is \, |Y + a|^2\}.$$

The next lemma will be called a lemma of symmetrization.

LEMMA 1.38. *If $\varsigma_1, \varsigma_2, \varsigma_3$ are independent* $\mathbf{H}$*-r.e., then*

$$|E \exp\{is \, |\varsigma_1 + \varsigma_2 + \varsigma_3|^2\}| \leq (E \exp\{is(\widetilde{\varsigma}_1, \widetilde{\varsigma}_2)\})^{1/4}, \tag{1.66}$$

where $\widetilde{\varsigma}_i$ is a symmetrization of ς_i, $i = 1, 2$.

Proof. By the Hölder inequality

$$|E \exp\{is \, |\varsigma_1 + \varsigma_2 + \varsigma_3|^2\}| = |EE_{\varsigma_1} \exp\{is \, |\varsigma_1 + \varsigma_2 + \varsigma_3|^2\}| \leq$$

$$\leq (E \, |E_{\varsigma_1} \exp\{is \, |\varsigma_1 + \varsigma_2 + \varsigma_3|^2\}|^2)^{1/2}, \tag{1.67}$$

where E_{ς_1} denotes that the mean is taken with respect to the variable ς_1 at fixed values of the rest. If $\hat{\varsigma}_1$ is an independent copy of r.e. ς_1 then for any $a \in \mathbf{H}$ r.v. $|\varsigma_1 + a|^2$ and $|\hat{\varsigma}_1 + a|^2$ are independent and identically distributed. Consequently,

$$|E \exp\{is \, |a + \varsigma_1|^2\}|^2 = E \exp\{is(|\varsigma_1 + a|^2 - |\hat{\varsigma}_1 + a|^2)\}$$

$$= E \exp\{is(2a + \varsigma_1 + \hat{\varsigma}_1, \widetilde{\varsigma}_1)\}.$$

Hence, also from (1.67), by applying the Hölder inequality again, we obtain

$$|E \exp\{is \, |\varsigma_1 + \varsigma_2 + \varsigma_3|^2\}| \leq (E \exp\{is(2\varsigma_2 + 2\varsigma_3 + \varsigma_1 + \hat{\varsigma}_1, \widetilde{\varsigma}_1)\})^{1/2}$$

$$\leq (E \, |E_{\varsigma_2} \exp\{2is(\varsigma_2, \widetilde{\varsigma}_1)\}|^2)^{1/4}$$

$$= (E \exp\{2is(\widetilde{\varsigma}_1, \widetilde{\varsigma}_2)\})^{1/4}. \; \square$$

LEMMA 1.39. *There exists an absolute constant C such that $\widetilde{K} \leq C/\lambda_5$.*

Proof. Let the r.v. γ be distributed normally with the parameters 0 and 1. It is known (see [158]) that if $\operatorname{Re} z \geq 0$, then

$$E \exp\{z\gamma^2\} = (1 - 2z)^{-1/2}. \tag{1.68}$$

By writing $\eta = (\eta_1 + \eta_2)/\sqrt{2}$, where η_1, η_2 are independent copies of η, and by applying the lemma on symmetrization, we obtain

$$|g(s)|^4 \leq E \exp\{is(\eta_1, \eta_2)\} = E \exp\left\{\frac{-s^2(T\eta, \eta)}{2}\right\}$$

$$= \prod_{j=1}^{\infty} (1 + \lambda_j^2 s^2)^{-1/2}.$$

Hence, by integrating $|g(t)|$ we obtain the necessary estimate of the quantity $\widetilde{K}$. $\square$

We shall now assume that all the random elements and random variables are independent in total.

Let the natural numbers k and m satisfy $k + m \leq [(n - 1)/8]$, where $[a]$ is the integer part of the number a. We shall define the quantities $\kappa(s)$ and $\kappa_g(s)$ assuming:

$$\kappa(s) = \inf_{k,m} \left(E \exp\left\{2i\left(\frac{s}{n}\right)(S, R)\right\}\right)^{1/4},$$

where

$$S = \sum_{j=1}^{k} \xi_j^{(t)},\ R = \sum_{j=k+1}^{k+m} \xi_j^{(t)}.$$

The quantity κ_g differs from $\kappa(s)$ only in the substitution of the summands $\xi_j^{(t)}$ for the Gaussian r.e. η_j (we also accept $\Sigma_{j\in\emptyset} = 0$) in the sums S and R.

LEMMA 1.40. *Let $k + m \leq n$. Then*

$$E\,|n^{-1/2}(\xi_1^{(t)} + \cdots + \xi_k^{(s)} + \eta_1 + \cdots + \eta_m)|^p \leq C(p).$$

Proof. If $m = 0$, then the estimate is easily obtained from the Bernstein inequality (see Theorem 1.3.12); besides, it must be taken into account that $E\,|\xi|^2 = 1$ and $E\xi = 0$. But if $m > 0$, then the application of the inequality $(a + b)^p \leq 2^{p-1}(a^p + b^p)$, $a, b \geq 0$, shows that it is sufficient to estimate $E\,|\eta|^p$. But $E\,|\eta|^2 = 1$, and the inequality $E\,|\eta|^p \leq C(p)$ follows from Proposition 1.2.3. □

LEMMA 1.41. *Let $\lambda_1 \geq \lambda_2 \geq \cdots$ be eigenvalues of the covariance operator of $\mathbf{H}$-r.e. ξ, and let the number $\delta > 0$, r.e. $W = \tilde{\xi}^{(t)}\chi\{|\tilde{\xi}^{(t)}| < \delta\}$, where $\tilde{\xi}^{(t)}$ is the symmetrization of r.e. $\xi^{(t)} = \xi\chi\{|\xi| < n^{1/2}\}$. If $8\Lambda_2 < \lambda_q$ and $\delta \geq 32n^{1/2}\tilde{\Lambda}_3/\lambda_q$, then the operator $R = R_W$ has no less than q eigenvalues (with regard for their multiplicity) exceeding λ_q.*

Proof. It is sufficient to check (see [100]) that

$$(Rz, z) \geq (2Tz, z) - \lambda_q\,|z|^2 \quad \text{for all } z \in \mathbf{H}. \tag{1.69}$$

We have

$$(Rz, z) = E(W, z)^2 = E(\tilde{\xi}^{(t)}, z)^2 - E(\tilde{\xi}^{(t)}, z)^2\chi\{|\tilde{\xi}^{(t)}| > \delta\}.$$

By applying the inequality $(a + b)^3 \leq 4(a^3 + b^3)$, $a, b \geq 0$, we obtain

$$E(\tilde{\xi}^{(t)}, z)^2\chi\{|\tilde{\xi}^{(t)}| > \delta\} \leq \delta^{-1}|z|^2E\,|\tilde{\xi}^{(t)}|^3 \leq 8\delta^{-1}|z|^2E\,|\xi^{(t)}|^3.$$

Since

$$|E(\xi^{(t)}, z)| = |E(\xi - \xi_{(t)}, z)| \leq \Lambda_2^{1/2}\,|z|$$

then

$$E(\tilde{\xi}^{(t)}, z)^2 = 2E(\xi^{(t)}, z)^2 - 2(E(\xi^{(t)}, z))^2 \geq 2(Tz, z) - 4\Lambda_2\,|z|^2.$$

Collecting the estimate and taking into account the condition of the lemma, we obtain (1.69). □

LEMMA 1.42. *The following estimate takes place:*

$$|f(s) - g(s)| \leq C(T)\ |(1 + s^2)(1 + |a|^3)(\Lambda_2 + \tilde{\Lambda}_3)(\kappa(s) + \kappa_g(s)). \tag{1.70}$$

Proof. In the course of the proof of the lemma we shall denote:

$$U = n^{-1/2}\xi,\ V = n^{-1/2}\eta,\ U_j^{(1)} = n^{-1/2}\xi_j^{(t)},$$
$$V_j = n^{-1/2}\eta_j,\ U^{(1)} = n^{-1/2}\xi^{(t)}.$$

Also we denote

$$W_k = U_1^{(1)} + \cdots + U_{k-1}^{(1)} + V_{k+1} + \cdots + V_n,$$

$$\phi(z) = \exp\{is\ |z + a|^2\},\ z \in \mathbf{H}.$$

Then

$$|f(s) - g(s)| \le \beta_1 + .. + \beta_n,$$

where $\beta_l = |E\phi(W_l + U^{(1)}) - E\phi(W_l + V)|$.

Let us expand the function ϕ into the Taylor series

$$\phi(z + h) = \phi(z) + \phi'(z)(h) + \frac{1}{2!}\phi'''(z)(h)^2 + \frac{1}{2}E(1 - \alpha)^2\phi'''(z + \alpha)(h)^3, \tag{1.71}$$

where the r.v. α is uniformly distributed at the interval [0, 1]. Assuming in (1.71) $z = W_l$, $h = U^{(1)}$, and then $h = V$, we obtain

$$\beta_l \le \Gamma_1 + \Gamma_2 + \Gamma_3 + \Gamma_4,$$

where

$$\Gamma_1 = |E\phi'(W_l)(U^{(1)}) - E\phi'(W_l)(V)| = |E\phi'(W_l)(U - U^{(1)})|$$

(since the means of r.e. U and V coincide),

$$\Gamma_2 = |E\phi''(W_l)(U - U^{(1)})^2|$$

(since the covariances of r.e. U and V coincide),

$$\Gamma_3 = |E(1 - \alpha)^2\phi'''(W_l + \alpha U^{(1)})(U^{(1)})^3|,$$
$$\Gamma_4 = |E(1 - \alpha)^2\phi'''(W_l + \alpha V)(V)^3|.$$

We shall prove that for $l \ge n/2$

$$\Gamma_j \le Cs^2(1 + |s|)\kappa(s)(1 + |a|^3)(n^{-3/2} + E\ |U - U^{(1)}|^3 + E\ |U^{(1)}|^3),$$
$$j = 1, 2, 3, 4. \tag{1.72}$$

At first we shall estimate Γ_3. It is not difficult to calculate that

$$\phi'''(z)(h)^3 = -4s^2(z + a, h)\phi(z)[3\,|h|^2 + 2is\,(z + a, h)].$$

Hence it follows that the quantity Γ_3 does not exceed the finite sum of summands having the form

$$E\ |E_{W_l}\Phi(W_l)^j\phi(W_l + \alpha U^{(1)})|,$$

where Φ is a j-linear, $j = 0, 1, 2, 3$, continuous form in the space **H** satisfying

$$\|\Phi\| \le Cs^2(1 + |s|)(1 + |a|^3)(|U^{(1)}|^3 + |U^{(1)}|^6),$$

where C is some absolute constant.

Since $l \ge n/2$, we can write $W_l = A_1 + \cdots + A_4$, where each of the sums A_i, $1 \le i \le 4$, is composed of $[(n-1)/8]$ independent copies of r.e. $U^{(1)}$. Because of the j-linearity of Φ, we have

$$|E_{W_l}\Phi(W_l)^j\phi(W_l + \alpha U^{(1)})| \le \sum_{i_1=1}^{4} \cdots \sum_{i_j=1}^{4} |E_{W_l}\Phi(A_{i_1}, ..., A_{i_j})\phi(W_l + \alpha U^{(1)})|.$$

Since $j \le 3$, then among $A_{i_1}, ..., A_{i_j}$ at least one of the sums $A_1, ..., A_4$, is absent. Let, for definiteness, the sum A_1 be absent. By applying the lemma on symmetrization and by breaking up the sum A_1 into two sums

$$A_1 = \sum_{i=1}^{k} U_i^{(1)} + \sum_{i=k+1}^{k+m} U_i^{(1)},$$

we obtain

$$|E_{W_l}\Phi(A_{i_1}, ..., A_{i_j})\phi(W_l + \alpha U^{(t)})|$$
$$\le E_{W_l}\|\Phi\| \, |A_{i_1}| \cdots |A_i| \, |E_{A_1}\phi(W_l + \alpha U^{(1)})|$$
$$\le \kappa(s)\|\Phi\| E\, |A_{i_1}| \cdots |A_{i_j}| \le C\kappa(s)\|\Phi\|$$

(the last inequality being obtained by means of Lemma 1.40). Collecting the estimates and noting that $E\,|U^{(1)}|^6 \le E\,|U^{(1)}|^3$, we make certain that the quantity Γ_3 satisfies estimate (1.72). The estimate of the quantity Γ_4 differs from the estimate Γ_3 only in the fact that the moments $E\,|U^{(1)}|^3$ and $E\,|U^{(1)}|^6$ are substituted for $E\,|V|^3$ and $E\,|V|^6$, respectively. However, by Lemma 1.40, $E\,|\eta|^p \le C(p)$, therefore $E\,|V|^3 + E\,|V|^6 \le Cn^{-3/2}$. Hence, Γ_4 also satisfies (1.72). The estimate of the quantities Γ_1 and Γ_2 also repeats the estimate of the quantity Γ_3. Besides, the sum $E\,|U - U^{(1)}|^2 + E\,|U^{(1)}|^3$ will be substituted for $E\,|U - U^{(1)}| + E\,|U - U^{(1)}|^2$. But, by the Chebyshev inequality, $E\,|U - U^{(1)}| \le E\,|U - U^{(1)}|^2$. Thus, estimate (1.72) is completely proved.

In the case $l < n/2$, in the sum W_l there will be present no less than $(n-1)/2$ independent copies of the Gaussian r.e. V. Therefore, acting similarly as in the case $l \ge n/2$, we also arrive at an estimate of the form (1.72), but the characteristic $\kappa(s)$ in this estimate will be substituted for $\kappa_g(s)$. Summing up estimate (1.72) over j and l, we obtain

$$|f(s) - g(s)| \le Cs^2(1 + |s|)(1 + |a|^2)(n^{-1/2} + \Lambda_2 + \tilde{\Lambda}_3)(\kappa(s) + \kappa_g(s)).$$

By the use of the fact that $E\,|\xi|^2 = 1$, it is not difficult to show that $n\ -1/2 \le (\Lambda_2 + \tilde{\Lambda}_3)$. □

LEMMA 1.43. *Let $n \ge 25$, the integer $l > 0$. Suppose that an infinite number of eigenvalues of the covariance operator T are positive. Then $\kappa_g(s) \le C(l, T)(1 + s^2)^{-l}$. Besides, there exists a constant $C_1 = C_1(l, T)$ such that conditions $\Lambda_2 + \tilde{\Lambda}_3 \le C_1$ and $|s| \le 1/\tilde{\Lambda}_3$ imply $\kappa(s) \le C(l, T)(1 + s^2)^{-l}$.*

Proof. Choosing the numbers $m = k = [(n+1)/8]/2$ in the definition of $\kappa_g(s)$ and reasoning similarly as in the proof of Lemma 1.42, we easily obtain the estimate for $\kappa_g(s)$. We shall proceed to the estimate of $\kappa(s)$.

Since $\kappa(s) = \kappa(-s)$, then we can assume $s \ge 0$. At first we shall show that if the real numbers k, m satisfy $k + m \le [(n+1)/8]$, and the numbers $\tau, \rho, \delta, \gamma \ge 0$ satisfy $\tau \le s$, $2s\rho\delta \le n$, $\tau\gamma\delta \le (n)^{1/2}$, then

$$\frac{\kappa^4(s)}{2} \le \phi(\tau) + \exp\left\{-\frac{\rho^2}{2k + 4\rho\delta}\right\} + \exp\left\{-\frac{\gamma^2}{8}\right\}, \tag{1.73}$$

where $\phi(\tau) = E\exp\{-km\,\tau^2(RZ, Z)/4n^2\}$, R is a covariance operator of r.e. $W := \tilde{\xi}^{(t)}\chi\{|\tilde{\xi}^{(t)}| < \delta\}$, and Z is a Gaussian $\mathbf{H}$-r.e. with a covariance R.

Note that for the symmetric r.e. U for all $x \in \mathbf{H}$ and $\delta > 0$

$$E\exp\{i(x, U)\} \le E\exp\{i(x, U\chi\{|U| < \delta\})\}. \tag{1.74}$$

Applying estimate (1.74) $k + m$ times, we obtain

$$\kappa^4(s) \le EE_W^m \exp\left\{\frac{2is(S, W)}{n}\right\},$$

where $S = W_1 + \cdots + W_k$, besides r.e. $W, W_1, W_2, \cdots$ are independent and identically distributed.

On the interval $|u| \le 1$ the inequality $\cos u \le 1 - u^2/4$ is satisfied. Noting again that $1 - u \le \exp\{-u\}, u \ge 0$, for $2s\rho\delta \le u$ we obtain

$$\kappa^4(s) \le P\{|S| > \rho\} + E\chi\{|S| \le \rho\}E_W^m \cos\left(\frac{2s(S, W)}{n}\right)$$

$$\le P\{|S| > \rho\} + E\chi\{|S| \le \rho\}\left[1 - \frac{s^2 E_W(S, W)^2}{n^2}\right]^m$$

$$\le P\{|S| > \rho\} + E\exp\left\{-\frac{\tau^2 m(RS, S)}{2n^2}\right\},$$

since $\tau \le s$. In order to estimate $P\{|S| > \rho\}$ we shall apply the Bernstein inequality (see Theorem 1.3.12). Since $EW = 0$ and for $j \ge 2$

$$E|W|^j \le \delta^{j-2}E|W|^2 \le \delta^{j-2}E|\tilde{\xi}^{(t)}|^2 \le 2\delta^{j-2}E|\xi|^2 = 2\delta^{j-2},$$

then

$$P\{|S| > \rho\} \le 2\exp\left\{-\frac{\rho^2}{2k + 4\rho\delta}\right\}.$$

Since $\operatorname{cov} Z = R$,

$$E\exp\left\{-\frac{\tau^2 m(RS, S)}{2n^2}\right\} = EE_W^k \exp\left\{\frac{i\tau m^{1/2}(Z, W)}{n}\right\}.$$

When estimating the last mathematical expectation we shall act as we did in the preceding one. Since the trace of the covariance operator R does not exceed 2, then (see [92])

$$P\{|Z| > \gamma\} \le 2\exp\left\{-\frac{\gamma^2}{8}\right\}.$$

Therefore, for $\tau(m)^{1/2}\gamma\delta \le n$,

$$EE_W^k \exp\left\{\frac{i\tau m^{1/2}(Z, W)}{n}\right\} \le \phi(\tau) + 2\exp\left\{-\frac{\gamma^2}{8}\right\}.$$

Having collected the estimates we obtain (1.73). Besides, the condition $\tau m^{1/2}\gamma\delta \le n$ easily follows from the inequalities $m \le n$, $\tau\gamma\delta \le n^{1/2}$. We shall choose $\delta = T_0^{-1}n^{1/2}$, $T_0 = \lambda_q/(32\tilde{\Lambda}_3)$. A sufficiently large natural number $q = q(l)$ will be chosen later. Further, we can assume that the numbers τ and T_0 are larger than an arbitrarily large (fixed) constant $C(l)$; otherwise the estimate of the quantity $\kappa(s)$ easily follows from the condition $\tilde{\Lambda}_3 \le 3C_1(l, T)$ and from the obvious estimate $\kappa(s) \le 1$. By the use of Lemma 1.41 and formula (1.68) we obtain

$$\phi(\tau) \le C(q, T)\left[1 + \frac{\tau^2 km}{n^2}\right]^{-q}. \tag{1.75}$$

Assume $\gamma = C_1(l) \ln^{1/2} T_0$. Then

$$\exp\left\{-\frac{\gamma^2}{8}\right\} \le T_0^{-C_1(l)/8} \le T_0^{-4l} \le C(l, T)(1 + s^2)^{-4l} \tag{1.76}$$

on the interval $s \le 1/\tilde{\Lambda}_3$, only if the number $C_1(l)$ chosen is sufficiently large.

We shall prove that $\kappa(s)$ satisfies the estimate from the condition of the lemma on the interval $C_1 s \ln^{1/2} T_0 \le T_0$. For this reason, in (1.73) we set $\tau = s$, $k = m = [[(n-1)/8]/2]$, $2\rho = C_1(l) n^{1/2} \ln^{1/2} T_0$. Besides, in (1.75) and (1.76), as well as after a simple estimate of the quantity $\exp\{-\rho^2/(2k + 4\rho\delta)\}$, due to the choice of the sufficiently large constant $C_1(l)$ we shall obtain the necessary estimate of the quantity $\kappa(s)$.

We shall now show that there exists a constant $C_2(l)$ such that on the interval $C_1^{-1}(l) T_0 \ln^{-1/2} T_0 \le s \le T_0^{4/3}/C_2(l)$ the quantity $\kappa(s)$ admits the estimate from the condition of the lemma. Therefore, in (1.73) we shall assume

$$\tau = C_1^{-1}(l) T_0 \ln^{-1/2} T_0, \quad \rho = C_2 n^{1/2} T_0^{-1/3},$$

$$k = [n T_0^{-4/3}], \quad k + m = \left[\frac{n-1}{8}\right],$$

and choosing a sufficiently large number $q = q(l)$ in (1.75), our arguments will be similar to the preceding ones. The constant $C_2(l)$ is also to be chosen sufficiently large. Note that in view of the condition $\tilde{\Lambda}_3 \le C_1(l, T)$ we can assume that $T_0^{4/3} C_2^{-1}(l) \ge T_0$. □

Proof of Theorem 1.35. In the course of the proof the theorem we can assume that an infinite number of eigenvalues of the operator T are positive. Otherwise the r.e. ξ with probability 1 is concentrated in some finite-dimensional subspace of the space **H**. Besides, one can assume that the covariance matrix of r.e. ξ is non-degenerate. Therefore, we can apply the well-known estimate in a finite-dimensional space (see, for instance, [31]) implying (1.65). Thus, let ξ be an infinite-dimensional r.e. By applying Lemmas 1.36 and 1.37 where $T_0 = 1/\tilde{\Lambda}_3$, and Lemmas 1.42, 1.43 and also 1.40 where $p = 3$, we obtain estimate (1.65) by means of elementary computations if only $n \ge 25$ and $\Lambda_2 + \tilde{\Lambda}_3 \le C_1(T)$. But if $n \le 25$ or $\Lambda_2 + \tilde{\Lambda}_3 \ge C_1(T)$, then estimate (1.65) obviously follows from the trivial inequality $\Delta_n(a) \le 2$.
□

5.2. Estimates of the Rate of Convergence in the Central Limit Theorem in Some Banach Spaces

Two aims are set in this section. Firstly, based on the examples of certain Banach spaces, in particular, of the spaces **C**[0, 1] and $\mathbf{c}_0$, we are to show the estimates of the rate of convergence in the central limit theorem that follow from the general results of the preceding section. Secondly, we are to construct the estimates of convergence on the balls by the use of special methods, taking into account the structure of the particular space.

We confine ourselves to estimates of the distance $\Delta_n(\xi, \eta, \mathscr{W}_f)$, where $\xi \in \mathscr{D}(\eta)$ and f is a positively homogeneous function satisfying the Lipschitz condition. Therefore, we shall assume that $|f(x) - f(y)| \le \|x - y\|$ and $f(tx) = tf(x)$ for all $t > 0$, with the elements x from the given space. In the case when $f(\cdot) = \|\cdot\|$, instead of $\Delta_n(\xi, \eta; \mathscr{W}_f)$; we shall write $\Delta_n(\xi, \eta)$.

Estimates in the Space C(S). First we consider the space C[0, 1] and the Gaussian r.e. η concentrated in the space $\mathrm{Lip}_\alpha[0, 1]$, $\alpha \in (0, 1]$. By $\mathscr{D}_{3,\alpha}(\eta)$ we shall denote the class of all C[0, 1]−r.e. ξ satisfying the conditions $E\xi = 0$,

$$\nu_{3,\alpha} = \nu_{3,\alpha}(\xi, \eta) := \int_{C[0,1]} \|x\|^3_{\mathrm{Lip}_\alpha[0,1]} \, |L(\xi) - L(\eta)| \, (dx) < \infty,$$

$E\xi(s)\xi(t) = E\eta(s)\eta(t)$ for all $s, t \in [0, 1]$.

THEOREM 2.1. *Let $P\{\eta \in \mathrm{Lip}_\alpha[0, 1]\} = 1$ and $\alpha \in (1/2, 1]$. If the r.v. $f(\eta)$ has a bounded density, then there exists a constant $C = C(\alpha, T_\eta)$ such that for all $\xi \in \mathscr{D}_{3,\alpha}(\eta)$ and $n \geq 1$ the following estimate takes place:*

$$\Delta_n(\xi, \eta; \mathscr{W}_f) \leq C\nu_{3,\alpha}^{1/3} n^{-1/6}. \tag{2.1}$$

Proof. Let the number $\beta \in (1/2, \alpha)$. If $\xi \in \mathscr{D}_{3,\alpha}(\eta)$ then

$$P\{\xi \in \mathrm{lip}_\beta[0, 1]\} = P\{\eta \in \mathrm{lip}_\beta[0, 1]\} = 1.$$

Assuming that ξ and η are the elements of the space $\mathbf{L}_0(\mathrm{lip}_\beta[0, 1])$ and denoting by i_β the operator of inclusion $\mathrm{lip}_\beta[0, 1]$ into C[0, 1], we have $\Delta_n(\xi, \eta; \mathscr{W}_f) = \Delta_{n,i_\beta}(\xi, \eta; \mathscr{W}_f)$. By Theorem 2.2.20, the class $\mathscr{W}_f$ is $(\mathbf{Q}^k_{i_\beta}, C\mathbf{K}_{1,k})$-uniform for each $k \geq 1$. There remains to prove that $E\xi(s)\xi(t) = E\eta(s)\eta(t)$ for all $s, t \in [0, 1]$ implies $ES(\xi)^2 = ES(\eta)^2$ for each symmetric operator $S \in \mathbf{L}^2(\mathrm{lip}_\beta[0, 1], \mathbf{R})$. By $\{\phi_i^*, i \in N\}$ we shall denote the coordinate functionals of the Schauder basis of the space $\mathrm{lip}_\beta[0, 1]$ (see [191]). It can easily be seen that the equality we need will be proved if we prove that

$$E\langle \xi, \phi_i^* \rangle \langle \xi, \phi_j^* \rangle = E\langle \eta, \phi_i^* \rangle \langle \eta, \phi_j^* \rangle \quad \text{for all } i, j \geq 1.$$

We can assume $\|\phi_i^*\| = 1$, $i \geq 1$. By the Krein–Milman theorem (see [85]), the set $V_1 := \{\phi^* \in \mathrm{lip}_\beta^*[0, 1] : \|\phi^*\| = 1\}$ is contained in a closed convex hull of its extreme points. Therefore it is sufficient to make sure that $E\langle \xi, \phi^* \rangle \langle \xi, \psi^* \rangle = E\langle \eta, \phi^* \rangle \times \langle \eta, \psi^* \rangle$ for the extreme points ϕ^*, ψ^* of the set V_1. It is known (see [125]) that the extreme points of the set coincide with the set of functionals $(\delta_s; \delta_{s,t} : s, t \in [0, 1]) \subset \mathrm{lip}_\beta^*[0, 1]$, where

$$\langle x, \delta_s \rangle = x(s), \quad \langle x, \delta_{s,t} \rangle = x(s) - x(t), \quad x \in \mathrm{lip}_\beta[0, 1].$$

Thus the necessary equality follows from the equalities

$$E\langle \xi, \delta_s \rangle \langle \xi, \delta_t \rangle = E\langle \eta, \delta_s \rangle \langle \eta, \delta_t \rangle,$$

$$\begin{aligned} E\langle \xi, \delta_{s,t} \rangle \langle \xi, \delta_{s',t'} \rangle &= E\xi(s)\xi(s') - E\xi(s)\xi(t') + E\xi(t)\xi(t') - E\xi(t)\xi(s') \\ &= E\langle \eta, \delta_{s,t} \rangle \langle \eta, \delta_{s',t'} \rangle. \end{aligned}$$

Now we can make use of Corollary 1.14. □

If $P\{\eta \in \mathrm{Lip}_\alpha[0, 1]\} = 1$ and $\alpha \in (0, 1/2]$, then to estimate the quantity $\Delta_n(\xi, \eta; \mathscr{W}_f)$ we shall apply Theorem 1.21. To this aim some auxiliary statements will be necessary. For technical reasons it is convenient to substitute the interval [0, 1] for the interval $[-\pi, \pi]$.

Let $\mathbf{W}_\beta$, $1/2 < \beta \leq \alpha$ stand for the Hilbert space defined in the proof of Theorem 2.2.20. We recall that $a_n(\phi)$, $n = 0, \pm 1, \pm 2, \cdots$, are Fourier coefficients of the function $\phi \in \mathbf{L}_2([\ \pi, \pi])$ with respect to the orthonormal basis

$$\{(2\pi)^{-1/2}, \pi^{-1/2} \sin nt, \pi^{-1/2} \cos nt, n = 1, 2, \cdots\}.$$

Then by $\tilde{\phi}$ we shall denote a periodical (with a period 2π) extension of the function $\phi \in \mathbf{L}_2[-\pi, \pi]$.

Consider the function $\psi: \mathbf{R}_1 \to \mathbf{R}_1$ from the class $\mathbf{C}^\infty(\mathbf{R}_1)$ with a bounded support, normed so that $\int_{-\infty}^{\infty} \psi(t)\,dt = 1$. For $\varepsilon > 0$ we assume $\psi_\varepsilon(t) = \varepsilon^{-1}\psi(\varepsilon^{-1}t)$ and for the function $\phi \in \mathrm{L}_2[-\pi, \pi]$

$$(A_\varepsilon\phi)(t) = \int_{-\infty}^{\infty} \tilde{\phi}(t - \tau)\psi_\varepsilon(\tau)\,d\tau. \tag{2.2}$$

It is obvious that $A_\varepsilon\phi$ is a function from the class $\mathbf{C}^\infty(\mathbf{R}_1)$; it is periodical together with all its derivatives, therefore the linear operator $A_\varepsilon : \mathrm{L}_2[-\pi, \pi] \to \mathbf{C}^\infty(\mathbf{R}_1)$ is correctly defined.

PROPOSITION 2.2. *If $\beta > \alpha$, $\alpha, \beta \in (0, 1]$, then, as defined by relation* (2.2), *the operator $A_\varepsilon \in \mathbf{L}(\mathrm{lip}_\alpha[-\pi, \pi], \mathbf{W}_\beta)$. Besides, there exists a constant $C(\alpha, \beta, \psi)$ such that*

$$\|A_\varepsilon\phi\|_{\mathbf{W}_\beta} \le C(\alpha, \beta, \psi)\max(1, \varepsilon^{\alpha-\beta})\|\phi\|_{\mathrm{lip}_\alpha}.$$

Proof. We shall estimate the Fourier coefficients $a_n(A_\varepsilon\phi)$. By changing the order of integration and by performing the substitution $t = z + \tau$, in the internal integral, we have

$$\begin{aligned}
&(2\pi)^{-1}\int_{-\pi}^{\pi} e^{-int}\int_{-\infty}^{\infty}\tilde{\phi}(t - \tau)\psi_\varepsilon(\tau)\,d\tau\,dt \\
&= (2\pi)^{-1}\int_{-\infty}^{\infty} e^{-int}\int_{-\pi}^{\pi}\tilde{\phi}(t - \tau)\psi_\varepsilon(\tau)\,dt\,d\tau \\
&= (2\pi)^{-1}\int_{-\pi}^{\pi} e^{int}\,\phi(t)\,dt\,\hat{\psi}(\varepsilon n),
\end{aligned}$$

where $\hat{\psi}$ is a Fourier transform of the function ψ. Using the relation obtained and noting that $\sup_t |t|^{2\beta-2\alpha}|\hat{\psi}(t)|^2 < \infty$, we obtain

$$\begin{aligned}
\|A_\varepsilon\phi\|^2_{\mathbf{W}_\beta} &= |a_0(\phi)|^2 + \sum_{n\ge1}(\varepsilon n)^{2\beta-2\alpha}|\hat{\psi}(\varepsilon n)|^2 n^{2\alpha}|a_n(\phi)|^2\varepsilon^{2\alpha-2\beta} \\
&\le C(\alpha, \beta, \psi)(1 + \varepsilon^{2\alpha-2\beta})\|\phi\|^2_{\mathrm{lip}_\alpha}. \square
\end{aligned}$$

Now by $i_{\mathbf{W}_\beta}$ we denote the inclusion operator $\mathbf{W}_\beta \to \mathbf{C}[-\pi, \pi]$.

PROPOSITION 2.3. *Let $\beta > \alpha$, $\alpha, \beta \in (0, 1]$. Then the following estimate takes place*

$$\|i_\alpha - i_{\mathbf{W}_\beta}\cdot A_\varepsilon\|_2 \le C(\alpha, \beta, \psi)\varepsilon^\alpha\max\left(1, \log\frac{1}{\varepsilon}\right).$$

Proof. To simplify the writing, we assume $T_\varepsilon = i_\alpha - i_{\mathbf{W}_\beta}\cdot A_\varepsilon$. If $t, \tau \in \mathbf{R}_1$, $\phi \in \mathrm{lip}_\alpha[-\pi, \pi]$, then $|\tilde{\phi}(t - \tau) - \tilde{\phi}(t)| \le \|\phi\|_{\mathrm{lip}_\alpha}|\tau|^\alpha$, therefore for $s \in [-\pi, \pi]$

$$\begin{aligned}
\|T_\varepsilon^*(\delta_s)\| &\le \sup_{\|\phi\|_{\mathrm{lip}_\alpha}\le1}|(T_\varepsilon\phi)(s)| \\
&\le \sup_{\|\phi\|_{\mathrm{lip}_\alpha}\le1}\left|\int_{-\infty}^{\infty}(\tilde{\phi}(s) - \tilde{\phi}(s - \tau))\psi_\varepsilon(\tau)\,d\tau\right| \le C(\psi)\varepsilon^\alpha.
\end{aligned}$$

Similarly, for $s, t \in [-\pi, \pi]$

$$\|T_\varepsilon^*(\delta_s) - T_\varepsilon^*(\delta_t)\| \le 2|s - t|^\alpha.$$

Applying Theorem 3.1.8 where the metric $\rho(s, t) = |s - t|^\alpha$ we obtain: for each $k \ge 1$

$$\|T_\varepsilon\|_2 \le C(\psi, \alpha)[\varepsilon^\alpha k^{1/2} + 2^{-k}k^{1/2}].$$

There remains to choose the number $k \ge 1$ appropriately. $\square$

THEOREM 2.4. *Let $P(\eta \in \mathrm{Lip}_\alpha[0, 1]) = 1$ and $\alpha \in (0, 1/2]$. If the r.v. $f(\eta)$ has a bounded* *density, then there exists a constant* $C = C(\alpha, T_\eta, E\|\xi\|^2)$ *such that for all* $n \geq 1$ *and for all* $\xi \in \mathscr{D}_{3,\alpha}(\eta)$ *the following estimate takes place:*

$$\Delta_n(\xi, \eta, \mathscr{W}_f) \leq C \max\{\nu_{3,\alpha}^{1/3} n^{-1/6};\ [\nu_{3,\alpha} n^{-1/2}]^{(4\alpha/6\alpha+3)-\varepsilon}\} \tag{2.3}$$

for each $\varepsilon > 0$.

Proof. Let the number $\alpha' < \alpha$ and $i_{\alpha'}$ be an inclusion operator $\mathrm{lip}_{\alpha'}[0, 1] \to \mathbf{C}[0, 1]$. As in the proof of the preceding theorem we have

$$\Delta_n(\xi, \eta, \mathscr{W}_f) = \Delta_{n,i_{\alpha'}}(\xi, \eta, \mathscr{W}_f),$$

but now we shall make use of Theorem 1.21.

Let the number $\beta \in (1/2; 1]$. From Theorem 2.2.20 and Proposition 2.2 and 2.3 we conclude that there exists a Hilbert space $\mathbf{H}_\beta \subset \mathbf{C}[0, 1]$ such that the class $\mathscr{W}_f$ is $(\mathbf{C}^k_{\mathbf{H}_\beta}, C\mathbf{K}_{1,k})$-uniform for each $k \geq 1$ and for each $\varepsilon > 0$ there exists an operator $u_\varepsilon \in \mathbf{L}(\mathrm{lip}_{\alpha'}[0, 1], \mathbf{H}_\beta)$ with the following properties: $\|i_\alpha - i_{\mathbf{H}_\beta} \cdot u_\varepsilon\|_2 \leq \varepsilon$, where $i_{\mathbf{H}_\beta}$ is the inclusion operator $\mathbf{H}_\beta \to \mathbf{C}[0, 1]$, and $\|u_\varepsilon\| \leq C(\alpha', \beta)\max(1, \varepsilon^{1-\beta/\alpha'})$ for each $0 < \alpha'' < \alpha'$. There remains to make use of Theorem 1.21 with $\gamma = (\beta/\alpha'') - 1$ and to take into account the arbitrariness of the numbers $\beta > 1/2$ and $\alpha'' < \alpha' < \alpha$. □

COROLLARY 2.5. *Let* $w(\cdot)$ *be a standard Wiener process on* [0, 1], *the process* $\xi(\cdot)$ *satisfies the conditions:*

$$E\xi(s)\xi(t) = Ew(s)w(t) \quad \text{for all} \quad s, t \in [0, 1];$$

$$\tilde{\nu}_3 := \int \|x\|^3_{\mathrm{Lip}_\alpha} |L(\xi) - L(w)|(dx) < \infty$$

for some $\alpha \in (0, 1/2]$. *If the r.v.* $f(w)$ *has a bounded density, then for each* $\varepsilon > 0$ *there exists a constant* $C = C(\varepsilon, T_w, E\|\xi\|^2_{\mathrm{Lip}_\alpha[0,1]})$ *such that for all* $n \geq 1$ *the following estimate takes place:*

$$\Delta_n(\xi, w, \mathscr{W}_f) \leq C(\tilde{\nu}_3 n^{-1/2})^{1/3-\varepsilon}.$$

This estimate, as well as estimate (2.3), can be improved if $f(\cdot) = \|\cdot\|$. Moreover, under the method applied one can consider the space $\mathbf{C}(S)$ for the arbitrary metric compact (S, d).

We recall that the notation $K(\rho, x)$ was introduced before Theorem 3.1.8.

The following elementary inequalities (for all $X \in \mathbf{L}_0(\mathbf{C}(S))$, $t \geq 0$, $x > 0$) underlie the method applied:

$$P\{\|X\| \leq t\} \leq P\{\sup_{s \in K(\rho,x)} |X(s)| \leq t\}, \tag{2.4}$$

$$P(\|X\| \leq t) \geq P\{\sup_{s \in K(\rho,x)} |X(s)| \leq t - \varepsilon\} - P\{\sup_{\rho(s,t) \leq x} |X(s) - X(t)| > \varepsilon\}. \tag{2.5}$$

By $\mathscr{D}_3(\eta)$ we shall denote the class of all $\mathbf{C}(S)$-r.e. $\xi \in \mathscr{D}(\eta)$ such that

$$\nu_3 \equiv \nu_3(\xi, \eta) := \int_{\mathbf{C}(S)} \|x\|^3 |L(\xi) - L(\eta)|(dx) < \infty.$$

Let ρ be a continuous semimetric, with respect to d, on S satisfying the condition $\int_0^a H^{1/2}(S, \rho, x)\,dx < \infty$ for some $a > 0$. Let us denote

$$H_m = H(S, \rho, 2^{-m}), \quad \kappa(m) = \sum_{i=m+1}^{\infty} 2^{-i} H_i^{1/2}.$$

THEOREM 2.6. *Let*

$$P\{\eta \in \mathrm{Lip}_\rho(S)\} = 1, \ E\|\xi\|^2_{\mathrm{Lip}_\rho(S)} < \infty, \ \xi \in \mathcal{D}_3(\eta).$$

If for all $\varepsilon > 0$

$$\sup_{r>0} P\{r \le \|\eta\| \le r + \varepsilon\} \le \phi(T_\eta)\varepsilon,$$

then there exists a constant $C_0 = C_0(T_\eta)$ *such that for all* $n \ge 1$ *the following estimate takes place:*

$$\Delta_n(\xi, \eta) \le C \max\{C_0(\nu_3 n^{-1/2})^{1/3}; \kappa^{2/3}(m(n))A_2^{1/3}\phi^{2/3}(T_\eta)\},$$

where

$$m(n) = m(n, C_1) := \sup\{m \ge 1 : H^3_m(\ln H_m)^9 \kappa^{-1}(m) \le C_1 n^{1/2} A_2(\phi(T_\eta))^{-1}\nu_3^{-1}\},$$

$$A_2 := E\|\xi\|^2_{\mathrm{Lip}_\rho(S)} + E\|\eta\|^2_{\mathrm{Lip}_\rho(S)},$$

and C, C_1 *are some absolute constants.*

Proof. Let us define the class $U_n(\eta)$ consisting of all possible vectors $(Z_1, \ldots, Z_n)$ of independent $\mathbf{C}(S)$ − r.e. $Z_1, \ldots, Z_n$ satisfying the conditions

$$Z_i \in \mathcal{D}(\eta), \ \nu_3(Z_i, \eta) \le \nu_3(\xi, \eta), \ E\|Z_i\|^2_{\mathrm{Lip}_\rho(S)} \le A.$$

We shall denote

$$\delta_n \equiv \delta_n(\xi, \eta) := \sup\left\{ \sup_t \ge 0 \left| P\left\{ \left\| \sum_{i=1}^n Z_i \right\| < t \right\} - P\left\{ \left\| \sum_{i=1}^n \eta_i \right\| < t \right\} \right| : (Z_1, \ldots, Z_n) \in U_n(\eta) \right\},$$

$$\delta_{n,m} \equiv \delta_{n,m}(\xi, \eta) := \sup\left\{ \sup_t \ge 0 \left| P\left\{ \sup_{s \in K(\rho, 2^{-m})} \left| \sum_{i=1}^n Z_i(s) \right| < t \right\} - P\left\{ \sup_{s \in K(\rho, 2^{-m})} \left| \sum_{i=1}^n \eta_i(s) \right| < t \right\} \right| : (Z_1, \ldots, Z_n) \in U_n(\eta) \right\}.$$

By means of inequalities (2.4), (2.5) and the condition of the theorem, we can show that for all $\varepsilon > 0$ the inequality

$$|\delta_n - \delta_{n,m}| \le I_n(\varepsilon) + \phi(T_\eta)\varepsilon n^{-1/2}, \tag{2.6}$$

takes place, where

$$I_n(\varepsilon) := \sup\left\{ P\left\{ \sup_{\rho(s,t) \le 2^{-m}} \left| \sum_{i=1}^n (Z_i(s) - Z_i(t)) \right| > \varepsilon \right\} : (Z_1, \ldots, Z_n) \in U_n(\eta) \right\}.$$

Let $\widetilde{Z}_i$ be a symmetrization of r.e. Z_i, $\{\varepsilon_i, i \ge 1\}$ be a Rademacher sequence. By the use of the inequality of symmetrization (Proposition 1.3.11) and applying the Chebyshev inequality, we obtain

$$P\left\{ \sup_{\rho(s,t) \le 2^{-m}} \left| \sum_{i=1}^n (Z_i(s) - Z_i(t)) \right| > \varepsilon \right\}$$

$$\leq C\varepsilon^{-2}E\left(E' \sup_{\rho(s,t)\leq 2^{-m}} \left| \sum_{i=1}^{n} (\widetilde{Z}_i(s) - \widetilde{Z}_i(t))\varepsilon_i \right|\right)^2, \tag{2.7}$$

where E' denotes the mean with respect to the probability space on which the r.e. ε_i, $i \geq 1$, are defined.

Reasoning the same way as in the proof of Theorem 3.1.8 (see the estimate of the quantity I_2 on page 49) we easily make sure that the mean E' in (2.7) does not exceed the quantity

$$C\kappa^2(m) \sum_{i=1}^{n} \|Z_i\|^2_{\mathrm{Lip}_\rho(S)},$$

Consequently,

$$I_n(\varepsilon) \leq C\kappa^2(m)nA_2\varepsilon^{-2}. \tag{2.8}$$

Assuming in (2.6)

$$\varepsilon = (n^{3/2}\phi^{-1}(T_\eta)A_2\kappa^2(m))^{1/3} \tag{2.9}$$

and taking into account estimate (2.8), we obtain

$$|\delta_n - \delta_{n,m}| \leq C\kappa^{2/3}(m)A_2^{1/3}\phi^{2/3}(T_\eta). \tag{2.10}$$

Note that here we could have used the known estimates in a finite-dimensional space and thus estimate $\delta_{n,m}$. All the estimates known, however, are rather badly dependent on the dimension of the space and, besides, have a rather complicated dependence on the covariance operator. Therefore, we shall estimate $\delta_{n,m}$.

Let $\mathbf{l}_\infty^k$ denote the space $\mathbf{R}_k$ with the norm $\|x\| = \max_{1\leq i\leq k} |x_i|$, $l_{\infty,\lambda}^k$ be the space $\mathbf{R}_k$ with the norm $\|x\| = \max_{1\leq i\leq k} |x_i| \lambda_i^{-1}$, where $\lambda = (\lambda_1, \lambda_2, \ldots) \in \mathbf{c}_0^+$, and let i_λ be an identical inclusion operator $\mathbf{l}_{\infty,\lambda}^k \to \mathbf{l}_\infty^k$. If for all $\delta > 0$,

$$\sum_{i=1}^{\infty} \exp\{-\delta(\lambda_i \ln^2 \lambda_i)^{-1}\} < \infty,$$

then it follows from Theorem 2.2.30 that the class $\mathbf{W}_0(\mathbf{l}_\infty^k)$ of the balls of the space $\mathbf{l}_\infty^k$ is $(\mathbf{Q}_{i_\lambda}^3, \mathbf{K})$-uniform, where $\mathbf{K}(\varepsilon) = (C_1\varepsilon^{-1}, C_2\varepsilon^{-2}, C_3\varepsilon^{-3})$. Assuming $\lambda_i = (\ln i \ln^3 \ln i)^{-1}$, $i > 2$, we can show that the class $\mathbf{W}_0(\mathbf{l}_\infty^k)$ is also $(\mathbf{Q}_{\mathbf{l}_\infty^k}^3, \mathbf{K}_k)$-uniform, where the function $\mathbf{K}_k(\varepsilon) = C(K_{k,1}(\varepsilon), K_{k,2}(\varepsilon), K_{k,3}(\varepsilon))$, C is an absolute constant, $K_{k,i}(\varepsilon) = \varepsilon^{-i} \ln^i k \ln^{3i} \ln k$.

Using this fact and estimating $\delta_{n,m}$, as in the proof of Lemma 1.8 (only slight changes of technical nature are required), we obtain the following recurrent inequality: for $n \geq 2$ and all $\varepsilon > 0$

$$\delta_{n,m} \leq C\Big\{\varepsilon^{-3}H_m^3 \ln^9 H_m n\nu_3\delta_{n-1,m} +$$

$$+ \max\Big(H_m^2 \ln^6 H_m\psi_1(\varepsilon, (n-1)T_\eta)\varepsilon^{-2}; \varepsilon^{-3}H_m^3 \ln^9 H_m \times$$

$$\times \psi\left(\frac{\varepsilon}{2}, (n-1)T_\eta\right)\Big)\Big\}n\nu_3 + \psi(\varepsilon, nT_\eta), \tag{2.11}$$

where

$$\psi(\varepsilon, kT_\eta) = \sup_{t\geq 0} P\left\{t \leq \sup_{s\in K(\rho,\alpha_m)} \left|\sum_{i=1}^{k} \eta_i(s)\right| \leq t + \varepsilon\right\},$$

$$\psi_1(\varepsilon, (n-1)T_\eta) = \sup_{r \geq \varepsilon/2} r^{-1}\psi(r, (n-1)T_\eta).$$

We shall use inequalities (2.4) and (2.5) again, but now we use them to estimate the quantity $\psi(\varepsilon, kT_\eta)$. Besides, taking into account (2.7) we obtain

$$\psi(\varepsilon, kT_\eta) \leq C[\phi^{2/3}(T_\eta)A_2^{1/3}(\kappa(m))^{2/3} + \varepsilon k^{-1/2}\phi(T_\eta)]. \tag{2.12}$$

Let the number ε be defined by relation (2.9). Then it follows from (2.12) that

$$\psi(\varepsilon, nT_\eta) \leq C(\kappa(m))^{2/3}A_2^{1/3}\phi^{2/3}(T_\eta),$$

$$\psi_1(\varepsilon, (n-1)T_\eta) \leq Cn^{-1/2}\phi(T_\eta).$$

By means of these estimates and relations (2.10) and (2.11) we obtain a recurrent inequality for δ_n at $n \geq 2$: for all $m \geq 1$

$$\begin{aligned}\delta_n \leq C[&(\kappa(m))^{-2}H_m^3 \ln^9 H_m\phi(T_\eta)A_2^{-1}\nu_3 n^{-1/2}\delta_{n-1} + \\ &+ (\kappa(m))^{-4/3}H_m^3 \ln^9 H_m\phi^{5/3}(T_\eta)A_2^{-2/3}n^{-1/2}\nu_3 + \\ &+ (\kappa(m))^{2/3}A_2^{1/3}\phi^{2/3}(T_\eta)].\end{aligned} \tag{2.13}$$

Now by induction with respect to $n \geq 1$ we shall show the estimate

$$\delta_n \leq \overline{C}V_n(C_1, C_0), \tag{2.14}$$

where

$$V_n(C_1, C_0) = \max(C_0(\nu_3 n^{-1/2})^{1/3}, A_2^{1/3}\phi^{2/3}(T_\eta)(\kappa(m(n, C_1)))^{2/3}).$$

In the case of $n = 1$, estimate (2.14) follows from Theorem 1.11. If $\delta_{n-1} \leq \overline{C}V_{n-1}(C_1, C_0)$, then for $n \geq 2$ we obtain from the recurrent inequality (2.13) the following estimate for δ_n:

$$\begin{aligned}\delta_n \leq C&(\kappa(m))^{-2}H_m^3 \ln^9 H_m\phi(T_n)A_2^{-1}n^{-1/2}\nu_3\overline{C}V_{n-1}(C_1, C_0) + \\ &+ (\kappa(m))^{-4/3}H_m^3 \ln^9 H_m\phi^{5/3}(T_\eta)A_2^{-2/3}n^{-1/2}\nu_3 + \\ &+ (\kappa(m))^{2/3}A_2^{1/3}\phi^{2/3}(T_\eta).\end{aligned}$$

Hence, observing that $V_{n-1}(C_1, C_0) \leq 2V_n(C_1, C_0)$ for each $m(n-1, C_1) \leq m < m(n, C_1)$, we obtain

$$\begin{aligned}\delta_n &\leq 2C\overline{C}C_1V_n(C_1, C_0) + CC_1(\kappa(m))^{2/3}\phi^{2/3}(T_\eta)A^{1/3}(T_\eta) + \\ &\quad + C(\kappa(m))^{2/3}A_2^{1/3}\phi^{2/3}(T_\eta) \\ &\leq \overline{C}\max(C_0(\nu_3 n^{-1/2})^{1/3}, A_2^{1/3}\phi^{2/3}(T_\eta)(\kappa(m))^{2/3}) \times \\ &\quad \times (2CC_1 + CC_1\overline{C}^{-1} + C\overline{C}^{-1}).\end{aligned}$$

Thus we complete the proof both of the estimate (2.14) and of the theorem since $\Delta_n(\xi, \eta) \leq \delta_n$. □

COROLLARY 2.7. *If the conditions of Theorems* 2.6 *are satisfied and the semimetric* ρ *is such that:*

(a) $N(S, \rho, x) \leq Cx^{-\alpha}$, $\alpha > 0$,

(b) $N(S, \rho, x) \leq C\exp\{x^{-1/\alpha}\}$, $\alpha > 1/2$,

then for all $n \geq 1$ *the following estimates take place: in the case of* (a)

$$\Delta_n(\xi, \eta) \leq C\nu_3^{1/3} n^{-1/6} \ln n,$$

in the case of (b)

$$\Delta_n(\xi, \eta) \leq C \max(\nu_3 n^{-1/2}, (\nu_3 n^{-1/2})^{(1-1/2\alpha)/3} (\ln n^{3(1-1/2\alpha)})),$$

where $C = C(T_\eta, A_2, \alpha)$.

We shall also formulate a corollary for the case $S = [0, 1]$ and for a limit Wiener process $\eta = w$. We recall that r.v. $\|w\|$ has a bounded density (see Supplement 4.4.6).

COROLLARY 2.8. *If* $\xi \in \mathcal{D}_3(w)$ *and* $A := E\|\xi\|^2_{\mathrm{lip}_\alpha[0,1]} < \infty$, $0 < \alpha \leq 1/2$, *then there exists a constant* $C = C(T_w, A)$ *such that for all* $n \geq 1$ *the following estimate takes place:*

$$\Delta_n(\xi, w) \leq C\nu_3^{1/3} n^{-1/6} \ln n.$$

Estimates in the space $\mathbf{c}_0$. Let the sequence $\lambda = (\lambda_1, \lambda_2, \ldots) \in \mathbf{c}_0^+$. By i_λ we denote the inclusion operator of the space $\mathbf{c}_{0,\lambda}$ into $\mathbf{c}_0$. If the Gaussian $\mathbf{c}_0$−r.e. η is concentrated in the space $\mathbf{c}_{0,\lambda}$ and $\xi \in \mathbf{L}_0(\mathbf{c}_0)$, then by $\nu_{3,\lambda} \equiv \nu_{3,\lambda}(\xi, \eta)$ we shall denote the pseudomoment

$$\int_{\mathbf{c}_0} \|x\|^3_{\mathbf{c}_{0,\lambda}} \, |\, \mathbf{L}(\xi) - \mathbf{L}(\eta) \,|\, (dx).$$

By $\mathcal{D}_{3,\lambda}(\eta)$ we denote the class of all $\mathbf{c}_0$−r.e. satisfying the conditions $E\xi = 0$, $E\xi_i\xi_j = E\eta_i\eta_j$ for all i, j, $\nu_{3,\lambda} < \infty$.

THEOREM 2.9. *Let the sequence* $\lambda \in \mathbf{c}_0^+$ *satisfy the condition* $\sup_i i^\beta \lambda_i < \infty$ *for* $\beta > 1/2$. *If* $P(\eta \in \mathbf{c}_{0,\lambda}) = 1$ *and r.v.* $f(\eta)$ *has a bounded density then for all* $n \geq 1$ *and for all* $\xi \in \mathcal{D}_{3,\lambda}(\eta)$ *the following estimate takes place:*

$$\Delta_n(\xi, \eta, \mathcal{W}_f) \leq C(n^{-1/2}\nu_{3,\lambda})^{1/3}. \tag{2.15}$$

Proof. We can assume that $\xi, \eta \in \mathbf{L}_0(\mathbf{c}_{0,\lambda})$. Then $\Delta_n(\xi, \eta, \mathcal{W}_f) = \Delta_{n,i_\lambda}(\xi, \eta, \mathcal{W}_f)$. Therefore estimate (2.15) follows from Theorem 2.4 and Theorem 2.2.19. □

By the use of Theorem 2.15 we can weaken the condition on the sequence λ; however, the estimate of the quantity $\Delta_n(\xi, \eta, \mathcal{W}_f)$ then sharply deteriorates. Better results are obtained in the case $f(\cdot) = \|\cdot\|$.

THEOREM 2.10. *For each* $\varepsilon > 0$ *let*

$$\sum_{i=1}^{\infty} \exp\{-\varepsilon\lambda_i^{-1}(\log \lambda_i)^{-2}\} < \infty.$$

If $P(\eta \in \mathbf{c}_{0,\lambda}) = 1$ *and r.v.* $\|\eta\|$ *has a bounded density then for all* $n \geq 1$ *and all* $\xi \in \mathcal{D}_{3,\lambda}(\eta)$ *the following estimate takes place:*

$$\Delta_n(\xi, \eta) \leq C(\nu_{3,\lambda} n^{-1/2})^{1/3},$$

where $C = C(T_\eta, \lambda)$.

Proof. As in the proof of the preceding result, we have the equality

$$\Delta_n(\xi, \eta) = \Delta_{n,i_\lambda}(\xi, \eta, \mathbf{W}_0).$$

In order to apply Theorem 1.13, we are but to note that the required uniformity of the class $\mathbf{W}_0(\mathbf{c}_0)$ is proved by Theorem 2.2.30. □

THEOREM 2.11. *Let $\lambda \in \mathbf{c}_0^+$ be a monotone sequence and for each $\varepsilon > 0$ let*

$$\sum_{i=1}^{\infty} \exp\{-\varepsilon\lambda_i^{-(1+\beta)}(\ln \lambda_i)^{-2}\} < \infty, \ \beta \in (0, 1). \tag{2.16}$$

If $P\{\eta \in \mathbf{c}_{0,\lambda}\} = 1$ and r.v. $\|\eta\|$ has a bounded density, then for all $n \geq 1$ and $\xi \in \mathcal{D}_{3,\lambda}(\eta)$ the following estimate takes place:

$$\Delta_n(\xi, \eta) \leq C(\nu_3 n^{-1/2})^{1/3-\beta/(3(2-\beta))} \tag{2.17}$$

where the constant $C = C(T_\eta, E\|\xi\|^2_{\mathbf{c}_{0,\lambda}}, \beta, \lambda)$.

Proof. Let $\mu_i = \lambda_i^{1+\beta}$, $i \geq 1$, and i_μ is an inclusion operator of $\mathbf{c}_{0,\mu}$ into $\mathbf{c}_0$. We shall define the operator $\mathbf{c}_{0,\lambda} \to \mathbf{c}_{0,\mu}$, for each $N \geq 1$ assuming $u_N(x) = (x_1, \ldots, x_N, 0, 0, \ldots)$. We estimate the norm of the operator u_N as

$$\|u_N\| = \sup_{\|x\|_{\mathbf{c}_{0,\lambda}} \leq 1} \|u_N(x)\|_{\mathbf{c}_{0,\mu}} \leq \lambda_N^{-\beta}. \tag{2.18}$$

Let $\delta > 0$ and the number $N = N(\delta)$ be such that $\lambda_i \leq \delta^{1/1-\beta}$ for $i \geq N+1$ and $\lambda_N > \delta^{1/1-\beta}$. If $\delta \in (0, 1]$, then from Theorem 3.1.12 we obtain the following estimate of the norm of type-2 operator $i_\lambda - i_\mu \cdot u_N$:

$$\begin{aligned} \|i_\lambda - i_\mu \cdot u_N\|_2 &\leq 2\delta + 2\delta \sum_{i \geq N+1, \lambda_i^2 < \delta} \exp\{-\delta\lambda_i^{-2}\} \\ &\leq 2\delta\left(1 + \sum_{i=N+1}^{\infty} \exp\{-\lambda_i^{-(1+\beta)}\}\right). \end{aligned} \tag{2.19}$$

But if $\delta > 1$, then

$$\|i_\lambda - i_\mu \cdot u_N\|_2 \leq 2\delta\left(1 + \sum_{i=N+1}^{\infty} \exp\{-\lambda_i^{-2}\}\right). \tag{2.20}$$

It follows from relations (2.18)–(2.20) that for each $\varepsilon > 0$ we can find an operator $u_\varepsilon \in \mathbf{L}(\mathbf{c}_{0,\lambda}, \mathbf{c}_{0,\mu})$ such that

$$\|i_\lambda - i_\mu \cdot u_\varepsilon\|_2 \leq \varepsilon, \ \|u_\varepsilon\| \leq C \max\{1, \varepsilon^{-\beta/(1-\beta)}\}.$$

Now we can apply Theorem 1.21 with $\gamma = \beta/(1-\beta)$. □

At the end of the section we shall present one result showing that when imposing very strong requirements on $\mathbf{c}_0$−r.e. ξ we can obtain an estimate of the quantity $\Delta_n(\xi, \eta)$ of the order $O(n^{-1/2+\varepsilon})$. As is known, the boundedness of r.e. ξ (in the norm of the space $\mathbf{c}_0$) in a general case does not even imply the CLT. But the existence of the moment $E\|\xi\|^3_{\mathbf{c}_{0,\lambda}}$, where $\lambda \in \Lambda_0$, provides with the rate of convergence of the order of $O(n^{-1/6})$. Now, roughly speaking we shall require boundedness of the quantity $\|\xi\|_{\mathbf{c}_{0,\lambda}}$ and a regular behaviour of λ.

Let $\xi = (\xi^1, \xi^2, \ldots)$ be $\mathbf{c}_0$−r.e. $E\xi = 0$, $T := \operatorname{cov}\xi$, $\sigma_i^2 = E(\xi^i)^2$. By $T_{m\sigma}$ we shall denote the covariance matrix of the random vector $(\xi^1\sigma_1^{-1}, \ldots, \xi^m\sigma_m^{-1})$, and by $\kappa_k^{(m)}$, $k = 1, 2, \ldots, m$, we denote the eigenvalues of this matrix. Without loss of generality we can assume that $\kappa_k^{(m)} \geq \kappa_{k+1}^{(m)}$ for all k and m. Let $\eta = (\eta^1, \eta^2, \ldots)$ denote a Gaussian $\mathbf{c}_0$−r.e. with a zero mean and a covariance operator T.

THEOREM 2.12. *Let the following conditions be satisfied:*

$$\sigma_i^2 = (\ln i)^{-(1+\delta)}, \ \delta > 0, \ i \geq 2, \tag{2.21}$$

$$|\xi^j| \le M\sigma_j,\ j \ge 1, \tag{2.22}$$

$$\kappa_m^{(m)} > (\ln m)^{-(\delta_1 - \delta)},\ \delta_1 > \delta. \tag{2.23}$$

Then for sufficiently large n

$$\Delta_n(\xi, \eta) = O(n^{-1/2+\nu}), \tag{2.24}$$

where $\nu > 0$ *and the constant in* (2.24) *infinitely increases if* $\nu \to 0$.

Before proving the theorem, we shall prove some auxiliary assertions that are the main steps of the proof. Up to the end of the section we shall use the following notations: if $x \in c_0$, then $\|x\|_m = \sup_{i<m} |x_i|$; $S_n(\xi_i)$ denotes the ith coordinate of the sum $S_n(\xi)$.

LEMMA 2.13. *For any integer* $m \ge 1$ *and* $\varepsilon > 0$ *the following inequality takes place:*

$$\Delta_n(\xi, \eta) \le C\Big(\Delta_{n,m}(\xi, \eta) + P\{\|S_n(\xi)\|_m > \varepsilon\} + P\{\|\eta\|_m > \varepsilon\} + P\{\|\eta\| < \varepsilon\}\Big), \tag{2.25}$$

where

$$\Delta_{n,m}(\xi, \eta) = \sup_{x>0} |P\{\max_{i\le m} |S_n(\xi^i)| < x\} - P\{\max_{i\le m} (\eta^i) < x\}|.$$

Proof. Since ξ and η are fixed everywhere below, then instead of $\Delta_n(\xi, \eta)$ and $\Delta_{n,m}(\xi, \eta)$ we will write Δ_n and $\Delta_{n,m}$. We denote

$$\Delta_n(x) = |P\{\|S_n(\xi)\| < x\} - P\{\|\eta\| < x\}|.$$

We have

$$\begin{aligned}\Delta_n &\le \max\{\sup_{0\le x\le\varepsilon} \Delta_n(x), \sup_{x>\varepsilon} \Delta_n(x)\} \\ &\le \max\{P\{\|S_n(\xi)\| < \varepsilon\} + P\{\|\eta\| < \varepsilon\}, \sup_{x>\varepsilon} \Delta_n(x)\} \\ &\le \max\{2P\{\|\eta\| < \varepsilon\} + \Delta_n(\varepsilon), \sup_{x>\varepsilon} \Delta_n(x)\} \\ &\le \sup_{x\ge\varepsilon} \Delta_n(x) + 2P\{\|\eta\| < \varepsilon\};\end{aligned} \tag{2.26}$$

$$\begin{aligned}\Delta_n(x) \le\ & |P\{\|S_n(\xi)\| < x\} - P\{\max_{j\le m} |S_n(\xi^j)| < x\}| + \\ &+ |P\{\max_{j\le m} |S_n(\xi^j)| < x\} - P\{\max |\eta^j| < x\}| + \\ &+ |P\{\|\eta\| < x\} - P\{\max |\eta^j| < x\}|.\end{aligned} \tag{2.27}$$

It is easily seen that

$$P\{\max_{j\le m} |S_n(\xi^j)| < x\} - P\{\|S_n(\xi)\| < x\} \le P\{\|S_n(\xi)\|_m > x\}, \tag{2.28}$$

$$P\{\max_{j\le m} |\eta^j| < x\} - P\{\|\eta\| < x\} \le P\{\|\eta\|_m > x\}. \tag{2.29}$$

(2.25) is proved by inequalities (2.26) – (2.29). □

LEMMA 2.14. *Let* η *be a Gaussian* c_0*–r.e. satisfying* (2.21). *If* $\varepsilon^2(\ln m)^\delta > 2$, *then*

$$P\{\|\eta\|_m > \varepsilon\}$$
$$\leq C\left[\left(1 + \varepsilon(ln\ m)^{(1+\delta)/2}\right)\left(\frac{\varepsilon^2}{2}(ln\ m)^\delta - 1\right)m^{2^{-1}\varepsilon^2 \ln^\delta m - 1}\right]^{-1}. \tag{2.30}$$

Proof. It is sufficient to use the following two elementary inequalities:

$$P\{\|\eta\|_m > \varepsilon\} \leq \sum_{j>m} P\{|\eta_j| > \varepsilon\},$$

$$\left(\frac{2}{\pi}\right)^{1/2} \int_t^\infty \exp\left\{-\frac{u^2}{2}\right\} du \leq \frac{4}{3}\left(\frac{2}{\pi}\right)^{1/2}(1+t)^{-1}\exp\left\{-\frac{t^2}{2}\right\}$$
$$\leq \frac{4}{3}\left(\frac{2}{\pi}\right)^{1/2}\exp\left\{-\frac{t^2}{2}\right\}$$

and to carry out simple calculations. □

LEMMA 2.15. *Let condition* (2.23) *be satisfied. Then for any* $\beta > 0$ *and* $\varepsilon \in (0, \varepsilon_0)$, *where*

$$\varepsilon_0 = \varepsilon_0(\beta, \delta_1) := \max\left\{\varepsilon : \ln\frac{1}{\varepsilon} \leq \frac{1}{\beta}\left(\sqrt{\frac{\pi}{2}}\,\varepsilon^{-1}\right)^{1/1+\delta_1}\right\},$$

the following estimate takes place:

$$P(\|\eta\| < \varepsilon) \leq C\exp\left\{-\frac{1}{2}\varepsilon^{-\beta}\ln\left(\sqrt{\frac{\pi}{2}}\frac{1}{\varepsilon}\right)\right\}. \tag{2.31}$$

Proof. For any $k \geq 1$ we have

$$P\{\|\eta\| < \varepsilon\} \leq P\{\max_{j\leq k} |\eta^j| < \varepsilon\} \leq (2\varepsilon)^k\left[(2\pi)^{k/2}(\det T_{k,\sigma})^{1/2}\prod_{j=1}^k \sigma_j\right]^{-1}$$
$$= \exp\left\{-k\ln\left(\frac{1}{\varepsilon}\sqrt{\frac{\pi}{2}}\right) - \sum_{j=1}^k \ln((\kappa_j^{(k)})^{1/2}\sigma_j)\right\}$$
$$\leq C\exp\left\{-k\ln\left(\frac{1}{\varepsilon}\sqrt{\frac{\pi}{2}}\right) + \frac{1+\delta_1}{2}\sum_{j=3}^k \ln\ln j\right\}.$$

Now we choose $k = [\varepsilon^{-\beta}] + 1$ and apply the estimate

$$\sum_{j=3}^k \ln\ln j \leq k\ln\ln k.$$

After simple calculations we shall obtain (2.31). □

In estimating the first two terms in (2.25), i.e. $\Delta_{n,m}$ and $P\{\|S_n(\xi)\|_m > \varepsilon\}$ we shall make use of the estimate (already presented in Lemma 1.25) of the remainder term in the CLT in a finite-dimensional space $\mathbf{R}_k$ and the following exponential inequalities of large deviations.

LEMMA 2.16 [141]. *Let* X_i, $i \geq 1$, *be i.i.d. r.v.,* $EX_1 = 0$, *and* $|X_1| \leq M$, *a.s.,* $Z_n := \Sigma_{i=1}^n X_i$. *If for any* $0 < p < 1$ *the condition*

$$P\{Z_n < 0\} \leq 1 - p,\ P\{Z_n > 0\} \leq 1 - p \tag{2.32}$$

is satisfied, then for all $t > 0$ *the following inequality takes place:*

$$P\{|Z_n| > tn^{1/2}\} \le (2e^3)^{1/2}p^{-1}\exp\{-t^2(16M^2)^{-1}\}. \tag{2.33}$$

Proof of Theorem 2.12. In (2.25) we assume $m = [n^\alpha]$, where α is some parameter whose choice is explained below. By using conditions (2.22), (2.23) estimate (1.39) and the elementary inequality

$$E\left(\sum_{i=1}^{m}(\xi^i)^2\right)^{3/2} \le m^{3/2}\max_{1\le i\le m}E|\xi^i|^3,$$

we obtain

$$\Delta_{n,m} = O(n^{-(1/2)+(5/2)\alpha}(\ln n)^{3(\delta_1-\delta)}). \tag{2.34}$$

To estimate the quantity $P\{\|S_n(\xi)\|_m > \varepsilon\}$ we shall apply Lemma 2.16. Since condition (2.22) is satisfied, then by the use of the classical Berry–Esseen estimate, we can assert that for $n > n_0$ inequalities (2.32) take place with $p = 1/4$. Therefore

$$P\{\|S_n(\xi)\|_m > \varepsilon\} \le \sum_{j=m+1}^{\infty}P\{|S_n(\xi^j)| > \varepsilon\}$$

$$\le C\sum_{j=m+1}^{\infty}\exp\{-\varepsilon^2(16M^2\sigma_j^2)^{-1}\}. \tag{2.35}$$

Let

$$\varepsilon = \frac{4M(\gamma+1)^{1/2}}{(\alpha\ln n)^{\delta/2}}, \tag{2.36}$$

where the parameter $\gamma > 0$ will be chosen below. From (2.35), recalling that $m = [n^\alpha]$, we obtain

$$P\{\|S_n(\xi)\|_m > \varepsilon\} \le C\left(\frac{\varepsilon^2}{16M^2}\ln^\delta m - 1\right)^{-1} m^{-((\varepsilon^2/16M^2)\ln^\delta m - 1)}$$

$$\le C\gamma^{-1}n^{-\alpha\gamma}.$$

From estimate (2.30), by substituting the chosen values of the quantities ε and m, we have

$$P\{\|\eta\|_m > \varepsilon\} \le C(\gamma, M)n^{-\alpha\gamma}. \tag{2.37}$$

In (2.31) we assume $\beta = 2\delta^{-1}$. Then, for $n > \tilde{n}_0$, where $\tilde{n}_0$ depends on M, α, γ and δ, we obtain

$$P\{\|\eta\| < \varepsilon\} \le C\exp\left\{-\frac{\alpha\ln n\,\ln(\sqrt{\pi/2}\,(\alpha\ln n)^{\delta/2}/4M(\gamma+1)^{1/2})}{2(4M(\gamma+1)^{1/2})^{2/\delta}}\right\}$$

$$\le C\exp\left\{-\frac{1}{2}C_1(M,\gamma,\alpha)\ln(C_2(M,\gamma,\alpha,\delta)\ln^{\delta/2}n)\ln n\right\} \le Cn^{-1/2}.$$

Now we have the estimates of all the terms present in inequality (2.25). By the chosen small number $\nu > 0$ we shall find $\tilde{\tilde{n}}_0$ and α such that for $n > \tilde{\tilde{n}}_0$ the inequality $n^{5/2\alpha}(\ln n)^{3(\delta_1+\delta)} < n^\nu$ holds. Assuming $\gamma = (2\alpha)^{-1}$, it follows from estimates (2.34)–(2.38) that, for $n > \max(n_0, \tilde{n}_0, \tilde{\tilde{n}}_0)$, (2.24) takes place. □

5.3. Estimates of the Rate of Convergence in the Prokhorov Metric and the Bounded Lipschitz Metric

Let η be a Gaussian $\mathbf{B}$-r.e. with a zero mean and a covariance operator $T = T_\eta$. In this section the estimates of the quantities $\pi(S_n(\xi), \eta)$ and $\rho_{\mathrm{BL}}(S_n(\xi), \eta)$, where $\xi \in \mathscr{D}(\eta)$, will be given. We shall consider both the case of a general Banach space $\mathbf{B}$ and particular spaces, e.g. C[0, 1] and $\mathbf{c}_0$. Like most of the estimates of Section 5.1, the estimates of the quantities written above differ significantly from the corresponding results in a finite-dimensional space, but the examples show that in a general case these estimates cannot be improved.

Some Inequalities. Let $\mathscr{F}$ be a class of measurable functions $f: \mathbf{B} \to \mathbf{R}$. The distances $L_{\mathscr{F}}$ and $\varsigma_{\mathscr{F}}$ in the set $\mathbf{L}_0(\mathbf{B})$, $(X_1, X_2 \in \mathbf{L}_0(\mathbf{B}))$ are defined as follows:

$$L_{\mathscr{F}}(X_1, X_2) := \sup_{f \in \mathscr{F}} L(f(X_1), f(X_2)),$$

$$\varsigma_{\mathscr{F}}(X_1, X_2) := \sup_{f \in \mathscr{F}} \left| \int_{\mathbf{B}} f(x)(L(X_1) - L(X_2))(\mathrm{d}x) \right|,$$

where the Levy metric between r.v. ς_1 and r.v. ς_2 is

$$L(\varsigma_1, \varsigma_2) := \inf\{\varepsilon > 0 : P\{\varsigma_1 < r\} < P\{\varsigma_2 < r + \varepsilon\} + \varepsilon,$$

$$P\{\varsigma_2 < r\} < P\{\varsigma_1 \le r + \varepsilon\} + \varepsilon, \text{ for all } r \in \mathbf{R}\}.$$

If $\mathscr{F} = \{f \in \mathrm{BL}(\mathbf{B}), \|f\|_{\mathrm{BL}} \le 1\}$, then $\varsigma_{\mathscr{F}}$ coincides with ρ_{BL}, and instead of $L_{\mathscr{F}}$ we shall write L_{BL}.

THEOREM 3.1. *For all $X_1, X_2 \in \mathbf{L}_0(\mathbf{B})$ the following inequalities take place:*

$$\pi(X_1, X_2) \le L_{\mathrm{BL}}(X_1, X_2) \le \rho_{\mathrm{BL}}^{1/2}(X_1, X_2). \tag{3.1}$$

Proof. First, we shall prove the left-hand inequality. Let $A \in \mathscr{B}(\mathbf{B})$ be a closed set,

$$d(A, x) = \inf\{\|x - y\| : y \in A\}, \; f(x) = \phi(d(A, x)),$$

where the continuous function $\phi: \mathbf{R} \to \mathbf{R}$ is linear in the interval [0, 1], $\phi(t) = 0$, if $t < 0$ and $\phi(t) = 1$, if $t > 1$. Since $f \in \mathrm{BL}(\mathbf{B})$ and $\|f\|_{\mathrm{BL}} \le 1$, then

$$\begin{aligned} L_{\mathrm{BL}}(X_1, X_2) \ge L(f(X_1), f(X_2)) &\ge \inf\{\varepsilon > 0 : P\{f(X_1) = 0\} \\ &\le P\{f(X_2) < \varepsilon\} + \varepsilon, P\{f(X_2) = 0\} \\ &\le P\{f(X_1) < \varepsilon\} + \varepsilon\} = \pi(X_1, X_2). \end{aligned}$$

Let, further, the function $f \in \mathrm{BL}(\mathbf{B})$, $\|f\|_{\mathrm{BL}} \le 1$, be such that $\varepsilon = L(f(X_1), f(X_2)) > 0$ (if for all f, $\varepsilon = 0$, then the right-hand inequality in (3.1) is obvious). By the definition of the Levy metric L, there exists a number $r \in \mathbf{R}$ such that

$$\varepsilon \le P\{f(X_1) \le r\} - P\{f(X_2) \le r + \varepsilon\} \tag{3.2}$$

or

$$\varepsilon \le P\{f(X_2) \le r\} - P\{f(X_1) \le r + \varepsilon\}.$$

Suppose $g(x) := 1 - \phi(\varepsilon^{-1}(f(x) - r))$, where the function ϕ is defined above. From (3.2) and the properties of the function g we obtain

$$\varepsilon \leq \left| \int_{\mathscr{B}} g(x)(L(X_1) - L(X_2))(dx) \right|. \tag{3.3}$$

Since $\varepsilon \leq 1$, consequently, $\|\varepsilon g\|_{\mathrm{BL}} \leq 1$ and (3.3) implies $\varepsilon \leq \varepsilon^{-1}\rho_{\mathrm{BL}}(X_1, X_2)$, which coincides with the right-hand inequality in (3.1). □

THEOREM 3.2. *For all $X_1, X_2 \in \mathbf{L}_0(\mathbf{B})$ the following inequality takes place:*

$$\rho_{\mathrm{BL}}(X_1, X_2) \leq 2\pi(X_1, X_2). \tag{3.4}$$

Proof. We shall make use of the following corollary of the Strassen theorem (see, for instance, [197], [203]): for arbitrary probability measures ν_1, ν_2 on $\mathscr{B}(\mathbf{B})$ there exists a probability space $(\Omega, \mathscr{A}, P)$ and $\mathbf{B}$-r.e. $\widetilde{X}_1, \widetilde{X}_2$ such that $L(\widetilde{X}_1) = \nu_1, L(\widetilde{X}_2) = \nu_2$ and

$$\inf\{t > 0 : P\{\|\widetilde{X}_1 - \widetilde{X}_2\| > t\} \leq t\} \leq \pi(\nu_1, \nu_2).$$

Having applied this corollary, with $\nu_1 = L(X_1)$, $\nu_2 = L(X_2)$ in the case $f \in \mathrm{BL}(\mathbf{B})$, $\|f\|_{\mathrm{BL}} \leq 1$, we obtain

$$\begin{aligned} &\left| \int_B f(x)(L(X_1) - L(X_2))(dx) \right| \\ &= |E(f(\widetilde{X}_1) - f(\widetilde{X}_2))| \leq E\,|f(\widetilde{X}_1) - f(\widetilde{X}_2)| \\ &\leq \sup_{x \neq y} \frac{|f(x) - f(y)|}{\|x - y\|} \int_{\|\widetilde{X}_1 - \widetilde{X}_2\| \leq \pi(X_1, X_2)} \|\widetilde{X}_1 - \widetilde{X}_2\|\, dP + \\ &\quad + 2 \sup_x |f(x)| \int_{\|\widetilde{X}_1 - \widetilde{X}_2\| > \pi(X_1, X_2)} dP \leq 2\pi(X_1, X_2). \square \end{aligned}$$

Next, we shall consider the relation between the metrics π, ρ_{BL}, L_{BL} and the distance ς_F, where $\mathscr{F}$ is some class of smooth functions. To this end we shall consider the operator $u \in \mathbf{L}(\mathbf{F}, \mathbf{B})$, where $\mathbf{F}$ is some Banach space and define some classes of u-differentiable functions.

Let $m \geq 1$ be an integer and let $\mathbf{K} = (K_0, K_1, \ldots, K_m)$ be a collection of non-negative numbers (it is possible that some of the numbers $K_0, \ldots, K_{m-1}$ are equal to $+\infty$).

Let the class $\mathscr{F} = \mathscr{F}(m, u, \mathbf{K})$ consist of measurable functions $f \in \mathbf{C}_u^m(\mathbf{B}, \mathbf{R})$ such that

$$\|f\|_\infty \leq K_0, \quad \|f\|_i \leq K_i, \quad i = 1, \ldots, m - 1, \quad \|f\|_{m-1,1} \leq K_m$$

(the seminorms $\|\cdot\|_i$, $\|\ \|_{m,\alpha}$ are defined on page 22). By $\varsigma_{m,u}^{\mathbf{K}}$ we define the distance $\varsigma_{\mathscr{F}(m,u,\mathbf{K})}$. In the case $\mathbf{K} = (\infty, \infty, \ldots, \infty, 1)$ instead of $\varsigma_{m,u}^{\mathbf{K}}$ we shall write $\varsigma_{m,u}$, and, in the case $\mathbf{K} = (1, \ldots, 1)$ – $\varsigma_{m,u}^1$. If $\mathbf{B} = \mathbf{F}$ and $u = I_{\mathbf{B}}$, then the index u will be omitted.

The interest for the equalities of the form $\pi \leq f(\varsigma_{m,u}^{\mathbf{K}})$ or $\rho_{\mathrm{BL}} \leq f(\varsigma_{m,u}^{\mathbf{K}})$, where the function $f(t) \to 0$ if $t \to 0$, was summoned by the following properties of the distance $\varsigma_{m,u}^{\mathbf{K}}$, very useful in studying the sums of independent r.e.

PROPOSITION 3.3. (a) *For each $c \in \mathbf{R}$,*

$$\varsigma_{m,u}^{\mathbf{K}}(cX_1, cX_2) \leq |c|^m \varsigma_{m,u}(X_1, X_2)K_m.$$

(b) *If r.e. Z does not depend on X_1, X_2, then*

$$\varsigma_{m,u}^{\mathbf{K}}(X_1 + Z, X_2 + Z) \leq \varsigma_{m,u}^{\mathbf{K}}(X_1, X_2).$$

(c) *For independent $\mathbf{B}$-r.e. $X_1, \ldots, X_n, Y_1, \ldots, Y_n$,*

$$\varsigma_{m,u}^{\mathbf{K}}\left(\sum_{i=1}^n X_1, \sum_{i=1}^n Y_i\right) \leq \sum_{i=1}^n \varsigma_{m,u}^{\mathbf{K}}(X_i, Y_i).$$

Proof. (a) and (b) follows from the definition of the distance $\varsigma_{\mathscr{F}}$, and (c) follows from (b) and the triangle inequality. □

THEOREM 3.4. *If the class $\mathscr{F}$ consists of continuous functions $f: \mathbf{B} \to \mathbf{R}$ such that the class $\{\mathscr{W}_f, f \in \mathscr{F}\}$ is $(\mathbf{Q}_u^{m-1,1}; \mathbf{K})$-uniform where $\mathbf{K}(\varepsilon) = (K_1\varepsilon^{-1}, ..., K_m\varepsilon^{-m})$, then for all $X_1, X_2 \in \mathbf{L}_0(\mathbf{B})$ the following inequality takes place:*

$$L_F(X_1, X_2) \leq (\varsigma_{m,u}^{\mathbf{K}'}(X_1, X_2))^{1/(m+1)}, \tag{3.5}$$

where $\mathbf{K}' = (1, K_1, ..., K_m)$.

Proof. Repeating the reasoning at the proof of the second inequality in (3.1) and noting that in (3.3) the function g can be taken from the class

$$\mathscr{F}(W_r(f), W_{r+\varepsilon}(f)) \cap (\mathbf{Q}_u^{m-1,1}; \mathbf{K}(\varepsilon)),$$

we easily obtain (3.5). □

THEOREM 3.5. *If the class $\mathscr{F}$ consists of continuous functions $f: \mathbf{B} \to \mathbf{R}$ such that the class $(\mathscr{W}_f, f \in \mathscr{F})$ is $(\mathbf{Q}_u^{m-1,1}; \mathbf{K})$-uniform, where $\mathbf{K}(\varepsilon) = (K_1\varepsilon^{-1}, ..., K_m\varepsilon^{-m})$, then there exist constants $\widetilde{K}_i = \widetilde{K}_i(K_1, ..., K_m)$, $i = 1, ..., m$, such that for all $X_1, X_2 \in \mathbf{L}_0(\mathbf{B})$ the following inequality takes place:*

$$\varsigma_{\mathscr{F}}(X_1, X_2) \leq$$
$$\leq 2 \max\{\varsigma_{m,u}^{\mathbf{K}}(X_1, X_2); [\|L(X_1) - L(X_2)\|^{m-1}\varsigma_{m,u}^{\mathbf{K}}(X_1, X_2)]^{1/m}\},$$

where $\mathbf{K} = (\infty, \widetilde{K}_1, ..., \widetilde{K}_m)$.

Proof. It follows from Theorem 2.2.8. that there exist constants $\widetilde{K}_i$, $i = 1, ..., m$, such that each function $f \in \mathscr{F}$ is uniformly $(\mathbf{Q}_u^{m-1,1}; \widetilde{\mathbf{K}})$-approximable where $\widetilde{\mathbf{K}}(\varepsilon) = (\widetilde{K}_1, \widetilde{K}_2\varepsilon^{+1}, ..., \widetilde{K}_m\varepsilon^{+(m+1)})$. Consequently, for each $\varepsilon > 0$ a function $g_\varepsilon \in (\mathbf{Q}_u^{m-1,1}, \widetilde{\mathbf{K}}(\varepsilon))$ can be found such that $\sup_x |f(x) - g_\varepsilon(x)| \leq \varepsilon$. By applying this inequality we obtain

$$\left| \int_{\mathbf{B}} f(x)(L(X_1) - L(X_2))(dx) \right|$$
$$\leq \varepsilon \|L(X_1) - L(X_2)\| + \max(\varepsilon^{-1}, \varepsilon^{-(m-1)})\varsigma_{m,u}^{\mathbf{K}}(X_1, X_2). \tag{3.6}$$

If $\varsigma_{m,u}^{\mathbf{K}}(X_1, X_2) \geq \|L(X_1) - L(X_2)\|$, then admitting $\varepsilon = 1$ in (3.6) we obtain

$$\varsigma_{\mathscr{F}}(X_1, X_2) \leq 2\varsigma_{m,u}^{\mathbf{K}}(X_1, X_2). \tag{3.7}$$

Otherwise, in (3.6) we assume

$$\varepsilon = (\varsigma_{m,u}^{\mathbf{K}}(X_1, X_2)\|L(X_1) - L(X_2)\|^{-1})^{1/m} < 1$$

and we obtain the inequality

$$\varsigma_{\mathscr{F}}(X_1, X_2) \leq 2(\varsigma_{m,u}^{\mathbf{K}}(X_1, X_2)\|L(X_1) - L(X_2)\|^{m-1})^{1/m}. \tag{3.8}$$

Relations (3.7) and (3.8) imply the assertion of the theorem. □

THEOREM 3.6. *If the operator $u \in \mathbf{L}(\mathbf{F}, \mathbf{H}) \cdot \pi_\gamma(\mathbf{H}, \mathbf{B})$, then for each $m \geq 1$ there exist constants $C_1 = C_1(m, u)$ and $C_2 = C_2(m, u)$ such that for all $X_1, X_2 \in \mathbf{L}_0(\mathbf{B})$ the following inequalities take place:*

$$\rho_{\mathrm{BL}}(X_1, X_2) \leq$$
$$\leq C_1 \max\{\varsigma^{\mathbf{K}}_{m,u}(X_1, X_2); (\varsigma^{\mathbf{K}}_{m,u}(X_1, X_2) \, \|L(X_1) - L(X_2)\|^{m-1})^{1/m}\}, \tag{3.9}$$

where $\mathbf{K} = (\infty, 1, ..., 1)$;

$$L_{\mathrm{BL}}(X_1, X_2) \leq C_2(\varsigma^1_{m,u}(X_1, X_2))^{1/m+1}, \tag{3.10}$$
$$\pi(X_1, X_2) \leq C_2(\varsigma^1_{m,u}(X_1, X_2))^{1/m+1}. \tag{3.11}$$

Proof. Inequalities (3.9) and (3.10) follow from Theorems 3.5, 3.4 and from Theorem 2.2.16 (also see Remark 2.2.5), and (3.11) follows from (3.10) and Theorem 3.1. □

Estimates of the Rate of Convergence. The following estimates are a direct corollary of Theorem 3.6 and Proposition 3.3.

THEOREM 3.7. *If* $u \in \mathbf{L}(\mathbf{F}, \mathbf{H}) \cdot \pi_\gamma(\mathbf{H}, \mathbf{B})$, *then for each* $m \geq 1$ *there exists a constant* $C = C(m, u)$ *such that for all* $n \geq 1$ *and* $\xi \in \mathscr{D}(\eta)$ *the following estimates take place:*

$$\rho_{\mathrm{BL}}(S_n(\xi), \eta) \leq C(\varsigma_{m,u}(\xi, \eta)n^{-(m-2)/2})^{1/m}, \tag{3.12}$$
$$\pi(S_n(\xi), \eta) \leq C(\varsigma_{m,u}(\xi, \eta)n^{-(m-2)/2})^{1/(m+1)}. \tag{3.13}$$

The following results shows, in particular, that in a general case it is impossible in (3.12) and (3.13) to substitute the characteristic $\varsigma_{m,u}(\xi, \eta)$ for $\varsigma_m(\xi, \eta)$ if $m > 2$, or for classical moment characteristics.

THEOREM 3.8. *For any monotone sequence* $(b_1, b_2, ..., b_n, ...) \in \mathbf{c}_0^+$, *there exists a Gaussian* $\mathbf{l}_2$*-r.e.* η *and* $\mathbf{l}_2$*-r.e.* $\xi \in \mathscr{D}(\eta)$ *such that* $P\{\|\xi\| \leq 1\} = 1$; $\varsigma_m(\xi, \eta) < \infty$ *for each* $m \leq 3$ *and*

$$\lim_{n\to\infty} \inf \pi(S_n(\xi, \eta)b_n^{-1} > 0, \tag{3.14}$$
$$\lim_{n\to\infty} \inf \rho_{\mathrm{BL}}(S_n(\xi), \eta)b_n^{-1} > 0. \tag{3.15}$$

Proof. Let $\xi = X$, $\eta = Y$, where $X, Y - \mathbf{l}_2$-r.e. defiined in proving Theorem 1.24 in the case $m = 2$. Having made use of the fact that r.v. $\|u(Y)\|$ has a bounded density, where the operator u has been also defined in the proof of Theorem 1.24, and by the use of the inequality of smoothing (1.1), we easily obtain:

$$\sup_{t\geq 0} |P\{\|S_n(u(X))\| < t\} - P\{\|u(Y)\| < t\} \leq C\rho^{1/2}_{\mathrm{BL}}(S_n(X), Y).$$

Combining this inequality and the assertion of Theorem 1.24 we obtain (3.14), and (3.15) follows from (3.14) and Theorem 3.1. By Proposition 3.9 presented below, $\varsigma_m(X, Y) < \infty$ for all $m \leq 3$. □

Estimates (3.12) and (3.13) are meaningful and ensure corresponding rates of convergence only under the condition $\varsigma_{m,u}(\xi, \eta) < \infty$ and $m > 2$. Therefore, we shall consider the conditions providing the finiteness of the quantity $\varsigma_{m,u}$.

PROPOSITION 3.9. *If* $\mathbf{F}$*-r.e.* X_1, X_2 *satisfy the condition: for each symmetric operator*

$$T \in \mathbf{L}^i(\mathbf{F}, \mathbf{R}) \quad ET(X_1)^i = ET(X_2)^i, \quad i = 1, ..., m-1,$$

then for each operator $u \in \mathbf{L}(\mathbf{F}, \mathbf{B})$ *the following inequalities are satisfied:*

$$\varsigma_{m,u}(u(X_1), u(X_2)) \le C(m)\nu_m(X_1, X_2), \tag{3.16}$$

$$\nu_m(X_1, X_2) := \int_{\mathbf{F}} \|x\|^m \, |L(X_1) - L(X_2)| \, (dx). \tag{3.17}$$

Proof. It is sufficient to apply the Taylor formula and to carry out some simple calculations. □

Note that the condition

$$ET(X_1)^i = ET(X_2)^i, \; i = 1, \ldots, m-1,$$

in the case of symmetric operators $T \in \mathbf{L}^i(\mathbf{F}, \mathbf{R})$ is necessary for the finiteness of the distance $\varsigma_{m,u}(u(X_1), u(X_2))$. Proposition 3.9 and Theorem 3.8 show the reason to apply the same scheme as in Section 5.1 when considering the rates of convergence in the CLT in the Prokhorov metric and in the bounded Lipschitz metric. Therefore, in the sequel we shall use the notations introduced in Section 5.1. We recall that Y stands for a Gaussian $\mathbf{F}$-r.v. with zero mean.

THEOREM 3.10. *If the operator* $u \in \mathbf{L}(\mathbf{F}, \mathbf{H}) \cdot \pi_\gamma(\mathbf{H}, \mathbf{B})$, *then there exists a constant* $C = C(u)$ *such that for all* $n \ge 1$ *and* $X \in \mathcal{D}^*(Y)$ *the following estimates take place:*

$$\rho_{\mathrm{BL}}(S_n(u(X)), u(Y)) \le C\nu_3^{1/3}(X, Y)n^{-1/6}, \tag{3.18}$$

$$\pi(S_n(u(X)), u(Y((\le C\nu_3^{1/4}(X, Y)n^{-1/8}. \tag{3.19}$$

Proof. Estimates (3.18) and (3.19) are corollaries of Theorem 3.7 and Proposition 3.9. □

We now turn to the consideration of examples showing that, generally, the exponents at n in estimates (3.18) and (3.19) are non-improvable.

Let the r.v. X be defined by the equalities

$$P\{X = a\} = P\{X = -a\} = p, \; P\{X = 0\} = 1 - 2p,$$

where $a > 0$, $p \le 1/2$. Then let $\sigma^2 = 2pa^2$ be a dispersion of r.v. X and γ be a Gaussian r.v. with a dispersion σ^2 and a zero mean. We shall define the function $f_{a,n} : \mathbf{R} \to \mathbf{R}$, assuming

$$f_{a,n}(x) = \begin{cases} x, & \text{if } x \in \left[0, \dfrac{a}{2n^{1/2}}\right], \\ \dfrac{a}{n^{1/2}} - x, & \text{if } x \in \left[\dfrac{a}{2n^{1/2}}, \dfrac{a}{n^{1/2}}\right], \\ 0, & \text{if } x \,\overline{\in}\, \left[0, \dfrac{a}{n^{1/2}}\right]. \end{cases}$$

It is obvious that $|f_{a,n}(x) - f_{a,n}(y)| \le |x - y|$ and $|f_{a,n}(x)| \le 1$, if $a < 2n^{1/2}$.

LEMMA 3.11. (a) $Ef_{a,n}(S_n(X)) = 0$, (b) $Ef_{a,n}(\gamma) \ge \pi^{-1/2}ap^{1/2}(1 - \exp\{-(1/16pn)\})$.

Proof. Statement (a) is obvious. We shall prove (b). We have

$$Ef_{a,n}(\gamma) \ge (2\pi)^{-1/2} \int_0^{a/2n^{1/2}} t \exp\left\{-\frac{t^2}{2\sigma^2}\right\}\sigma^{-1}\, dt$$

$$= (2\pi)^{-1/2}\sigma\left(1 - \exp\left\{-\frac{a^2}{8n\sigma^2}\right\}\right),$$

what coincides with (b). □

LEMMA 3.12. *If* $\delta_n := P\{S_n(X) = 0\}$ *and* $pn < 1/2$, *then*

$$1 - 2pn \le \delta_n \le 1 - 2pn + 10(pn)^2.$$

Proof. It can be easily seen that

$$\delta_n = \sum_{m=1}^{[n/2]} \begin{bmatrix} n \\ 2m \end{bmatrix} \begin{bmatrix} 2m \\ m \end{bmatrix} p^{2m}(1 - 2p)^{n-2m}.$$

Since

$$\begin{bmatrix} n \\ 2m \end{bmatrix} \begin{bmatrix} 2m \\ m \end{bmatrix} \le n^m,$$

then

$$\delta_n \le (1 - 2p)^n + \sum_{m=1}^{\infty} (pn)^{2m} \le (1 - 2p)^n + (pn)^2(1 - (pn)^2)^{-1}$$

$$\le (1 - 2p)^n + 2(pn)^2 \le 1 - 2pn + \sum_{k=2}^{\infty} (2pn)^k + 2(pn)^2$$

$$\le 1 - 2pn + 10(pn)^2.$$

Now we shall estimate δ_n from below:

$$P\{S_n(X) = 0\} \ge (1 - 2p)^n \ge 1 - 2pn,$$

since $(1 - t)^n \ge 1 - tn$, if $0 < t < 1$. □

LEMMA 3.13. *If* $pn < 1/200$, *then*

$$L(S_n(X), \gamma) \ge 2^{-1} \min(an^{-1/2}, pn).$$

Proof. By the definition of the Levy metric, it is sufficient to show that for each $\varepsilon < 2^{-1} \min(an^{-1/2}, pn)$ the following inequality takes place:

$$P\{S_n(X) \le -an^{-1/2}\} > P\{\gamma \le -an^{-1/2} + \varepsilon\} + \varepsilon. \tag{3.20}$$

Since $\varepsilon < a/2n^{-1/2}$, then

$$P\{\gamma \le -an^{-1/2} + \varepsilon\} < P\{\gamma < -an^{-1/2}\}$$

and (3.20) follows from the inequality

$$1 - P\{S_n(X) =)\} > P\left\{|\gamma| > \frac{a}{2n^{1/2}}\right\} + 2\varepsilon.$$

When estimating $P\{S_n(X) = 0\}$ by means of Lemma 3.12 we note that it is sufficient to prove the estimate

$$1 - 2pn > P\{|\gamma_0| > z\} + pn,$$

where γ_0 is a standard Gaussian r.v., $z = (8pn)^{-1/2}$. This estimate follows from the inequality

$$z^2P\{\,|\gamma_0|\,<z\}>\frac{3}{(2\sqrt{2})^2},$$

which is true if $z>200$. □

Let $\lambda=(\lambda_1,\lambda_2,...)$ be a sequence of positive numbers, $\sup_{i\geq 1}\lambda_i^{-1}<\infty$. By i_λ we denote the inclusion operator $\mathbf{l}_{2,\lambda}\to\mathbf{l}_2$.

THEOREM 3.14. *For each monotonic sequence $(b_n)_{n\in N}\in\mathbf{c}_0^+$ there exists a Gaussian $\mathbf{l}_{2,\lambda}$-r.e. η, a symmetric $\mathbf{l}_{2,\lambda}$-r.e. $\xi\in\mathcal{D}(\eta)$, $E\|\xi\|_{\mathbf{l}_{2,\lambda}}^\beta<\infty$, $\beta>2$ and a sequence $n_s\uparrow\infty$ such that for all $s\geq 1$*

$$\rho_{\mathrm{BL}}(S_n(i_\lambda\xi),\,i_\lambda\eta)\geq Cb_{n_s}n_s^{1/\beta-1/2}. \tag{3.21}$$

Proof. We shall define the distance $\kappa_0(\cdot,\cdot)$ on $\mathbf{L}_0(\mathbf{l}_2)$ assuming

$$\kappa_0(X_1,X_2):=\sup\{\,|Ef((a,X_1))-Ef((a,X_2))|\,:a\in\mathbf{l}_2,$$
$$\|a\|\leq 1,\ f\in\mathrm{BL}(\mathbf{R}_1),\ \|f\|_{\mathrm{BL}}<1\}.$$

It is obvious that

$$\rho_{\mathrm{BL}}(X_1,X_2)\geq\kappa_0(X_1,X_2),$$

therefore, it is sufficient to estimate the distance κ_0 from below. We choose the subsequence $n_s\uparrow\infty$ such, that

$$2\sum_{s=1}^{\infty}n_s^{-1}\leq 1,\ \sum_{s=1}^{\infty}\lambda_s^{\beta/2}b_{n_s}^\beta<\infty,\ n_s^{1/\beta-1/2}b_{n_s}\leq 1.$$

We shall assume $p_s=n_s^{-1}$, $a_s=n_s^{1/\beta}b_{n_s}$ and define $\mathbf{l}_{2,\lambda}$-r.e. ξ:

$$P\{\xi=a_se_s\}=P\{\xi=-a_se_s\}=p_s,\ P\{\xi=0\}=1-2\sum_{s=1}^{\infty}p_s.$$

Then

$$E\|\xi\|_{\mathbf{l}_{2,\lambda}}^\beta=2\sum_{s=1}^{\infty}(\lambda_s^{1/2}a_s)^\beta p_s=2\sum_{s=1}^{\infty}\lambda_s^{\beta/2}n_sb_{n_s}^\beta n_s^{-1}$$
$$=2\sum_{s=1}^{\infty}\lambda_s^{\beta/2}b_{n_s}^\beta<\infty.$$

Consider the function $f(x)=f_{a_s,n_s}(x_s)$ where x_s denotes the sth coordinate of the vector $x\in\mathbf{l}_2$. It is clear that

$$|f(x)|<a_sn_s^{-1/2}=n_s^{1/\beta-1/2}b_{n_s}\leq 1.$$

Besides $f(x)=g((x,e_s))$, where the function g satisfies the conditions $|g(t)-g(s)|\leq|t-s|$, $|g(t)|\leq 1$.

Therefore, applying estimate (b) of Lemma 3.11, we obtain

$$\kappa_0(S_{n_s}(i_\lambda\xi),\,i_\lambda\eta)\geq|Ef(S_{n_s}(\xi))-Ef(\eta)|$$
$$=Ef(\eta_s)\geq\pi^{-1/2}a_sp_s^{1/2}\left(1-\exp\left\{-\frac{1}{16}p_sn_s\right\}\right)$$

$$= \pi^{-1/2} n_s^{1/\beta - 1/2} b_{n_s}. \square$$

THEOREM 3.15. *For any monotonic sequence $(b_n)_{n-N} \in \mathbf{c}_0^+$ there exist a Gaussian $\mathbf{l}_{2,\lambda}$-r.e. η, a symmetric $\mathbf{l}_{2,\lambda}$-r.e. $\xi \in \mathscr{D}(\eta)$, $E\|\xi\|^\beta_{\mathbf{l}_{2,\lambda}} < \infty$, $\beta > 2$ and a sequence $n_s \uparrow \infty$ such that*

$$\pi(S_{n_s}(i_\lambda \xi), i_\lambda \eta) \geq b_{n_s} n_s^{-(\beta-2)/2(\beta+1)}.$$

Proof. Let π_0 be a Prokhorov metric restricted to the class of all half-spaces. Obviously, it is sufficient to prove the estimate from below for the metric π_0.

We choose a sequence $n_s \uparrow \infty$ such that

$$2 \sum_{s=1}^{\infty} \lambda_s^{\beta/2} b_{n_s}^{\beta+1} < \infty, \quad 2 \sum_{s=1}^{\infty} n_s^{-3\beta/2(\beta+1)} b_{n_s} \leq 1.$$

Assume $a_s = n_s^{3/2(\beta+1)} b_{n_s}, p_s = n_s^{-3\beta/2(\beta+1)} b_{n_s}$. We shall define $\mathbf{l}_{2,\lambda}$-r.e. ξ assuming

$$P\{\xi = a_s e_s\} = P\{\xi = -a_s e_s\} = p_s, \quad P\{\xi = 0\} = 1 - 2 \sum_{s=1}^{\infty} p_s.$$

It is obvious that $a_s p_s \to 0$ as $s \to \infty$. We have

$$E\|\xi\|^\beta_{\mathbf{l}_{2,\lambda}} = 2 \sum_{s=1}^{\infty} \lambda_s^{\beta/2} a_s^\beta p_s = 2 \sum_{s=1}^{\infty} \lambda_s^{\beta/2} b_{n_s}^{\beta+1} < \infty.$$

Noting that $\pi_0(S_{n_s}(i_\lambda \xi), i_\lambda \eta) \geq L(S_{n_s}(\xi_s), \eta_s)$ and applying Lemma 3.13, we obtain

$$\pi_0(S_{n_s}(i_\lambda \xi), i_\lambda \eta) \geq 2^{-1} \min(a_s n_s^{-1/2}; p_s n_s) = n_s^{(2-\beta)/2(\beta+1)} b_{n_s}. \square$$

In order to make sure of the optimality of the power indices at n in estimates (3.18) and (3.19), there remains to note that the sequence $\lambda = (\lambda_1, \lambda_2, \ldots)$ can be chosen such that $i_\lambda \in \pi_\gamma(\mathbf{l}_{2,\lambda}, \mathbf{l}_2)$ and to assume $\beta = 3$ in Theorems 3.14 and 3.15.

REMARK 3.16. Theorems 3.14 and 3.15 also show that the exponents in equalities (3.10), (3.11) are non-improvable.

Next we shall consider the possibility of extending the class of operators $u \in \mathbf{L}(\mathbf{F}, \mathbf{B})$ for which we can estimate the rate of convergence to zero of distances $\pi(S_n(u(X)), u(Y))$ and $\rho_{\mathrm{BL}}(S_n(u(X)), u(Y))$.

We denote by $\mathbf{H_B} \subset \mathbf{B}$ a Hilbert space such that the inclusion operator $i_{\mathbf{H_B}} : \mathbf{H_B} \to \mathbf{B}$ is γ-radonifying and $\|i_{\mathbf{H_B}}\| \leq 1$. We also denote

$$A_2 \equiv A_2(X, Y) := (E\|X\|^2 + E\|Y\|^2)^{1/2}.$$

THEOREM 3.17. *Let $u \in \mathbf{L}(\mathbf{F}, \mathbf{B})$. If for each $\varepsilon > 0$ there exists an operator $u_\varepsilon \in \mathbf{L}(\mathbf{F}, \mathbf{H_B})$ such that*

(1) $\|u_\varepsilon\| \leq C \max(1, \varepsilon^{-\gamma}), \gamma \geq 0$,

(2) $\|u - i_{\mathbf{H_B}} \cdot u_\varepsilon\|_2 \leq \varepsilon$,

then there exists a constant $C = C(\gamma, u)$ such that for all $X \in \mathscr{D}^*(Y)$ and for all $n \geq 1$ the following estimates take place:

$$\pi(S_n(u(X)), u(Y)) \leq C \max(n^{-1/8} \zeta_3^{1/4}(X, Y), (n^{-1/2} \zeta_3(X, Y) A_2^{3\gamma})^{2/8+9\gamma}), \tag{3.22}$$

$$\rho_{\mathrm{BL}}(S_n(u(X)), u(Y))$$

$$\leq C\max\{(n^{-1/2}\varsigma_3(X, Y))^{1/3}, (n^{-1/2}\varsigma_3(X, Y)A_2^{3\gamma})^{1/3+\gamma}\}. \tag{3.23}$$

Proof. Let the number $\delta > 0$ and operator $u_\delta \in \mathbf{L}(\mathbf{F}, \mathbf{H_B})$ satisfy the conditions of theorem. We denote:

$$X_\delta = i_{\mathbf{H_B}} \cdot u_\delta(X),\ X_\delta = i_{\mathbf{H_B}}\, u_\delta(Y),\ \psi(t) = C\max(1, t^{-\gamma}).$$

First, we shall consider the metric π. Let

$$f \in \mathrm{BL}(\mathbf{B}),\ \|f\|_{\mathrm{BL}} \leq 1 \text{ and } \varepsilon = L(f(S_n(u(X))), f(u(Y))).$$

There exists a number $r - \mathbf{R}$ such that

$$\varepsilon \leq P\{f(S_n(u(X))) \leq r\} - P\{f(u(Y)) \leq r + \varepsilon\}$$

or

$$\varepsilon \leq -P\{f(S_n(u(X))) \leq r + \varepsilon\} + P\{f(u(Y)) \leq r\}.$$

Let us assume for definiteness that the first of the above inequalities holds. By the use of inequalities (1.29) and (1.30) from Section 5.1, we obtain

$$\varepsilon \leq I_1 + I_2, \tag{3.24}$$

where

$$I_1 = P\left\{f(S_n(X_\delta)) \leq r + \frac{\varepsilon}{4}\right\} - P\left\{f(Y_\delta) \leq r + \frac{3\varepsilon}{4}\right\},$$

$$I_2 = P\left\{\|S_n(u(X)) - X_\delta\| > \frac{\varepsilon}{4}\right\} + P\left\{\|u(Y) - Y_\delta\| > \frac{\varepsilon}{4}\right\}.$$

From condition (2) and the Chebyshev inequality there easily follows the estimate

$$I_2 \leq C\delta^2\varepsilon^{-2}A_s^2. \tag{3.25}$$

By Theorem 2.2.16 there exists a function

$$g_\varepsilon \in \mathcal{F}(W_{r+\varepsilon/4}(f), W_{r+3\varepsilon/4}(f)) \cap (\mathbf{Q}^3_{\mathbf{H_B}}, \mathbf{K}(\varepsilon)),$$

where $\mathbf{K}(\varepsilon) = C(\varepsilon^{-1}, \varepsilon^{-2}, \varepsilon^{-3})$. By means of this function we estimate I_1 as follows

$$I_1 \leq \left| \int_{\mathbf{B}} g_\varepsilon(x)(L(S_n(X_\delta)) - L(Y_\delta))(dx) \right| \leq \varsigma^{\overline{\mathbf{K}}(\varepsilon)}_{3, i_{\mathbf{H_B}}}(S_n(X_\delta), Y_\delta),$$

where

$$\overline{\mathbf{K}}(\varepsilon) = (+\infty, C\varepsilon^{-1}, C\varepsilon^{-2}, C\varepsilon^{-3}).$$

From the properties of the distance $\varsigma_{\mathcal{F}}$ given in Proposition 3.3 we obtain the estimate

$$I_1 \leq C\psi^3(\delta)\varepsilon^{-3}n^{-1/2}\varsigma_3(X, Y). \tag{3.26}$$

Relations (3.24)-(3.26) imply the inequality

$$\varepsilon \leq C(\delta^2\varepsilon^{-2}A_2^2 + \psi^3(\delta)\varepsilon^{-3}n^{-1/2}\varsigma_3(X, Y)).$$

Assuming $\delta = C\varepsilon^{3/2}A_2^{-1}$, we obtain

$$\varepsilon \leq C\psi^3(\varepsilon^{3/2}A_2^{-1})\varepsilon^{-3}n^{-1/2}\varsigma_3(X, Y), \tag{3.27}$$

whence estimate (3.22) is easily derived.

We shall now estimate the distance $\rho_{\mathrm{BL}}(S_n(u(X)), n(Y))$. Let $f \in \mathrm{BL}(\mathbf{B})$, $\|f\|_{\mathrm{BL}} \leq 1$. By Theorem 2.2.17, for each $\varepsilon > 0$ there exists a function $g_\varepsilon \in (\mathbf{Q}^3_{i_{\mathbf{H_B}}}, \mathbf{K}(\varepsilon))$, where

$K(\varepsilon) = C(1, \varepsilon^{-1}, \varepsilon^{-2})$, such that $\sup_{x\in \mathbf{B}} |f(x) - g_\varepsilon(x)| \le \varepsilon$.

As in the estimation of the metric π, we have

$$\left| \int_{\mathbf{B}} f(x)(L(S_n(u(X))) - L(u(Y)))(dx) \right| \le 2\varepsilon +$$

$$+ \left| \int_{\mathbf{B}} g_\varepsilon(x)(L(S_n(X_\delta)) - L(Y_\delta))(dx) \right| + E\|S_n(u(X) - X_\delta)\| +$$

$$+ E\|Y_\delta - u(Y)\| \le C(\psi^3(\delta)\max(1, \varepsilon^{-2})\varsigma_3(X, Y)n^{-1/2} + A_2\delta + \varepsilon).$$

Now assuming $A_2\delta = \varepsilon$, we obtain

$$\rho_{BL}(S_n(u(X)), u(Y))$$
$$\le C(\psi^3(\delta)\max(1, \delta^{-2}A_2^{-2})\varsigma_3(X, Y)n^{-1/2} + A_2\delta), \tag{3.28}$$

and there remains to choose the number $\delta > 0$ appropriately. □

REMARK 3.18. If instead of condition (2) of Theorem 3.17 the inequality

$$\|u_\varepsilon\| \le C\max(1, (\ln \varepsilon^{-1})^\gamma), \quad \gamma \ge 0,$$

holds, then for $X \in \mathscr{D}^*(Y)$ such that $\varsigma_3(X, Y) < \infty$ and for all $n \ge 1$ we obtain

$$\pi(S_n(u(X)), u(Y)) = O(n^{-1/8}(\ln n)^{3\gamma/4}), \quad \rho_{BL}(S_n(u(X)), u)Y)) = O(n^{-1/6}\ln^\gamma n).$$

This easily follows from relations (3.27) and (3.28). But if the norm $\|u_\varepsilon\|$ is growing exponentially when $\varepsilon \to 0$, then the corresponding rate of convergence to zero of the distances in the Prokhorov metric and in the metric ρ_{BL} will have a logarithmic order.

Estimates in the spaces C[0, 1] and $\mathbf{c}_0$. This section will be completed by the estimate of the distances $\pi(S_n(\xi), \eta)$ and $\rho_{BL}(S_n(\xi), \eta)$ in the case when the random elements take the values in one of the spaces C[0, 1], $\mathbf{c}_0$. We shall show through an example of the space $\mathbf{c}_0$ that in a particular space, based on its structure, one can obtain more exact results than those which follow from the general estimates presented above. Below, we shall use the notations introduced in Section 2.2. First, we shall consider the space C[0, 1]. Let η be a Gaussian C[0, 1]−r.e.

THEOREM 3.19. *If* $P\{\eta \in \mathrm{Lip}_\alpha[0, 1]\} = 1$ *and* $\xi \in \mathscr{D}_{3,\alpha}(\eta)$, *then for all* $n \ge 1$ *the following estimates take place:*

(a) *If* $\alpha \in (1/2, 1]$, *then*

$$\pi(S_n(\xi), \eta) \le C\nu_{3,\alpha}^{1/4}n^{-1/8}, \tag{3.29}$$

$$\rho_{BL}(S_n(\xi), \eta) \le C\nu_{3,\alpha}^{1/3}n^{-1/6}; \tag{3.30}$$

(b) *If* $\alpha \in (0, 1/2]$, *then for each epsilon* > 0

$$\pi(S_n(\xi), \eta) \le C\max(\nu_{3,\alpha}^{1/4}n^{-1/8}, (\nu_{3,\alpha}n^{-1/2})^{(4\alpha/(9-2\alpha))-\varepsilon}). \tag{3.31}$$

$$\rho_{BL}(S_n(\xi), \eta) \le C\max((\nu_{3,\alpha}n^{-1/2})^{1/3}, (\nu_{3,\alpha}n^{-1/2})^{(2\alpha/3)-\varepsilon}). \tag{3.32}$$

Proof. Let the number $\alpha' < \alpha$. Then in the case $\alpha' \in (1/2, 1]$, as has been shown in Theorem 2.2.20, there exists a Hilbert space **H** such that

$$i_\alpha \in \mathbf{L}(\mathrm{lip}_{\alpha'}, \mathbf{H}) \cdot \pi_\gamma(\mathbf{H}, \mathrm{C}[0, 1]).$$

Assuming $\xi, \eta \in \mathrm{L}_0(\mathrm{lip}_\alpha[0, 1])$ and noting that

$$\pi(S_n(\xi), \eta) = \pi(S_n(i_{\alpha'}\xi), i_{\alpha'}\eta),$$

estimate (3.29) is obtained from Theorem 3.10. Similarly we obtain estimate (3.30).

In the case $\alpha \in (0, 1/2]$, having made use of Proposition 2.3, we conclude that there exists a Hilbert space **H** and operators $\nu \in \pi_\gamma(\mathbf{H}, \mathbf{C}[0, 1])$, $u_\varepsilon \in \mathbf{L}(\mathrm{lip}_{\alpha'}[0, 1], \mathbf{H})$ such that

$$\|i_{\alpha'} - \nu \cdot u_\varepsilon\|_2 \leq \varepsilon, \quad \|u_\varepsilon\| \leq C \max(1, \varepsilon^{1-\beta/\alpha''}),$$

where the number $\beta > 1/2$, $\alpha'' \in (0, \alpha')$. Now estimates (3.31), (3.32) easily follow from Theorem 3.17. □

Let $\eta, \xi \in \mathbf{L}_0(\mathbf{c}_0)$, a sequence $\lambda \in \mathbf{c}_0^+$.

THEOREM 3.20. *Let* $\lambda_i = i^{-\beta}$, $\beta > 0$ *and* $P\{\eta \in \mathbf{c}_{0,\lambda}\} = 1$, $\xi \in D_{3,\lambda}(\eta)$. *For all* $n \geq 1$ *the following estimates take place:*

(a) *In the case* $\beta > 1/2$,

$$\pi(S_n(\xi), \eta) \leq C(\nu_{3,\lambda} n^{-1/2})^{1/4}, \tag{3.33}$$

$$\rho_{\mathrm{BL}}(S_n(\xi), \eta) \leq C(\nu_{3,\lambda} n^{-1/2})^{1/3}. \tag{3.34}$$

(b) *In the case* $\beta \in (0, 1/2]$ *for each* $\varepsilon > 0$

$$\pi(S_n(\xi), \eta) \leq C \max\{(\nu_{3,\lambda} n^{-1/2})^{1/4}, (\nu_{3,\lambda} n^{-1/2})^{(4/\beta/(9-2\beta))-\varepsilon}\},$$

$$\rho_{\mathrm{BL}}(S_n(\xi), \eta) \leq C \max\{(\nu_{3,\lambda} n^{-1/2})^{1/3}, (\nu_{3,\lambda} n^{-1/2})^{(2\beta/3)-\varepsilon}\}.$$

Proof. In the case $\beta > 1/2$ the operator $i_\lambda \in \mathbf{L}(\mathbf{c}_{0,\lambda}, \mathbf{H}) \cdot \pi_\gamma(\mathbf{H}, \mathbf{c}_0)$ and estimates (3.33) and (3.34) follow from Theorem 3.11.

Let $\beta \leq 1/2$, $\mu_i = (\log i)^{-2}$, $i > 1$. Then the inclusion operator $i_\mu : \mathbf{l}_{2,\mu} \to \mathbf{c}_0$ is γ-radonifying. Defining for $N \geq 1$ the operator

$$u_N : \mathbf{c}_{0,\lambda} \to \mathbf{l}_{2,\mu}, \; u_N(x) = (x_1, \ldots, x_N, 0, 0, \ldots),$$

we have

$$\|u_N\| \leq CN^{(1-2beta)/2},$$

$$\|(i_\lambda - i_\mu \cdot u_N)^* e_i\| \leq \begin{cases} i^{-\beta}, & \text{if } i > N, \\ 0, & \text{if } i \leq N. \end{cases}$$

Having applied Theorem 3.1.12 to estimate the norm of the type-2 operator $i_\lambda - i_\mu \cdot u_N$, we shall easily obtain $\|i_\lambda - i_\mu \cdot u_N\|_2 \leq CN^{-\beta/\alpha}$ for each $\alpha > 1$. Therefore, we can apply Theorem 3.17 where $\gamma = \alpha(1 - 2\beta)(2\beta)^{-1}$. □

Estimates (3.33) and (3.34) also take place when $P\{\eta \in \mathbf{l}_{2,\lambda}\} = P\{\xi \in \mathbf{l}_{2,\lambda}\} = 1$, where the sequence $\lambda \in \Lambda_0$ since in this case the inclusion operator $i_\lambda \in \pi_\gamma(\mathbf{l}_{2,\lambda}, \mathbf{c}_0)$.

It turns out that estimate (3.33) also takes place when $P(\eta \in \mathbf{l}_2) = 1$. We shall present this result whih shows, in particular, that the condition $u \in \mathbf{L}(\mathbf{F}, \mathbf{H}) \cdot \pi_\gamma(\mathbf{H}, \mathbf{B})$ in Theorem 3.10 is not necessary. Note that in the proof an approximating function constructed by means of non-Gaussian measure is employed. We shall first prove two auxiliary lemmas intended for the construction of this function.

LEMMA 3.21. *There exists an infinitely differentiable function* $\phi : \mathbf{R} \to \mathbf{R}$ *such that* $\phi(x) > 0$ *for* $x \in (-1, 1)$, $\phi(x) = 0$ *for* $x \overline{\in} (-1, 1)$, $\int_{-\infty}^{\infty} \phi(x)\, dx = 1$. *Besides the functions*

$$\psi_{k,l}(x) = \begin{cases} \phi^{(k)}(x)\phi^{(l)}(x)(phi(x))^{-1} & \text{for } x \in (-1, 1), \\ 0 & \text{for } x \overline{\in} (-1, 1), \end{cases} \quad k,l = 0, 1, \cdots$$

are continuous and bounded.

Proof. Let us assume

$$\phi(x) = \begin{cases} C \exp\{-(1 - x^2)^{-2}\}, & x \in (-1, 1), \\ 0 & x \overline{\in} (-1, 1), \end{cases}$$

where the constant C is chosen such that

$$\int_{-\infty}^{\infty} \phi(x)\, dx = 1.$$

The continuity of the function $\psi_{k,l}$ follows from the relations

$$\lim_{x \uparrow 1} \psi_{k,l}(x) = 0, \quad \lim_{x \downarrow -1} \psi_{k,l}(x) = 0,$$

and can be easily checked by the use of L'Hopital's rule. For instance,

$$\lim_{x \uparrow 1} \frac{\phi'(x)\phi''(x)}{\phi(x)} = \lim_{x \uparrow 1} \frac{\phi''(x)\phi''(x) + \phi'(x)\phi'''(x)}{\phi'(x)} = \lim_{x \uparrow 1} \frac{2\phi''(x)\phi'''(x)}{\phi''(x)} = 0. \square$$

In the space $\mathbf{l}_2^k$ we shall consider a probability measure μ having the density

$$p(x) = \phi(x_1) \cdots \phi(x_k), \ x = (x_1, \ldots, x_k) \in \mathbf{l}_2^k,$$

with respect to the standard Lebesgue measure. It is easy to verify that the density p is infinitely differentiable. We shall define the s-linear mapping $q_s(x) \in \mathbf{L}^s(\mathbf{l}_2^k, \mathbf{R})$, setting

$$q_s(x)(h_1, \ldots, h_s) := \frac{p^{(s)}(x)(h_1, \ldots, h_s)}{p(x)} \quad \left(\frac{0}{0} := 0\right).$$

LEMMA 3.22. *Uniformly for all $k \geq 1$ the following estimates take place:*

$$\int_{\mathbf{l}_2^k} (q_s(x)(h)^s)^2 \mu(dx) \leq C(s) \,|\, h \,|^{2s}.$$

Proof. It is sufficient to check that for $|h| < 1$ the following estimate takes place:

$$I := \int_{\mathbf{l}_2^k} (q_s(x)(h)^s)^2 \mu(dx) \leq C(s).$$

Let $h = (h_1, \ldots, h_k)$ and let ∂_i denote differentiation along the direction $e_i(0, \ldots, 0, 1, 0, \ldots, 0)$ (the unit is on the ith place). Then

$$I = \Sigma_i \Sigma_j h_{i_1} \cdots h_{i_s} h_{j_1} \cdots h_{j_s} I_{i,j}, \tag{3.35}$$

where

$$I_{i,j} = \int_{\mathbf{l}_2^k} \partial_{i_1} \cdots \partial_{i_s} p(x) \partial_{j_1} \cdots \partial_{j_s} p(x) \frac{dx}{p(x)},$$

and the summation is performed over all possible multi-indices $i = (i_1, \ldots, i_s)$, $j = (j_1, \ldots, j_s)$ such that $1 \leq i_1, \ldots, i_s, j_1, \ldots, j_s \leq k$. For the multi-indices $i = (i_1, \ldots, i_s)$ and $j = (j_1, \ldots, j_s)$, we shall consider the sets

$$A(i) = \{i_1\} \cup \{i_2\} \cup \cdots \cup \{i_s\}, \ A(j) = \{j_1\} \cup \cdots \cup \{j_s\},$$

and let the elements of these set be the integers $1 \leq \alpha_1 < \cdots < \alpha_m \leq k$ and $1 \leq \beta_1 < \cdots < \beta_l \leq k$, $m,l \leq s$, respectively. Let the number α_1 (the numbers $\alpha_2, \ldots, \alpha_m$) be present in the collection $(i_1, i_2, ..., i_s)\gamma_1$ times exactly (respectively, $\gamma_2, ..., \gamma_m$ times). Similarly, let the number β_1 (the numbers $\beta_2, ..., \beta_l$) be present in the collection $(j_1, ..., j_s)\delta_1$ times exactly (respectively, $\delta_2, ..., \delta_l$ times). Since $p(x) = \phi(x_1) \cdots \phi(x_k)$, then

$$I_{i,j} = \int_{\mathbf{l}_2^k} \prod_{r \in \{\alpha_1, \ldots, \alpha_m\}} \phi(x_r)\phi^{(\gamma_1)}(x_{\alpha_1}) \cdots \phi^{(\gamma_m)}(x_{\alpha_m}) \times$$

$$\times \frac{\phi^{(\delta_1)}(x_{\beta_1})}{\phi(x_{\beta_1})} \cdots \frac{\phi^{(\delta_1)}(x_{\beta_l})}{\phi(x_{\beta_l})} \mathrm{d}x_1 \cdots \mathrm{d}x_k.$$

We shall prove that

$$|I_{i,j}| \leq \begin{cases} C(s), & \text{if } A(i) = A(j), \\ 0, & \text{if } A(i) \neq A(j). \end{cases} \tag{3.36}$$

Let $A(i) \neq A(j)$. Suppose for definiteness that $\alpha_i \in A(j)$. Then I_{ij} can be presented as the product of integrals among which there will be

$$\int_{-\infty}^{\infty} \phi^{(\gamma_1)}(x_{\alpha_1})\, \mathrm{d}x_{\alpha_1} = 0$$

(since $\gamma_1 \geq 1$, and the function ϕ has a compact support), therefore, $I_{i,j} = 0$.

But if $A(i) = A(j)$, then, taking into account that $\int_{-\infty}^{\infty} \phi(x)\, \mathrm{d}x = 1$, we obtain

$$|I_{i,j}| \leq \int_{-\infty}^{\infty} |\phi^{(\gamma_1)}(t)| \frac{\mathrm{d}t}{\phi(t)} \cdots \int_{-\infty}^{\infty} |\phi^{(\gamma_l)}(t)\phi^{(\delta_l)}(t)| \frac{\mathrm{d}t}{\phi(t)}.$$

Since $\gamma_1, ..., \gamma_l, \delta_1, ..., \delta_l \geq 1$ and $\gamma_1 + \cdots + \gamma_l = \delta_1 + \cdots + \delta_l = s$, then having applied Lemma 3.22 we obtain $|I_{i,j}| \leq C(s)$. Thus (3.36) is proved. From (3.36) and (3.35) it follows that

$$I \leq C(s) \sum_i \sum_{j:A(i)=A(j)} |h_{\alpha_1}|^{\gamma_1+\delta_1} \cdots |h_{\alpha_l}|^{\gamma_l+\delta_l}. \tag{3.37}$$

Since $\gamma_1, \delta_1, ..., \gamma_l, \delta_l \geq 1$ and by the assumption $|h|_2 \leq 1$ it means that $|h_1|, ..., |h_k| \leq 1$; also, therefore

$$|h_{\alpha_1}|^{\gamma_1+\delta_1} \cdots |h_{\alpha_l}^{\gamma_l+\delta_l} \leq |h_{\alpha_1}|^2 \cdots |h_{\alpha_l}|^2.$$

By grouping the summands in (3.37) correspondingly, we obtain

$$I \leq C(s) \sum_{l=1}^{s} \sum_{1 \leq \alpha_1 < \ldots < \alpha_l \leq k} \sum_{i:A(i)=\{\alpha_1, \ldots, \alpha_l\}} \sum_{j:A(j)=\{\alpha_1, \ldots, \alpha_l\}} |h_{\alpha_1}|^2 \cdots |h_{\alpha_l}|^2.$$

If the collection $1 \leq \alpha_1 < \cdots < \alpha_l \leq k$, $l = 1, 2, ..., s$ is fixed, then the number of multi-indices $i = (i_1, ..., i_s)$ such that $A(i) = (\alpha_1, ..., \alpha_l)$ does not exceed some constant $C(l, s) \leq C(s)$. A similar remark also takes place in the case of multi-indices $j = (j_1, ..., j_s)$. Therefore,

$$I \leq C(s) \sum_{l=1}^{s} \sum_{\alpha_1=1}^{k} \cdots \sum_{\alpha_l=1}^{k} |h_{\alpha_1}|^2 \cdots |h_{\alpha_l}|^2 \leq C(S). \square$$

LEMMA 3.23. *Let the class $\mathscr{F}(\mathbf{l}_\infty^k)$ consist of functions $f: \mathbf{l}_\infty^k \to \mathbf{R}$, satisfying the Lipschitz condition $|f(x) - f(y)| \leq \|x - y\|$. Then the class $\mathscr{W}_F = \{\mathscr{W}_f, f \in \mathscr{F}(\mathbf{l}_\infty^k)\}$ is $(\mathbf{Q}_{\mathbf{l}_2^k}^m, \mathbf{K})$-uniform for each $m \geq 1$, where the function $\mathbf{K}(\varepsilon) = (K_1\varepsilon^{-1}, ..., K_m\varepsilon^{-m})$. Besides, the function K is independent of k.*

Proof. For the function $f \in \mathrm{BL}(\mathbf{l}_\infty^k)$, such that $\|f\|_{\mathrm{BL}} \leq 1$, we denote

$$f_{t,k}(x) = \int_{l_2^k} f(x+y)\mu_t(dy),$$

where the measure $\mu_t(A) = \mu(t^{-1}A)$, and the measure μ was defined before Lemma 3.22. Then we have

$$|f_{t,k}(x) - f(x)|$$
$$\leq \int_{l_2^k} \sup_{i\leq k} |x_i| \mu_t(dx) = t\int_{l_2^k} \sup_{i\leq k} |x_i| p(x)\, dx \leq Ct, \tag{3.38}$$

and the constant C, as it can easily be seen, is independent of k.

Thus, it follows from Lemma 3.22 and relation (3.38) that the function f is uniformly $(\mathbf{Q}_{l_2^k}^m, \mathbf{K})$-approximable and, the function $\mathbf{K}$ is independent of k. Next, the reasonings of the proof of Theorem 2.2.15 must be repeated. □

THEOREM 3.24. *Let* $\eta \in \mathbf{L}_0(\mathbf{c}_0)$, $\xi \in \mathscr{D}(\eta)$. *If*

$$P\{\eta \in l_2\} = 1 \text{ and } \nu_3 := \int_{\mathbf{c}_0} \|x\|_{l_2}^2\, |L(\xi) - L(\eta)|(dx) < \infty,$$

then there exists an absolute constant $C = C(T_\eta)$ *such that for all* $n > 1$ *the following estimate takes place:*

$$\pi(S_n(\xi), \eta) \leq C(\nu_3 n^{-1/2})^{1/4}. \tag{3.39}$$

Proof. Let $P_k : \mathbf{c}_0 \to \mathbf{l}_\infty^k$ be a projection on the first k-coordinates. It follows from Lemma 3.23 and Theorem 3.4 that

$$\pi(S_n(P_k\xi), P_k\eta) \leq C(\varsigma_{3,i_k}(S_n(P_k\xi), P_k\eta))^{1/4};$$

besides, the constant C is independent of k, $i_k \equiv I_{l_\infty^k}$. Hence, it is easy to obtain the estimate

$$\pi(S_n(P_k(\xi), P_k\eta) \leq C(\nu_3 n^{-1/2})^{1/4}. \tag{3.40}$$

Passing to the limit as $k \to \infty$ we obtain (3.39) from (3.40). □

5.4. Supplements

5.4.1. THEOREM [177]. *Let* $(\beta_n) \in \mathbf{c}_0^+$. *If for each centred bounded* $\mathbf{B}$*-r.e.* $\xi \in \mathscr{D}(\eta)$ *the inequality* $\Delta_n(\xi, \eta; \mathbf{W}_0(\mathbf{B})) \leq \beta_n$, *holds, then* $\mathbf{B}$ *is a space of type-2.*

5.4.2. THEOREM [178]. *Let* $(\beta_n) \in \mathbf{c}_0^+$, *the number* $\varepsilon > 0$. *In the space* $\mathbf{H}$ *there exists a norm* ϕ, *a Gaussian* $\mathbf{H}$*-r.e.* η *and a bounded* $\mathbf{H}$*-r.e.* $\xi \in \mathscr{D}(\eta)$ *such that*

(a) *for all* $x \in \mathbf{H}$

$$(1-\varepsilon)\phi(x) \leq |x| \leq (1+\varepsilon)\phi(x);$$

(b) $\phi \in \mathbf{C}^\infty(\mathbf{H}\ \{0\}, \mathbf{R})$ *and for all* $k \geq 1$

$$\sup_{|x|=1} \|\phi^{(k)}(x)\| < \infty;$$

(c) *for infinitely many* $n \geq 1$

$$\Delta_n(\xi, \eta; \mathscr{W}_\phi) \geq \beta_n.$$

Let η be a Gaussian $\mathbf{H}$-r.e.

$$E\eta = 0,\ \sigma^2 := E\,|\eta|^2,\ \sigma_1^2 \geq \sigma_2^2 \geq \cdots$$

be eigenvalues of a covariance operator $T_\eta = cov\ \eta$. Suppose

$$\psi_j = \left(\sum_{i\geq j} \sigma_i^4\right)^{1/4},\ j \geq 1.$$

5.4.3. THEOREM [225]. *Let $\xi \in \mathscr{D}(\eta)$, $E\,|\xi|^3 < \infty$. If $\sigma_i^2 > 0$ for an infinite number of indices $i \in N$, then $\Delta_n(\xi, \eta; \mathbf{W}_0(\mathbf{H})) = o(n^{-1/2})$ for $n \to \infty$.*

5.4.4. THEOREM [129]. *Let $\xi \in \mathscr{D}(\eta)$, $\beta_3 := E\,|\xi|^3 < \infty$. There exist an absolute constant $C > 0$ such that for all $n \geq 1$, $a \in \mathbf{H}$,*

$$\Delta_n(\xi, \eta; \mathbf{W}_a(\mathbf{H})) \leq C\beta_3\sigma^2(\sigma^3 + |a|^3)\left(\sum_{k=1}^{8} \sigma_k\right)^{-1} n^{-1/2}.$$

5.4.5. THEOREM [190]. *Let $\xi \in \mathscr{D}(\eta)$, $E\,|\xi|^3 < \infty$, $\sigma = 1$. There exists an absolute constant $C > 0$ such that for all $n \geq 1$, $a \in \mathbf{H}$, and $r \geq 0$*

$$|P\{|S_n(\xi) + a| < r\} - P\{|\eta + a| < r\}|$$
$$\leq C \min\left\{\prod_{j=1}^{7} \sigma_j^{-6/7} + (\sigma_1\sigma_2\sigma_3\sigma_7^2)^{-1};\ \prod_{j=1}^{7} \psi_j^{-6/7} + \sigma_1^2(\psi_1\psi_2\psi_3\psi_7^4)^{-1}\right\} \times$$
$$\times \left\{E\,|\xi|^3 + \min\left(1; \frac{r^3}{|a|^3}\right) E\,|(\xi, a)|^3\right\}(1 + |r - |a||)^{-3} n^{-1/2}.$$

THEOREM [190]. *Let $\xi \in \mathscr{D}(\eta)$, $E\,|\xi|^4 < \infty$, $\sigma_l^2 > 0$, $\sigma = 1$. For each $\varepsilon > 0$ there exists a constant $C(\varepsilon) > 0$ such that for all $n \geq 1$ and $r \geq 0$*

$$|P\{|S_n(\xi)| < r\} - P\{|\eta| < r\}|$$
$$\leq C\left\{\prod_{j=1}^{13} \sigma_j^{-12/13}(E\,|\xi|^4 + (E\,|\xi|^3)^2)n^{-1} + \right.$$
$$\left. + C(\varepsilon)\left(\prod_{j=1}^{l} \sigma_j^{-4/(l+2)} + \prod_{j=1}^{9} \sigma_j^{-8/9}\right)\left(\frac{L^2}{n}\right)^{l-2/l+\varepsilon}\right\}(1 + r)^{-4},$$

where C is an absolute constant and the number $L \geq 0$ is such that $E\,|\xi|^2\chi(|E| \geq L) \leq \sigma_l^2/4$.

5.4.6. Let $A: \mathbf{H} \to \mathbf{H}$ be a bounded symmetric operator, $u \in \mathbf{H}$, function $\omega(x) = (Ax, x) + (u, x)$ a set $V_{a,r} = \{x \in \mathbf{H} : \omega(a + x) < r\}$, $a \in \mathbf{H}$ and let ρ be a distance from zero to the set $\partial V_{r,a}$.

THEOREM [26]. *Let the image of the operator $(AT_\eta)^2$ be infinite-dimensional, the numbers $q \geq 2$, $m \geq 1$. We denote $x = r - (Aa, a) - u(a)$. Let $\xi \in D(\eta)$. Then*

$$P\{S_n(\xi) \in V_{r,a}\} - P\{\eta \in V_{r,a}\} = R_1 + R_2,$$

where the remainders R_1, R_2 admit the estimates

$$|R_1| \leq 2nP\{|\xi| > n^{1/2}(1 + \rho)\} +$$
$$+ C(q)(1 + \rho)^{-q} n^{1-q/2} E\,|\xi|^q \chi\{n^{1/2} \leq |\xi|\ M n^{1/2}(1 + \rho)\};$$

$$|R_2| \leq C(m, A, T_\eta)(1 + |x|)^{-m}(E|\xi|^2\chi\{|\xi| > n^{1/2}\} +$$
$$+ n^{-1/2}E|\xi|^3\chi\{|\xi| \leq n^{1/2}\}(1 + |2Aa + u|^{3+m})\left(1 + \frac{|2Aa + u|}{1 + |x|}\right).$$

Let η be a Gaussian $\mathbf{B}$-r.e., $E\eta = 0$, $T_\eta = cov\ \eta$, η_1 be an independent copy of r.e. η. For the function $f \in \mathbf{D}^2(\mathbf{B}, \mathbf{R})$ we assume:

$$\beta_1(x) = \langle T_\eta f'(x), f'(x)\rangle, \quad \beta_2(x, y) = \langle T_\eta f''(x)(y), f''(x)(y)\rangle.$$

5.4.7. THEOREM [75]. *Let the following conditions be satisfied:*

1. *The function* $f \in \mathbf{C}^4(\mathbf{B}, \mathbf{R})$ *and*

(a) $\|f^{(k)}(x)\| \leq C(1 + \|x\|^r), \ 1 \leq k \leq 4;$

(b) $\|f^{(4)}(x) - f^{(4)}(y)\| \leq C(1 + \|x\| + \|y\|)^r \|x - y\|^\beta$

for all $x, y \in \mathbf{B}$ *and some* $C > 0, r \geq 0, 0 < \beta \leq 1$.

2. $P\{\beta_1(\eta) \leq \varepsilon\} = O(\varepsilon^k)$ *when* $\varepsilon \to 0$, *where* $k \geq 340/\beta$. *If* $\xi \in \mathcal{D}(\eta)$ *and* $E\|\xi\|^4 < \infty$, *then*

$$\Delta_n(\xi, \eta; \mathcal{W}_f) = O(n^{-1/2}).$$

5.4.8. THEOREM [217].

1. *The function* $f \in \mathbf{C}^7(\mathbf{B}, \mathbf{R})$ *and*

$$\|f^{(k)}(x)\| \leq C(1 + \|x\|^r)$$

for $1 \leq k \leq 7$ *and for any* $x \in \mathbf{B}$ *and some* $C > 0, r > 0$.

2. *For any* $L > 0$ *there can be found a constant* $C(L) > 0$ *such that for all* $s > 1$

(a) $E\ \exp\{-s\beta_1(\eta)\} \leq C(L)s^{-L}$;

(b) $E\ \exp\{-s\beta_2(\eta, \eta_1)\} \leq C(L)s^{-L}$.

If $\xi \in \mathcal{D}(\eta)$ *and* $E\|\xi\|^4 < \infty$, *then for each* $\varepsilon > 0$ *there exists a constant* $C(\varepsilon) = C(\varepsilon, T_\eta) > 0$ *such that*

$$\Delta_n(\xi, \eta; \mathcal{W}_f) \leq C(\varepsilon)n^{-1+\varepsilon}.$$

5.4.9. THEOREM [96]. *Let* $k \in N$, η *be a Gaussian* $\mathbf{l}_{2k}$*-r.e. not concentrated in any finite-dimensional space,* $E\eta = 0$. *If* $\xi \in D(\eta)$ *and* $\beta_3 := E\|\xi\|^3 < \infty$, *then for all* $n \neq 1$ *and* $a \in \mathbf{l}_{2k}$

$$\Delta_n(\xi, \eta; \mathbf{W}_a(\mathbf{l}_{2k})) \leq C\beta_3(1 + \|a\|^{3(2k-1)})n^{-1/2}$$

with some constant $C = C(T_\eta)$.

5.4.10. Let $\{e_i\}_{i\geq 1}$ be a standard basis of the space $\mathbf{l}_q$, and let η be a Gaussian $\mathbf{l}_p$-r.e. $1 \leq p < \infty$. Suppose that $E\langle \eta, e_i\rangle\langle \eta, e_j\rangle = 0, i \neq j$;

$$t_i := E\langle\eta, e_i\rangle^2 \sim i^{-\gamma}$$

for some $\gamma > 2/p$.

THEOREM [148]. 1. *If* $\xi \in \mathcal{D}(\eta)$ *and*

$$E\left(\sum_{i=1}^{\infty} <\xi, e_i>^2 t_i^{-\delta}\right)^{3/2} < \infty, \ \delta > 0,$$

then

$$\Delta_n(\xi, \eta; \mathbf{W}_0(\mathbf{l}_p)) = O(n^{-g_1(\gamma, p, \delta)}),$$

where

$$g_1(\gamma, p, \delta) = \frac{\frac{1}{2}p(\gamma - 1)}{(4 - 3\delta)\gamma p + p + 3\gamma(1 - \delta) + 2}.$$

2. *If* $\xi \in \mathscr{D}(\eta)$ *and*

$$E\left(\sum_{i=1}^{\infty} <\xi, e_i>^2 t_i^{-(1-(1/\gamma)-\varepsilon)}\right)^{3/2} < \infty,$$

then

$$\Delta_n(\xi, \eta; \mathbf{W}_0(\mathbf{l}_p)) = O(n^{-g_2(\gamma, p, \delta)}),$$

where

$$g_2(\gamma, p, \delta) = \frac{\frac{1}{2}p(\gamma - 1)}{p(\gamma + 1) + 3(1 + \varepsilon\gamma)(p + 1) + 2}.$$

5.4.11. A Banach space belongs to class $\mathscr{C}^m$, if the norm $\phi(\cdot) = \|\cdot\| \in \mathbf{C}^m(\mathbf{B}\backslash\{0\}, \mathbf{R})$ and

$$\sup_{\|x\|=1} \|\phi^{(i)}(x)\| < \infty, \ i = 1, ..., m.$$

We shall consider a Gaussian $\mathbf{B}$-r.e. η, $E\eta = 0$, which is not concentrated in any finite-dimensional subspace of the space $\mathbf{B}$.

THEOREM [17]. *Let the numbers* $2 < p < 4$, $0 < q \leq p$, $\mathbf{B}$-*r.e.* $\xi \in \mathscr{D}(\eta)$, $E\|\xi\|^p < \infty$. *If* $\mathbf{B} \in \mathscr{C}^6$, *then*

$$E\|S_n(\xi)\|^q - E\|\eta\|^q = o(n^{-(p-2)/2}).$$

THEOREM [17]. *Let the numbers* $p \geq 4$, $0 < q \leq p$, $\mathbf{B}$-*r.e.* $\xi \in \mathscr{D}(\eta)$, $E\|\xi\|^p < \infty$. *If either* $q \leq 6$ *and* $\mathbf{B} \in \mathscr{C}^9$ *or* $q > 6$ *and* $\mathbf{B} \in \mathscr{C}^6$, *then*

$$E\|S_n(\xi)\|^q - E\|\eta\|^q = O(n^{-1}).$$

Bibliographical Notes

Chapter 2

The classical material presented at the beginning of Section 2 – i.e. Gateaux and Fréchet differentiation – can be found, for instance, in books [45] and [100]. Derivatives in the directions from a subspace were considered in the works of V.I. Averbuch, O.G. Smol'janov and S.V. Fomin [5]–[7], I. Gross [79] and other authors. The classical definition of a variational derivative can be also interpreted as one of the definitions of a derivative along a subspace. The differentiation with respect to an operator (Definition 1.16) was suggested by A. Račkauskas. Theorem 1.20, concerning the equivalence of the definition of differentiation with respect to an operator and along a subspace, was proved by V.Yu. Bentkus and A. Račkauskas.

The functions $f : \mathbf{B} \to \mathbf{R}$, differentiable along a Hilbert subspace $\mathbf{H}$ reproducing a Gaussian measure on $\mathbf{B}$, were considered in the works of L. Gross [79], M.A. Piech [159], V.Yu. Bentkus [11] and others. In particular, the proof of Theorem 1.23 can be also found in [159] and [7].

Section 2.2 presents the results of a non-linear analysis in Banach spaces applied in estimating the rate of convergence in the CLT.

J. Keulbs and T. Kurtz, in their estimates of the rates of convergence in the CLT in a Hilbert space [103], were the first to apply smooth functions approximating the indicator functions of balls. V. Paulauskas [143], [145], [152] has shown that the estimates of convergence rate on sets in the CLT in Banach spaces can be reduced to two problems: the problem of constructing smooth in the sense of Fréchet functions approximating indicator functions; and the problem of estimating a Gaussian measure of an ε-strip of the boundary of sets. In particular, Theorem 2.22 and a variant of Theorem 2.21 in the case of Banach spaces with a sufficiently smooth norm have been obtained in [152] (see also [156]). In [152], Proposition 2.26 is proved, and in [154] the proof of a variant of Theorem 2.29 is presented. V.Yu. Bentkus and A. Račkauskas [22]–[25] were first to apply differentiation along the directions from a subspace in estimating the rates of convergence in the CLT. This enabled them to construct estimates in arbitrary Banach spaces without any restrictions on the smoothness of a norm. It was noted in [24] that the problem of approximation of indicator functions of sets can be reduced to a problem of the estimate of the best uniform approximation of functions by differentiable functions.

The main results presented in Section 2.2 (Theorems 2.8, 2.9, 2.15, 2.20, 2.29, 2.31) have been taken from [24]–[25].

The method of constructing smooth functions, making use of differentiable measures (see Supplement 2.3.7), has been considered in detail by V.Yu. Bentkus [11].

Chapter 3

The notions of spaces of type and cotype and those of operators of type and stable type, have been introduced in the works of B. Maurey, G. Pisier and J. Hoffmann–Jorgensen (see [86], [124], [160] and [161]).

Theorems 1.3, 1.4 and Proposition 1.5 can be found in [86]. Lemma 1.6 is contained in the works of G. Pisier [163]. The proof of Theorem 1.8 is presented in [163]. Corollary 1.9 is proved by J. Zinn [226] and this result was one of the first examples of type-2 operators. Theorem 1.12 is a corollary of some more general results of A. Račkauskas [174]. Corollary 1.13 (with the class Λ_1 instead of Λ_0) was proved by V. Paulauskas, A. Račkauskas and V. Sakalauskas [157], and the second proof of this corollary can be found in the paper of K. Liubinskas [121].

Different generalizations of the concept of the operators of type, cotype and stable type as well as their relations to limit theorems have been studied by Z. Jurek and K. Urbanik [91], R. Norvaiša [133], V. Paulauskas and A. Račkauskas [155], W.A. Woyczynski [220], M. Ledoux [114], D.Kh. Mushtari [128], A.N. Chuprunov [47] and others.

The work by Yu.V. Prokhorov [167] made a greatest influence on the subjects of limit theorems of probability theory in Banach spaces.

Many works were devoted to the central limit theorem in Banach spaces. Among those works of S.R.S. Varadhan [216], L. Le Cam [111], R.M. Dudley and V. Strassen [59], N.N. Vakhaniya [213], E. Gine [61], N.C. Jain and M. Marcus [90], J. Hoffmann-Jorgensen and G. Pisier [87] and S.A. Chobanian and V.I. Tarieladze [49] must be mentioned. The present list, of course, is not intended to be complete. The historical data on the studies of the CLT can be found in the monograph by A. Araujo and E. Gine [3].

The first example of a pre-Gaussian bounded random element, not satisfying the CLT, was constructed in the space C[0, 1] by R.M. Dudley and V. Strassen [59]. Example 2.3 is a particular case of the result of S.A. Chobanian and V.I. Tarieladze [48], [49]. Different examples concerning the CLT in Banach spaces can be found in the works of G. Pisier, E. Gine and J. Zinn [165], [70] and M. Ledoux [112]. Theorems 2.4 and 2.7 are available in [3]. Theorem 2.5 belongs to J. Hoffmann-Jorgensen and G. Pisier [87]. Based on B. Maurey's ideas, Theorems 2.8 and 2.11 were independently proved by G. Pisier [162] and S.A. Chobanian and V.I. Tarieladze [49].

Proposition 2.9 belongs to N.S. Jain [89]. Theorem 2.14 was proved in the work of N.S. Jain and M. Marcus [90] which completed a series of studies on the CLT in the space $C(S)$ (see [58], [59], [68]).

V. Paulauskas was the first to consider the CLT in the space $\mathbf{c}_0$ [153]. In this paper there is a proof of a weaker variant of Theorem 2.16 which, in its turn, was proved by V. Paulauskas and A. Račkauskas in [157], as well as by B. Heinkel, independently, by using a different method [82].

Chapter 4

A particularly important theorem, Theorem 1.2 (see also Supplement 4.3.1) was proved by B.S. Tsirel'son [207]. Some refinements of exponential estimate were obtained by V.A. Dmitrovskii [55].

Proposition 1.1, demonstrating an essential difference in the problem of describing uniform classes in an infinite-dimensional and a finite-dimensional space, has been proved by V. Paulauskas [143].

The boundedness of the distribution density of the norm of a Gaussian random element in a Hilbert space was proved by N.N. Vakhanija [213]. This result was independently obtained by J. Kuelbs and T. Kurtz [103].

The boundedness of the distribution density of the norm of a Gaussian $\mathbf{l}_p$-r.e., $1 \leq p < \infty$, under some assumptions about the independence of coordinates was proved by V. Paulauskas [145]. He revealed the dependence of the estimate of density on the shifts (Theorem 2.3 and

Proposition 2.4). Theorem 2.1 (the general result) was proved by Yu.A. Davydov by the fibring method. The results obtained by this method are available in a review article by Yu.A. Davydov and M.A. Lifshits [52] (see also the papers by M.A. Lifshits [118]–[120]).

V. Paulauskas [149]–[151] constructed the first example of a Gaussian **B**–r.e. with an unbounded distribution density of the norm (Theorem 3.3 in the case $\psi(t) = t$). A similar example was independently presented by M.A. Lifshits [52]. The idea of the construction of these examples belongs to B.S. Tsirel'son. Similar examples in Banach spaces with a smooth norm were presented by W.S. Rhee and M. Talagrand [176], [178] (see Supplement 4.4.8). Gy. Pap [138] has proved a particular case of Theorem 3.3 (with $\psi(t) = t^\alpha$, $1/2 < \alpha \leq 1$). V.Yu. Bentkus has suggested Lemma 3.4, which resulted in the variant of Theorem 3.1 presented in the text. Theorems 3.6 and 3.7 were proved by V. Paulauskas [151]. Recently a number of statements have been obtained on the distribution of functionals of stable random elements generalizing the part of results of Chapter 4. The analogue of Tsirel'son's theorem has been proved by R. Sztencel, T. Byczkowski, K. Samotij and T. Żak [44], [201], [232]. The first result on the boundedness of the distribution density of the norm of a stable **H**–r.e. has been obtained by V. Paulauskas [154], [156]. V.Yu. Bentkus and A. Račkauskas [22] have proved a similar result for spherically symmetric stable random elements of a Banach space. The boundedness of the distribution density of the norm of a stable **B**–r.e. in the case of a uniformly smooth and uniformly convex space **B** has been proved by M.A. Lifshits and V.N. Smorodina (the paper will be published in the journal *The Theory of Probability and Its Applications*). For the cases of a Hilbert space **H**, this result had been obtained earlier by Gy. Pap and V.Yu. Bentkus [20].

Chapter 5

The first estimates of the rate of convergence on the balls in the CLT in a Hilbert space (the estimates of the distance $\Delta_n := \Delta_n(\xi, \eta, \mathbf{W}_0(\mathbf{H}))$), constructed by N.N. Vakhanija and N.P. Kandelaki [214], [97] had a logarithmic order of decrease with respect to the number of summands n. For **H**–r.e. ξ with a specific structure (considering the rates of convergence in ω^2–criteria) V.V. Sazonov [187] has proved the estimate $\Delta_n = O(n^{-1/6})$. The same order in the case of **H**–r.e. ξ with a finite moment $E\,|\,\xi\,|^{7/2}$ was obtained by J. Kuelbs and T. Kurtz [103]. The moment condition has been reduced to $E\,|\,\xi\,|^3 < \infty$ by V. Paulauskas [145] who has also obtained the first non-uniform estimates on the balls [142] and estimates in the case of a stable limit law [156], [154]. Pseudomoments were applied in these works for the first time.

Assuming the coordinates of **H**–r.e. ξ to be independent and $E\,|\,\xi\,|^3 < \infty$, S.V. Nagayev and V.I. Chebotarev [131] have proved the estimate $\Delta_n = O(n^{-1/2})$. The condition of independence of coordinates has been weakened in the papers of Yu.V. Borovskich and A. Račkauskas [36], [172].

A considerable advance in the estimate of Δ_n has been made by F. Götze [74]. In this work the estimate $\Delta_n = O(n^{-1/2})$ under the condition $E\,|\,\xi\,|^6 < \infty$ has been proved. V.V. Jurinskii [95] has proved that $\Delta_n = O(n^{-1/2})$ if $E\,|\,\xi\,|^3 < \infty$. L.V. Osipov and V.I. Rotar' [134] by a different method obtained $\Delta_n = O(n^{-1/2}, \ln^\varepsilon n)$. Subsequently, the estimate $\Delta_n = O(n^{-1/2})$ was being refined in different directions: the dependence of the estimate on the covariance operator T_η was studied by S.V. Nagaev [129] (one particular result is formulated in Supplement 5.4.4); the estimate $\Delta_n = o(n^{-(p-2)/2})$ under the condition $E\,|\,\xi\,|^p < \infty$, $2 < p < 4$ was obtained by B.A. Zalesskii [225] (this estimate has been refined by V.V. Sazonov and B.A. Zalesskii [189], V.Yu. Bentkus [18], Yu.V. Borovskich [101]); non-uniform estimates on the balls have been constructed by V.V. Sazonov and B.A. Zalesskii [189], V.Yu.

Bentkus and B.A. Zalesskii [26] (one of the results of this paper is presented in Supplement 5.4.6), V.V. Sazonov, V.V. Ul'yanov and B.A. Zalesskii [190] (see Supplement 5.4.5) and V.V. Ul'yanov [208].

Asymptotic expansions completing, in a sense, the analysis of the accuracy of approximation on the balls in the CLT in a Hilbert space, have been constructed by F. Götze [76], V.Yu. Bentkus [18], [19], V.Yu. Bentkus and B.A. Zalesskii [26] and S.V. Nagaev and V.I. Chebotarev [130].

The proof of Theorem 1.35 presented in the text was suggested by V.Yu. Bentkus [18] and that is a slight modification of the proofs from the works of F. Götze [74] and V.V. Jurinskii [95].

The first estimates of the rate of convergence in the CLT on sets in Banach spaces have been obtained by V. Paulauskas [145] and V.M. Zolotarev [228]. In [145], Banach spaces with a sufficiently smooth norm have been considered. In particular, for $\mathbf{B}$-r.e. $\xi \in D(\eta)$, $E\|\xi\|^3 < \infty$, under some assumptions on the Gaussian measure $L(\eta)$ of the shifts of ε-strip of a ball, the estimate $\Delta_n := \Delta_n(\xi, \eta; \mathbf{W}_a(\mathbf{B})) = O(n^{-1/6})$ has been proved. In [145], [228], in the case of balls Theorem 1.3 has been proved. Different generalizations of the estimate $\Delta_n = O(n^{-1/6})$ have been obtained by V. Bernotas [27]–[29], V. Bernotas and V. Paulauskas [30] and V.V. Ul'yanov [209], [210].

The subsequent studies on the rate of convergence in the CLT on the sets can be subdivided into two directions: estimates on smooth sets and those on sets without assumption on smoothness of their boundary. It is to the former that the works by F. Götze [75] (see Supplement 5.4.7), B.A. Zalesskii and V.V. Sazonov [224], Yu.V. Borovskich [37] and T.R. Vinogradova [217] (see Supplement 5.4.8) refer. The second one has found its development in the papers of V.Yu. Bentkus and A. Račkauskas [22]–[24]. In particular, in [24] a variant of Theorem 1.7 with some additional restrictions on Gaussian measure of the shifts of the ε-strip of the set under consideration was proved.

Theorem 1.7 has been proved by A. Račkauskas [173]. Theorems 1.13, 1.18, 1.21 and Corollaries 1.14, 1.15, 1.19 are published for the first time. The proof of Theorem 1.11 was taken from [188]. In the case of a Hilbert space, this result has been obtained in [150].

The estimates from below of the rate of convergence in the CLT have been obtained by V.Yu. Bentkus [14], W.S. Rhee and M. Talagrand [178], J.S. Borisov [35]. Theorem 1.23 was taken from [16], and the result of [178] is presented in Supplement 5.4.2. Theorem 1.24 showing unimprovability of the order $O(n^{-1/6})$ in a general case is proved by V.Yu. Bentkus [14].

The estimates of the rate of convergence on sets of distributions of sums of dependent $\mathbf{B}$-r.e. to a Gaussian distribution are available in the works by P.L. Butzer, L. Hahn and M.Th. Roeckerath [43], W.S. Rhee and M. Talagrand [180] and P.P. Gudynas [80].

E. Gine [69] and V. Paulauskas [146] were the first to obtain the estimates of the rate of convergence on the balls in the CLT in a space $\mathbf{C}(S)$. The estimates in a particular case of the limit Wiener process had an order close to $O(n^{-1/21})$. The above results were generalized and refined in the papers of V. Sakalauskas [185], E.I. Ostrovskii [136]. V.Yu. Bentkus and A. Račkauskas [24] have proved the estimate

$$\Delta_n(\xi, \eta: \mathbf{W}_0(\mathbf{C}[0, 1])) = O(n^{-1/8})$$

in the case when r.e. ξ and η are concentrated in $\mathrm{Lip}_\alpha[0, 1]$, $1/2 < \alpha \leq 1$. V.Yu. Bentkus [12] has also considered $0 < \alpha \leq 1/2$ and, in the case of a limit Wiener process, has obtained the estimate of the order $O(n^{-1/8+\varepsilon})$. Theorem 2.6 has been proved by A. Račkauskas and is published for the first time. Propositions 2.2 and 2.3 are taken from V.Yu. Bentkus [12]. Theorem 2.12 has been proved by V. Paulauskas [147], [153]. A similar result under other restrictions has been obtained by A.V. Asrijev and V.J. Rotar [4].

Some papers are devoted to the estimates of the rate of convergence on the balls in the CLT in $\mathbf{l}_p$. V.V. Jurinskii [96] has obtained the estimate of the order $O(n^{-1/2})$ in $\mathbf{l}_{2k}$, $k \in N$ (see Supplement 5.4.9). In the other spaces $\mathbf{l}_p$ there are the results of V.J. Chebotarev [46], A. Račkauskas [171] (under some conditions of independence of coordinates) and V. Paulauskas [148] (see Supplement 5.4.10).

The papers of V.M. Zolotarev [227]–[230] developed a metric approach to the problems of estimating the proximity of distributions of sums of independent r.e. in general spaces. In particular, in [228] estimates were obtained of the rate of convergence in the CLT in the Prokhorov metric on the balls in Banach spaces with a sufficiently smooth norm.

Theorems 3.1 and 3.2 are available in the paper by R.M. Dudley [56]. Theorems 3.4 and 3.5 were proved by V.Yu. Bentkus and A. Račkauskas [23], [25]. Some relations of the Levy–Prokhorov type metrics with the metrics of functional type were established in the works of V.M. Zolotarev [227]–[230] and V.V. Senatov [192].

The estimates of the rate of convergence in the CLT in the Prokhorov metric have been obtained by V.V. Jurinskii [98] (in the case of a Hilbert space), R. Lapinskas [110] (in Banach spaces with the Schauder basis under the assumption of the independence of coordinates) and A.G. Kukush [104]–[105] (in a Hilbert space).

The estimates in the Prokhorov metric, as well as in the toplogically equivalent bounded Lipschitz metric in Banach spaces having in a general case the right order $O(n^{-1/8})$ (respectively, $O(n^{-1/6})$) have been proved by V.Yu. Bentkus and A. Račkauskas [23], [25]. Theorems 3.6 and 3.10 were taken from these papers. Theorems 3.14 and 3.15 were proved by V.Yu. Bentkus [14]. Theorem 3.8 has been also proved by V.V. Senatov [194] by the use of different method. Theorem 3.24 belongs to V.Yu. Bentkus [25].

Among other characteristics of closeness of distribution of the sum $\mathbf{B}$–r.e. to a Gaussian distribution, we shall mention the distance of a functional type $|Ef(S_n(\xi) - Ef(\eta)|$, where $f: \mathbf{B} \to \mathbf{R}$ is some function. The estimates of the rate of convergence to zero of this distance in the case of the functions f of different smoothness were considered by V. Paulauskas [144], V.M. Zolotarev [228], [229] and V.Yu. Bentkus and A. Račkauskas [22]. Some generalizations for the case of dependent summands can be found in the papers of P.P. Gudynas [80] and P.L. Butzer and D. Schulz [42]. Asymptotic expansions have been constructed by F. Götze [77] and V.Yu. Bentkus [13].

The estimates of the rate of convergence of moments in the CLT in a Hilbert space have been obtained by B.A. Zalesskii and V.V. Sazonov [223]. The estimate and asymptotic expansions in the case of Banach spaces have been constructed by V.Yu. Bentkus [17] (see Supplement 5.4.11).

References

[1] Acosta de, A. and Gine, E., 'Convergence of moments and related functionals in the general central limit theorem in Banach spaces', *Z. Wahrsch.verw.Geb.* **48,** (1979), 213–231.

[2] Andersen, N.T., Gine, E., Ossiander, M. and Zinn, J., 'The central limit theorem and the law of the iterated logarithm for empirical processes under local conditions', Preprint, 1986.

[3] Araujo, A. and Gine, E., *The Central Limit Theorem for the Real and Banach Valued Random Variables,* Wiley, New York, 1980, 233 pp.

[4] Asrijev, A.V. and Rotar, V.J., 'On the rate of convergence in the infinite–dimensional central limit theorem for the probability of hitting a parallelepiped', *Teor.veroj. primen.* **30,** No. 4 (1985), 652–661 (in Russian).

[5] Averbuch, V.I. and Smol'janov, O.G., 'Differentiation theory in topological spaces', *Uspechi mat.nauk* **22,** (1967), 257–281 (in Russian).

[6] Averbuch, V.I. and Smol'janov, O.G., 'Different definitions of derivative in linear toplogical spaces', *Uspechi mat.nauk.* **23,** No. 4 (1968), 67–117 (in Russian).

[7] Averbuch, V.I., Smol'janov, O.G. and Fomin, S.V., 'Generalized functions and differential equations in linear spaces. I: Differentiable measures', *Trudy Moskov.Mat.Obsc.* **21,** (1971), 133–174 (in Russian).

[8] Badrikian, A. and Chevat, S., 'Méasures cylindriques. Espaces de Wiener et fonctions aléatoires Gaussiennes', *Lect.Notes Math.* **379,** (1974).

[9] Bakštys, G. and Paulauskas, V.J., 'On approach of distributions of sums of Banach spaces valued random elements with infinitely divisible laws. I, II', *Lit.Matem.Sb.* **26,** (1986), 403–414; **27,** 2 (1987), 224–235 (in Russian).

[10] Barsov, S.S. and Ul'yanov, V.V., 'Estimates of the proximity of Gaussian measures', *Soviet.Math.Dokl.,* **34,** (1987), 462–466.

[11] Bentkus, V.Yu., 'Analytically of Gaussian measures', *Theor.Probab.Appl.* **27,** No. 1 (1982), 155–162.

[12] Bentkus, V.Yu., 'Estimates of the proximity of sums of independent random elements in the space **C**[0, 1]', *Lithuan.Math.J.* **23,** 1 (1983), 1–8.

[13] Bentkus, V.Yu., 'Asymptotes of the remainder term in the central limit theorem in Hilbert space', *Lithuan.Math.J.,* **24,** 1 (1984), 2–5.

[14] Bentkus, V.Yu., 'Lower bounds for the rate of convergence in the central limit theorem in Banach spaces', *Lithuan.Math.J.,* **25,** 4 (1986), 312–319.

[15] Bentkus, V.Yu., 'Differentiable functions defined in the spaces $\mathbf{c}_0$ and $\mathbf{R}^k$', *Lithuan.Math.J.,* **23,** 2 (1984), 146–154.

[16] Bentkus, V.Yu., 'Lower bounds for the sharpness of a normal approximation in Banach spaces', *Lithuan.Math.J.,* **24,** 1 (1984), 6–10.

[17] Bentkus, V.Yu., 'Asymptotes of moments in the central limit theorem in Banach spaces', *Lithuan.Math.J.,* **24,** 2 (1984), 113–124.

[18] Bentkus, V.Yu., 'Asymptotic expansions in the central limit theorem in Hilbert space', *Lithuan.Math.J.* **24,** 3 (1984), 210–224.

[19] Bentkus, V.Yu., 'Asymptotic expansions for distributions of sums of independent random elements of a Hilbert space', *Lithuan.Math.J.,* **24,** 4 (1984), 305–319.

[20] Bentkus, V.Yu. and Pap, Gy., 'Distribution of the norm of a stable random vector in a

Hilbert space', *Lithuan.Math.J.*, **26,** (1986), 114 – 120.

[21] Bentkus, V.Yu. and Paulauskas, V., 'Rate of convergence in the central limit theorem for Gaussian mixtures in infinite-dimensional spaces', *Lithuan.Math.J.* **23,** 1 (1983), 9 – 18.

[22] Bentkus, V.Yu. and Račkauskas, A., 'Rate of convergence in the central limit theorem in infinite – dimensional spaces', *Lithuan.Math.J.*, **21,** 4 (1982), 271 – 276.

[23] Bentkus, V.Yu. and Račkauskas, A., 'Estimates of the reate of convergence in the central limit theorem in the Prokhorov metric', *Soviet.Math.Dokl.*, **25,** No. 3 (1982), 652 – 656.

[24] Bentkus, V.Yu. and Račkauskas, A., 'Estimates of the rate of approximation of independent random variables in a Banach space. I, II', *Lithuan.Math.J.*, **23,** 3 (1983), 223 – 234; **22,** 4, 344 – 352.

[25] Bentkus, V.Yu. and Račkauskas, A., 'Estimates of the distance between sums of independent random elements in Banach spaces', *Theor.Probab.Appl.*, **29,** 1 (1985), 50 – 65.

[26] Bentkus, V.Yu and Zalesskii, B., 'Asymptotic expansions with nonuniform remainders in the central limit theorem in Hilbert space', *Lithuan.Math.J.*, **25,** 3 (1986), 199 – 208.

[27] Bernotas, V., 'Closeness of the distributions of two sums of independent random variables with values in certain Banach spaces', *Lithuan.Math.J.*, **18,** 4 (1979), 453 – 457.

[28] Bernotas, V., 'Uniform and nonuniform proximity bounds for the distributions of two normalized sums of independent random variables with values in Banach spaces', *Lithuan.Math.J.*, **19,** 4 (1980), 482 – 489.

[29] Bernotas, V., 'Sums of a random number of random variables with values in Hilbert space', *Lithuan.Math.J.*, **17,** 3 (1978), 297 – 304.

[30] Bernotas, V. and Paulauskas, V., 'A nonuniform estimate in the central limit theorem in some Banach spaces', *Lithuan.Math.J.*, **19,** 2 (1980), 177 – 190.

[31] Bhattacharya, R.N. and Ranga, Rao R., *Normal Approximation and Asymptotic Expansions,* Wiley, New York, 1976.

[32] Billingsley, P., *Convergence of Probability Measures,* Wiley, New York, 1968.

[33] Bonic, R. and Frampton, J., 'Smooth functions on Banach manifolds', *J.Math.Mech.,* **15,** (1966), 877 – 898.

[34] Borisov, J.S., 'On the rate of convergence of distributions of functionals of integral type', *Theor.Probab.Appl.*, **21,** 2 (1976), 283 – 299.

[35] Borisov, J.S., 'Remark on the convergence rate in the central limit theorem in Banach spaces', *Sibirsk.Mat.Z.*, **26,** 2 (1985), 29 – 35 (in Russian).

[36] Borovskich, Yu.V. and Rackauskas, A., 'Asymptotics of distributions in Banach spaces', *Lithuan.Math.J.*, **19,** 4 (1980), 472 – 481.

[37] Borovskich, Yu.V., 'On normal approximation in infinite-dimensional spaces', *Sov.Math.Dokl.*, **39,** (1984), 548 – 553.

[38] Buldygin, V.V., *Convergence of Random Series in Topological Spaces,* Naukova dumka, Kijev, 1980 (in Russian).

[39] Buldygin, V.V. and Solntsev, S.A., 'Oscilatory properties of Gaussian sequences'. In *Probabilistic Infinite-dimensional Analysis,* Math.Inst. AN Ukr. SSR, Kijev, 1981, pp. 15 – 29 (in Russian).

[40] Buldygin, V.V. and Solntsev, S.A., 'On asymptotic behaviour of a Gaussian – Markovian sequence'. In *Some Problems of Theory of Random Processes,* Math.Inst. AN Ukr. SSR, Kijev, 1982, pp. 24 – 33 (in Russian).

[41] Burysek, S., 'Eigenvalue problem, bifurcations and equations with analytic operators in Banach spaces'. In *Theory of Non-Linear Operators,* Schrift. Z.Math.Mech., DDR,

1975, **20,** 1–15.

[42] Butzer, P.L. and Schulz, D., 'On o–rate of closeness of two weighted random sums of dependent Banach-valued random variables', *Liet.mat.rink.,* **25,** 2 (1985), 40–52.

[43] Butzer, P.L., Hahn, L. and Roeckerath, M.Th., 'Central limit theorem and weak law of large numbers with rates for martingales in Banach spaces', *J. Multivar.Anal.,* **13,** No. 12 (1983), 287–301.

[44] Byczkowski, T. and Samotij, K., 'Absolute continuity of stable seminorms', *Ann.Probab.,* **13,** No. 2 (1985), 299–312.

[45] Cartan, A., *Differential Calculus, Differential Forms,* Mir, Moscow, 1971 (in Russian).

[46] Chebotarev, V.J., 'An estimate of the rate of convergence in the central limit theorem in $\mathbf{l}_p$', *Sibirsk.Mat.Z.,* **20,** No. 5 (1979), 1099–1116 (in Russian).

[47] Chuprunov, A.N., 'On a generalization of spaces of type p', *Abst. Commun. IV Vilnius Int. Conf. on Probab. and Math.Stat.,* 1985, Vol. 3, pp. 302–304 (in Russian).

[48] Chobanian, S.A. and Tarieladze, V.I., 'A counterexample concerning CLT in Banach spaces', *Lect.Notes Math.,* **656,** (1978), 25–30.

[49] Chobanian, S.A. and Tarieladze, V.I., 'Gaussian characterization of certain Banach spaces', *J.Multivar.Anal.,* **7,** No. 1 (1977), 183–203.

[50] Creekmore, J., 'Type and cotype of Lorentz $\mathbf{L}_{pq}$ spaces', *Proc.Kon. Ned.akad.wetensch.,* **84,** No. 2 (1981), 145–152.

[51] Davydov, Yu.A., 'On strong convergence of distributions of functionals of random processes. I, II', *Theor.Probab.Appl.,* **25,** No. 4 (1981), 772–789; **26,** No. 2 (1981), 258–278.

[52] Davydov, Yu.A. and Lifshits, M.A., 'Fibering method in some probabilistic problems', *J.Sov.Math.,* **31,** (1985), 2796–2858.

[53] Daletskii, Yu.L. and Fomin, S.V., *Measures and Differential Equations in Infinite–Dimensional Spaces,* Nauka, Moscow, 1983 (in Russian).

[54] Diestel, J., 'Geometry of Banach spaces – Selected topics', *Lect. Notes Math.,* **485,** (1975).

[55] Dmitrovskii, V.A., 'Estimates of the distribution of the maxima of Gaussian field'. In *Random processes and Fields,* Moscow State Univ., 1979, pp. 22–31.

[56] Dudley, R.M., 'Distance of probability measures and random variables', *Ann.Math.Stat.,* **39,** (1968), 1563–1572.

[57] Dudley, R.M., 'Sample functions of Gaussian process', *Ann.Probab.,* **1,** No. 1 (1973), 66–103.

[58] Dudley, R.M., 'Metric entropy and the central limit theorem in $\mathbf{C}(S)$', *Ann.Inst.Fourier.* **24,** (1974), 49–60.

[59] Dudley, R.M. and Strassen, V., 'The central limit theorem and entropy', *Lect. Notes Math.,* **89,** (1969), 224–231.

[60] Dudley, R.M., Hoffmann–Jorgensen, J. and Shepp, L.A., 'On the lower tail of Gaussian seminorms', *Ann.Probab.,* **7,** (1979), 319–342.

[61] Ehrhard, A., 'Sur la densité du maximum d'une fonction aléatoire Gaussienne', *Lect. Notes Math.,* **920,** (1980).

[62] Fabian, M., Whitfield, J.H.N. and Zizler, V., 'Norms with locally Lipschitzian derivatives', *Israel J. Math.,* **44,** No. 3 (1983), 262–276.

[63] Fernique, X., 'Intégrabilité des vecteurs Gaussiens', *C.R. Acad. Sci. Paris,* Ser. A (1970), 1698–1699.

[64] Feller, W., *An Introduction to Probability Theory and its Applications,* Vol. II, Wiley, New York, 1971.

[65] Fomin, S.V., 'Differentiable measures in linear spaces'. *Uspechi Mat.Nauk,* **23,** No. 1 (1968), 221–222 (in Russian).

[66] Gikhman, J.J. and Skorohod, A.V., *Theory of Random Processes,* Vol. I, Nauka, Moscow, 1971 (in Russian).

[67] Gine, E., 'On the central limit theorem for sample continuous processes', *Ann.Probab.,* **2,** (1974), 619–641.

[68] Gine, E., 'Some remarks on the central limit theorem in $\mathbf{C}(S)$', *Lect. Notes Math.,* **526,** (1976), 101–111.

[69] Gine, E., 'Bounds for the speed of convergence in the central limit theorem in $\mathbf{C}(S)$', *Z. Wahrsch.verw.Geb.,* **36,** (1976), 317–331.

[70] Gine, E. and Zinn, J., 'Central limit theorems and weak laws of large numbers in certain Banach spaces', *Z.Wahrsch.verw.Geb.,* **62,** (1983), 323–354.

[71] Gine, E and Zinn, J., 'Some limit theorems for empirical processes', *Ann.Probab.,* **12,** No. 4 (1984), 929–989.

[72] Goodman, V., Kuelbs, J. and Zinn, J., 'Some results on the LIL in Banach space with applications to weighted empirical processes', *Ann.Probab.,* **9,** No. 5 (1981), 713–752.

[73] Gorgadze, Z.G. and Tarieladze, V.I., 'On geometry of Orlicz spaces', *Lect. Notes Math.,* **828,** (1980), 47–52.

[74] Götze, F., 'Asymptotic expansions for bivariate von Mises functionals', *Z.Wahrsch.verw.Geb.,* **50,** 3 (1979), 333–355.

[75] Götze, F., 'On the rate of convergence in the central limit theorem in Banach spaces', *Ann.Probab.,* **14,** No. 3 (1986), 922–942.

[76] Götze, F., 'On Edgeworth expansions in Banach spaces', *Ann.Probab.,* **9,** No. 5 (1981), 852–859.

[77] Götze, F., 'Asymptotic expansions in functional limit theorem', *J.Multivar.Anal.,* **16,** (1985), 1–20.

[78] Götze, F., 'Expansions for von Mises functionals', *Z.Wahrsch.verw.Geb.,* **65,** (1984), 599–625.

[79] Gross, L., 'Potential theory on Hilbert space', *J.Funct.Anal.,* **1,** No. 2 (1967), 123–181.

[80] Gudynas, P.P., 'Approximation of distributions of sums of dependent random variables with values in a Banach space', *Lithuan.Math.J.,* **23,** 3 (1984), 251–262.

[81] Hahn, M., 'Conditions for sample continuity and the central limit theorems', *Ann. Probab.* **5,** No. 2 (1977), 351–360.

[82] Heinkel, B., 'Majorizing measures and limit theorems for $\mathbf{c}_0$-valued random variables', *Lect. Notes Math.,* **990,** (1983), 136–149.

[83] Heinkel, B., 'Some exponential inequalities with applications to the central limit theorem in $\mathbf{C}[0, 1]$', Université Louis Pasteur, preprint 325/p. 178, 1986.

[84] Hennequin, P. and Tortrat, A., *Theory of Probability and some its Applications,* Nauka, Moscow, 1974 (in Russian).

[85] Hille, E. and Phillips, R.S., *Functional Analysis and Semi-groups,* Amer.Math.Soc., 1957.

[86] Hoffmann-Jorgensen, J., 'Sums of independent Banach space-valued random variables', *Studia Math.,* **52,** No. 2, 159–189.

[87] Hoffmann-Jorgensen, J. and Pisier, G., 'The law of large numbers and central limit theorem in Banach spaces', *Ann.Probab.,* **4,** (1976), 587–599.

[88] Ibragimov, I.A., 'On probability of hitting a ball of small radius by a Gaussian vector with values in Hilbert space', *Zap.Naucn.Sem.Leningrad Otdel Math.Inst. Steklov,* **85,** (1979), 75–93 (in Russian).

[89] Jain, N.C., 'Central limit theorem and related questions in Banach spaces'. In *Proc.Symp. in Pure Math.,* 1977, Amer.Math.Soc., Providence, Vol. 31, pp. 55–65.

[90] Jain, N.C. and Marcus, M., 'Central limit theorems for $\mathbf{C}(S)$-valued random variables', *J.Funct.Anal.,* **19,** No. 3 (1975), 216–231.

[91] Jurek, Z. and Urbanik, K., 'Remarks on stable measures on Banach spaces', *Collog.Math.*, **38,** No. 2 (1978), 269–276.
[92] Jurinskii, V.V., 'Exponential inequalities for sums of random vectors', *J.Multivar.Anal.*, **6,** No. 4 (1976), 473–499.
[93] Jurinskii, V.V., 'A smoothing inequality for estimates of the Levy–Prokhorov distance', *Theor.Probab.Appl.*, **20,** 1 (1975), 1–10.
[94] Jurinskii, V.V., 'On the error of the Gaussian approximation for covolutions', *Theor.Probab.Appl*, **22,** 2 (1978), 236–247.
[95] Jurinskii, V.V., 'On the accuracy of normal approximation of the probability of hitting a ball', *Theor.Probab.Appl*, **27,** 2 (1983), 280–289.
[96] Jurinskii, V.V., 'An error of normal approximation', *Sibirsk.Mat.Z.*, **24,** No. 6 (1983), 188–199 (in Russian).
[97] Kandelaki, N.P., 'On limit theorem in Hilbert space', *Trudy Vicisl.Centra.Akad.Nauk.Gruzin. SSR*, **11,** (1965), 46–55 (in Russian).
[98] Kantorovich, L.V. and Akilov, G.P., *Functional Analysis*, Nauka, Moscow, 1977 (in Russian).
[99] Kahane, J.P., *Some Random Series of Functions*, Heath, Lexington, Massachusetts, 1968.
[100] Kolmogorov, A.N. and Fomin, S.V., *Elements of Theory of Functions and Functional Analysis*, Nauka, Moscow, 1968 (in Russian).
[101] Koroljuk, S.V. and Borovskich, Yu.V., *Asymptotic Analysis of Distributions of Statistics*, Naukova Dumka, Kijev, 1984 (in Russian).
[102] Kruglov, V.M., *Supplementary Chapters of Probability Theory*, Vyssaja Skola, Moscow, 1984 (in Russian).
[103] Kuelbs, J. and Kurtz, T., 'Berry–Esseen estimates in Hilbert space and an application to the law of the iterated logarithm', *Ann.Probab.* **2,** No. 3 (1974), 387–407.
[104] Kukush, A.G., 'Weak convergence of measures and convergence of semi invariants'. In *Teor.Veroj.Mat.Stat.*, Kijev, 1980, Vol. 23, pp. 74–80 (in Russian).
[105] Kukush, A.G., 'Central limit theorem in Hilbert space in terms of Levy–Prokhorov metric'. In *Teor.Veroj.Mat.Stat.*, Kijev, 1980, Vol. 25, pp. 55–63 (in Russian).
[106] Kuo, H.H., 'Gaussian measures in Banach spaces', *Lect. Notes Math.*, **463,** (1975).
[107] Kutateladze, S.S., *Foundations of Functional Analysis*, Nauka, Novosibirsk, 1983 (in Russian).
[108] Kuhn, T., 'γ–Summing operators in Banach spaces of type p, $1 < p \leq 2$ and cotype q, $2 \leq q < \infty$', *Theor.Probab.Appl.*, **26,** 1 (1981), 118–129.
[109] Kwapien, S., 'Isomorphic characterizations of inner product spaces by orthogonal series with vector-valued coefficients', *Studia Math.*, **44,** (1972), 583–585.
[110] Lapinskas, R., 'Approximation of partial sums in certain Banach spaces', *Lithuan.Math.J.*, **18,** 4 (1979), 494–498.
[111] Le Cam, L., 'Remarques sur le théorème limite central dans les éspaces localement convexes', *Probab. sur les Stuct.Algebr.*, C.N.R.S., Paris, 1970, pp. 233–249.
[112] Ledoux, M., 'A remark on the central limit theorem in Banach spaces', *Lect. Notes Math.*, **1080,** (1984), 144–151.
[113] Ledoux, M., 'Théorème limite central dans les éspaces $\mathbf{l}_p(\mathbf{B})$ $(I \leq p < \infty)$', *Ann.Inst. H. Poincaré*, **19,** (1983), 393–411.
[114] Ledoux, M., 'Sur une inéqualité de H.P. Rosenthal et le théorème limite central dans les éspaces de Banach', *Izrael J. Math.*, **50,** No. 4 (1985), 290–318.
[115] Ledoux, M., 'On the small balls condition in the central limit theorem in uniformly convex spaces', *Geometrical and Statistical Aspects of Probab. in Banach Spaces, Lecture Notes Math.* **1193,** (1986), 44–52.

[116] Ledoux, M. and Talagrad, M., 'Un critère sur les petites boules dans le théorème limite central', *Probab.Theor.Rel. Fields,* **77,** 1 (1988), 29–47.

[117] Linde, W. and Pietsch, A., 'Mappings of Gaussian cylindrical measures in Banach spaces', *Theor.Probab.Appl,* **19,** 3 (1975), 445–460.

[118] Lifshits, M., 'Fibering method and its application to the study of functionals of stochastic processes', *Theor.Probab.Appl.,* **27,** 1 (1982), 69–83.

[119] Lifshits, M., 'On the absolute continuity of distributions of functionals of random processes', *Theor.Probab.Appl.,* **27,** 3 (1983), 600–606.

[120] Lifshits, M., 'Density of the maxima of a Gaussian process', *Teor.veroj.primen.,* **29,** No. 4 (1984), 814–815 (in Russian).

[121] Liubinskas, K., 'On the central limit theorem in the space $\mathbf{c}_0$', *Abst. 24th conf. Lithuanian Math.Soc., Vilnius,* 1983, p. 117 (in Russian).

[122] Liusternik, L.A. and Sobolev, V.J., *Elements of Functional Analysis,* Nauka, Moscow, 1965 (in Russian).

[123] Maurey, B., 'Théorèmes de factorisation pour les opératoeurs linéaires a valeurs dans les éspaces L^p', *Asterisque,* **11,** (1974), 1–163.

[124] Maurey, B., 'Type et cotype dans les éspaces munis de structure locales inconditionnelles', *Sem.Maurey–Schwartz,* 1973–1974, exp. 14–15.

[125] Mayer–Wolf, E., 'Isometries between Banach spaces of Lipschitz functions', *Israel J.Math.,* **38,** No. 1–2 (1981), 58–74.

[126] Meschkov, V.Z., 'Smoothness properties in Banach spaces', *Studia Math.,* **43,** No. 2 (1978), 111–123.

[127] Moulis, N., 'Approximations de fonctions differentielles sur certain éspaces de Banach', *Ann.Inst. Fourier,* **21,** (1971), 293–345.

[128] Mushtari, D.Kh., 'Spaces of cotype p $(0 \leq p \leq 2)$', *Theor.Probab.Appl.,* **25,** 1 (1980), 105–117.

[129] Nagaev, S.V., 'On accuracy of normal approximation for distribution of sum of independent Hilbert space-valued random variables', *Lect. Notes Math.,* **1021,** (1982), 461–473.

[130] Chebotarev, V.I., 'On accuracy of the Gaussian approximation for distributions of sums of independent Hilbert space-valued random variables', *Abst.Commun. 4th Vilnius Conf. on Probab.Theor. and Math.Stat, Vilnius,* 1985, Vol. 4, pp. 208–210.

[131] Nagaev, S.V. and Chebotarev, V.I., 'Estimates of the rate of convergence in the central limit theorem in the space $\mathbf{l}_2$'. In *Mathematical Analysis and Related Problems of Mathematics,* Novosibirsk, 1978, 153–182 (in Russian).

[132] Nemirovskii, A.S. and Semenov, S.A., 'On polynomial approximation of functions on Hilbert space', *Math. USSR, Sbornik,* **21,** (1973), pp. 255–277 (1974).

[133] Norvaiša, R., 'Law of large numbers for identically distributed Banach-valued random variables', *Lithuan.Math.J.,* **24,** 4 (1985), 370–382.

[134] Osipov, L.V. and Rotar, V.I., 'On an infinite-dimensional central limit theorem', *Theor.Probab.Appl,* **29,** 2 (1985), 375–382.

[135] Ostrovskii, E.I., 'Gaussian semigroup in a Banach space', *Theor.veroj.primen.,* **20,** No. 2 (1975), 447–450 (in Russian).

[136] Ostrovskii, E.I., 'Nonuniform estimate of the rate of convergence in the central limit theorem in Banach spaces', *Abst. III Int. Vilnius Conf. on Probab.Theor. and Math.Stat., Vilnius,* 1981, Vol. 2, pp. 99–100 (in Russian).

[137] Pap, Gy., 'Dependence of Gaussian measure on covariance in Hilbert space', *Lect. Notes Math.,* **1080,** (1984), 188–194.

[138] Pap, Gy., 'Distribution of the norm of a Gaussian vector in a Banach space', *Lithuan.Math.J.,* **24,** 1 (1984), 50–52.

[139] Pap, Gy. and Paulauskas, V., 'Remark on the distribution of the norm of a Gaussian vector in the Banach space $\mathbf{c}_0$', *Lithuan.Math.J.*, **25,** 4 (1986), 359–362.

[140] Paulauskas, V., 'On sum of a random number of multivariate random vectors', *Lithuan.Math.J.*, **12,** No. 2 (1972), 109–131 (in Russian).

[141] Paulauskas, V., 'A multivariate inequality for large deviations', *Lithuan.Math.J.*, **12,** No. 1 (1972), 207–212 (in Russian).

[142] Paulauskas, V., 'A nonuniform estimate in the central limit theorem in Hilbert space', *Lithuan.Math.J.*, **15,** 4 (1976), 649–158.

[143] Paulauskas, V., 'On the closeness of the distributions of two sums of independent random variables with values in Hilbert space', *Lithuan.Math.J.*, **15,** 3 (1976), 494–509.

[144] Paulauskas, V., 'Convergence of some functionals of sums of independent random variables in a Banach space', *Lithuan.Math.J.*, **16,** (1976), 385–399 (1977).

[145] Paulauskas, V., 'On the rate of convergence in the central limit theorem in certain Banach spaes', *Theor.Probab.Appl.*, **21,** 4 (1977), 754–769.

[146] Paulauskas, V., 'Estimates of convergence rate in the central limit theorem in $\mathbf{C}(S)$', *Lithuan.Math.J.*, **16,** (1976), 587–611 (1977).

[147] Paulauskas, V., 'On the central limit theorem in the Banach space $\mathbf{c}_0$', *Soviet.Math.Dokl.*, **22,** No. 2 (1980), 349–351.

[148] Paulauskas, V., 'Estimate of the rate of convergence in the central limit theorem in $\mathbf{l}_p$ spaces', *Lithuan.Math.J.*, **21,** 1 (1981), 55–62.

[149] Paulauskas, V., 'On the distribution density of the norm of a Gaussian vector in a Banach space', *Soviet.Math.Dokl.*, **26,** No. 2 (1982), 499–500.

[150] Paulauskas, V., 'Variation distance and pseudomoments in Hilbert space', *Lithuan.Math.J.*, **18,** 1 (1978), 94–99.

[151] Paulauskas, V., 'On the density of the norm of Gaussian vector in Banach spaces', *Lect. Notes Math.*, **990,** (1982), 179–197.

[152] Paulauskas, V., 'On the approximation of indicator functions by smooth functions in Banach spaces'. In *Funct.Anal.Approx.* Birkhauser Verlag, Basel, 1981, pp. 385–394.

[153] Paulauskas, V., 'On the central limit theorem in Banach space $\mathbf{c}_0$', *Probab.Math.Stat.*, **3,** No. 2 (1984), 127–141.

[154] Paulauskas, V., 'The rate of convergence to stable laws and the law of itereated logarithm in Hilbert space', Univ. of Goteborg Dep. Math., 1977, Preprint No. 5.

[155] Paulauskas, V. and Račkauskas, A., 'Operators of stable type', *Lithuan.Math.J.*, **24,** 2 (1985), 160–171.

[156] Paulauskas, V. and Račkauskas, A., 'Infinitely divisible and stable laws in separable Banach spaces. II', *Lithuan.Math.J.*, **20,** 4 (1981), 305–316.

[157] Paulauskas, V., Račkauskas, A. and Sakalauskas, V., 'Central limit theorem in the space of squences converging to zero', *Lithuan.Math.J.*, **23,** 1 (1983), 87–97.

[158] Petrov, V.V., *Sums of Independent Random Variables.* Nauka, Moscow, 1972 (in Russian).

[159] Piech, M.A., 'Smooth functions on Banach spaces', *J.Math.Anal.Appl.*, **57,** No. 1 (1977), 56–67.

[160] Pisier, G., 'Type des éspaces normés', *C.R. Acad.Sci. Paris*, **276,** (1973), 1673–1676.

[161] Pisier, G., 'Type des éspaces normés', *Sem.Maurey–Schwartz*, 1973–1974, exp. 3.

[162] Pisier, G., 'Le théorme limite central et la loi du logarithm itérée dans les éspaces de Banach', *Sem.Maurey–Schwartz*, 1974–1975, exp. 3–4.

[163] Pisier, G., 'Some applications of the metric entropy conditions to harmonic analysis', *Lect. Notes Math.*, **995,** (1983), 123–155.

[164] Pisier, G., 'Martingales with values in uniformly convex spaces', *Israel J.Math.*, **20,** No.

3-4 (1975), 326-350.

[165] Pisier, G. and Zinn, J., 'On the limit theorems for random variable with values in $\mathbf{L}_p$, $p \geq 2$', *Z. Wahrsch. verw. Geb.*, **41**, (1978), 289-304.

[166] Prokhorov, Yu.V., 'The method of characteristic functions', *Proc. 4th Berkeley Symp. on Math. Stat. Probabl., Berkeley*, 1960, Vol. 2, pp. 403-419.

[167] Prokhorov, Yu.V., 'Convergence of random processes and limit theorems of probability theory', *Teor. veroj. primen.*, **1**, No. 2 (1956), 177-238 (in Russian).

[168] Prokhorov, Yu.V. and Sazonov, V.V., 'On the estimates of a rate of convergence in the central limit theorem in infinite-dimensional case', *Abst. first Soviet-Japan Symp. Probability Theory*, Nauka, Novosibirsk, 1969, Vol. 1, pp. 223-230.

[169] Qualtierotti, A.F., 'Some remarks on spherically invariant distributions', *J. Multivar. Anal.*, **4**, No. 3 (1974), 347-349.

[170] Qualtierotti, A.F., 'Mixtures of Gaussian cylindrical set measures and abstract Wiener spaces as models for detection of signals imbedded in noise', *Lect. Notes Math.*, **656**, (1978), 45-57.

[171] Račkauskas, A., 'On the rate of convergence in the central limit theorem in spaces $\mathbf{l}_p$', *Lithuan. Math. J.*, **20**, No. 3 (1980), p. 165 (in Russian).

[172] Račkauskas, A., 'Approximation in the uniform metric of sums of independent random variables with values in Hilbert space', *Lithuan. math. J.*, **21**, 3 (1982), 258-262.

[173] Račkauskas, A., 'On the rate of convergence in the central limit theorem', *Abst. 25th Conf. Lithuan. Math. Soc., Vilnius*, 1984, 239-240 (in Russian).

[174] Račkauskas, A., 'Operator ideal of type (p, q)', *Lithuan. Math. J.*, **24**, 4 (1984), 383-392.

[175] Račkauskas, A., 'On the smoothness of operators', *FSU, Jena, Forschungsergebnisse*, No. 17 (1985), 1-21.

[176] Rhee, W.S., 'On the distribution of the norm of a Gaussian measure', *Ann. Inst. H. Poincaré*, **20**, No. 3 (1984), 277-286.

[177] Rhee, W.S., 'On the rate of convergence in the central limit theorem and the type of the Banach space', *Stat. and Probab. Letters*, **2**, (1984), 59-62.

[178] Rhee, W.S. and Talagrand, M., 'Bad rates of convergence for the central limit theorem in Hilbert space', *Ann. Probab.*, **12**, No. 3 (1984), 843-850.

[179] Rhee, W.S. and Talagrand, M., 'Uniform convexity and the distribution of the norm for a Gaussian measure', *Probab. Theor. Rel. Fields*, **71**, No. 1 (1986), 59-68.

[180] Rhee, W.S. and Talagrand, M., 'On Berry-Esseen type bounds for m-dependent random variables valued in certain Banach spaces', *Z. Wahrsch. verw. Geb.*, **58**, (1981), 433-451.

[181] Rhee, W.S. and Talagrand, M., 'Uniform bound in the central limit theorem for Banach space-valued dependent random variables', *J. Multivar. Anal.* **20**, (1986), 303-320.

[182] Rolewicz, S., *Metric Linear Spaces*, PWN, Warschawa, 1972.

[183] Rozanov, Yu.V., *Gaussian Infinite-Dimensional Distributions*, Trudy Math.Inst.Stekl., Nauka, Moscow, 1968, Vol. CVII (in Russian).

[184] Sachanenko, A.I., 'An estimate of the density of the distribution of integral type functionals', *Trudy Inst. Math. Sibirs. Otd. AN SSSR*, **1**, (1982), 180-186 (in Russian).

[185] Sakalauskas, V., 'Estimation of the rate of approximation to stable laws in $\mathbf{C}(S)$', *Lithuan. Math. J.*, **22**, 1 (1982), 74-82.

[186] Sazonov, V.V., 'On w^2-criterion', *Sankhya*, Ser. A, **30**, No. 2 (1968), 205-209.

[187] Sazonov, V.V., 'An improvement of one estimate of the rate of convergence', *Teor. veroj. primen.*, **14**, No. 4 (1969), 667-678 (in Russian).

[188] Sazonov, V.V., 'Normal approximation. Some recent advances', *Lect. Notes Math.*, **879,** (1981), 1–105.

[189] Sazonov, V.V. and Zalesskii, B.A., 'On the central limit theorem in Hilbert space', Technical Report Center, Stochastic Process, Univ. North Carolina, Chapel Hill, 1983, No. 35, p. 32.

[190] Sazonov, V.V., Ul'yanov, V.V. and Zalesskii, B.A., 'Normal approximation in Hilbert space', *Abst.Commun. 4th Conf. on Probab.Theor. and Math.Stat., Vilnius,* 1985, Vol. 4, pp. 270–272.

[191] Semadeni, Z., 'Shauder bases in Banach spaces of continuous functions', *Lect. Notes Math.,* **918,** (1982), p. 135.

[192] Senatov, V.V., 'On a relationship between Levy–Prokhorov metrics and ideal metrics', *Lect. Notes Math.,* **900,** (1983), 193–198.

[193] Senatov, V.V. 'Several uniform estimates of the rate of convergence in the multidimensional central limit theorem', *Theor.Probab.Appl.,* **25,** (1981), 745–759.

[194] Senatov, V.V. 'Some lower estimates for the rate of convergence in the central limit theorem in a Hilbert space', *Sov.Math.Dokl.,* **23,** (1981), 188–192.

[195] Skorohod, A.V., *Integration in a Hilbert Space,* Nauka, Moscow, 1975 (in Russian).

[196] Solntsev, S.A., 'On tight oscillation of a Gaussian–Markovian sequence', *Teor.Veroj.Mat.Stat.,* **28** (1983), 131–132.

[197] Strassen, V., 'The existence of probability measures with given marginals', *Ann.Math.Stat.,* **36,** No. 2 (1965), 423–439.

[198] Sundaresan, L., 'Geometry of Lebesgue–Bochner function spaces – smoothness', *Trans.Amer.Math.Soc.,* **198,** (1974), 229–251.

[199] Sundaresan, L., 'Geometry and non-linear analysis in Banach spaces', *Pacific J. math.,* **102,** No. 2 (1982), 487–498.

[200] Sundaresan, L. and Swaminathan, S., 'Geometry and non-linear analysis in Banach spaces', *Lect. Notes Math.,* **1131,** (1985).

[201] Sztencel, R., 'On the lower tail of stable measures', *Bull.Acad.Pol.Sci.,* **32,** No. 11–12 (1984), 715–719.

[202] Szigeti, F., 'Differential topology in smooth Banach spaces'. In *Topics in Topology, Proc.Collog.Math.Soc.Janos Bolyai,* 1974, Vol. 8, pp. 589–608.

[203] Szulga, A., 'On minimal metrics in the space of random variables', *Teor.veroj.primen.,* **27.** No. 2 (1982), 401–405.

[204] Sytaja, G.N., 'On small balls of Gaussian measures'. In *Probabilistic Distributions in Infinite–Dimensional Spaces,* Kijev, 1978, pp. 154–170 (in Russian).

[205] Sytaja, G.N., 'On asymptotics of the measure of small balls for Gaussian processes equivalent to Brownian process', *Teor.Veroj.Mat.Stat.,* **19,** (1978), 128–133 (in Russian).

[206] Tomczak–Jaegermann, N., 'On the differentiability of the norm in trace classes s_p', *Sem.Maurey–Schwartz,* 1975/75, exp. 22.

[207] Tsirel'son, B.S., 'The density of the distribution of the maximum of a Gaussian process', *Theor.Probab.Appl.* **20,** 4 (1975), 847–855.

[208] Ul'yanov, V.V., 'Central limit theorem in Hilbert space, nonidentically distributed random variables case', *Abst.Commun. 4th Vilnius Conf. on Probab.Theor. and Math.Stat., Vilnius,* 1985, Vol. 4, pp. 316–317.

[209] Ul'yanov, V.V., 'An estimate for the rate of convergence in the central limit theorem in a real separable Hilbert space', *Math. Notes,* **29,** (1981), 78–82.

[210] Ul'yanov, V.V., 'Refinement of proximity estimates for two distributions of sums of random variables with values in Banach spaces', *Moscow.Univ.Comp.Math.Cybern.,* No. 1 (1981), 60–67.

[211] Ul'yanov, V.V., 'Estimates of the density of the norm of a Gaussian random variable in H', *Proc. of Conf. of Young Scientists of the Faculty of Computing Mathematics and Cybernetics,* Moscow Univ., 1984 (in Russian).
[212] Vainberg, M.M., *Variational Methods for Investigating Non-linear Operators,* Mir, Moscow, 1956 (in Russian).
[213] Vakhanija, N.N., *Probabilistic Distributions in Linear Spaces,* Mecniereba, Tbilisi, 1971 (in Russian).
[214] Vakhanija, N.N. and Kandelaki, N.P., 'On the estimate of the rate of convergence in central limit theorem in Hilbert space', *Trudy Vycisl. Centra Akad.Nauk.Gruzin. SSR,* No. 9 (1969), 150 – 160 (in Russian).
[215] Vakhanija, N.N., Tarieladze, V.I. and Cobanian, S.A., *Probabilistic Distribution in Banach Spaces,* Nauka, Moscow, 1985 (in Russian).
[216] Varadhan, S.R.S., 'Limit theorems for sums of independent random variables with values in a Hilbert space', *Sankhya,* **24,** (1962), 213 – 238.
[217] Vinogradova, T.R., 'On the accuracy of normal approximation on sets defined by a smooth functions. I', *Theor.Probab.Appl.,* **30,** (1986), 235 – 246.
[218] Wells, J., 'Differentiable functions on $\mathbf{c}_0$', *Bull.Amer.Math.Soc.,* **75,** (1969), 117 – 118.
[219] Whithield, J.H.M., 'Differentiable functions with bounded nonempty support on Banach spaces', *Bull.Amer.Math.Soc.,* **77,** (1966), 145 – 146.
[220] Woczynski, W.A., 'Geometry and martingales in Banach spaces, II. Independent increments', *Adv.Probab.Rel.Topics,* No. 4 (1978), 267 – 517.
[221] Ylvisaker, N.D., 'The expected number of zeros of a stationary Gaussian process', *Ann.Math.Stat.,* **36,** No. 3 (1965), 1043 – 1046.
[222] Zalesskii, B.A., 'Estimate of the accuracy of normal approximation in Hilbert space', *Theor.Probab.Appl,* **37,** 2 (1982), 290 – 298.
[223] Zalesskii, B.A. and Sazonov, V.V., 'Closeness of moments for normal approximation in a Hilbert space', *Theor.Probab.Appl.,* **28,** 2, (1984), 263 – 277.
[224] Zalesskii, B.A., 'On the rate of convergence on some class of sets in the central limit theorem in Hilbert space', *Teor.veroj.primen.,* **30,** No. 4 (1985), 662 – 670 (in Russian).
[225] Zalesskii, B.A., 'On an estimate of the rate of convergence in the central limit theorem in a Hilbert space', *Sov.Math.Dokl.,* **26,** (1982), 575 – 578.
[226] Zinn, J., 'A note on the central limit theorem in Banach spaces', *Ann.Probab.,* **5,** No. 1 (1977), 282 – 289.
[227] Zolotarev, V.M., 'Metric distances in spaces of random variables and their distributions', *Math. USSR, Sbornik,* **30,** (1976), 373 – 401.
[228] Zolotarev, V.M., 'Approximation of distributions of sums of independent random variables with values in infinite-dimennsional spaces', *Theor.Probab.Appl.,* **21,** 4 (1977), 721 – 737.
[229] Zolotarev, V.M., 'Ideal metrics in the problem of approximating distributions of sums of independent random variables', *Theor.Probab.Appl.,* **22,** 3 (1978), 433 – 450.
[230] Zolotarev, V.M., 'Probability metrics', *Theor.Probab.Appl,* **28,** 2 (1984), 278 – 302.
[231] Zygmund, A., *Trigonometrical Series,* Warsaw, 1935.
[232] Żak, T., 'On the continuity of the distribution of a seminorm of stable random vectors', *Bull.Acad.Pol.Sci.,* **32,** Nos 7 – 8 (1984), 519 – 521.

Index

absorbing class of sets, 83

Borel σ-algebra, 1
bounded Lipschitzian metric, 7

characteristic functional, 4
central limit theorem, 52
continuity set, 7
covariance, 3
covariance function, 4
covariance operator, 4
cotype of a Banach space, 55
cylindrical set, 1

differentiable function in a sense of Frechet, 11
- - in a sense of Gateaux, 12
- - with respect to an operator, 18
distribution of a random element, 2

estimate in central limit theorem of order $n^{-1/8}$, 83
- - - - of order $n^{-1/6}$, 84
- - - - in Hilbert space, 104

Fréchet derivative, 11

Gateaux derivative, 13
Gaussian random element, 5
- covariance operator, 6

Levy inequality, 9
lemma of symmetrization, 105

mathematical expectation in the sense of Gelfand-Pettis, 2
- - in the sense of Bochner, 3
measurable linear space, 1
method of covolutions, 80
- of approximation, 91
- of characteristic functions, 104
metric entropy, 6
mixing measure, 99
mixture of Gaussian distributions, 99
- continuous function, 7
- continuity set, 7
- uniform class of sets, 8

mixture of Gaussian distributions (contd.)
- uniform class of functions, 8

oscillation of a function, 8

pre-Gaussian random element, 6
Prokhorov metric, 7
Prokhorov theorem, 8

Rademacher sequence, 9
radonifying operator, 6
random element in a Banach space, 2

smoothing inequality, 80
symmetric random element, 2
symmetrization of random element, 2

Taylor formula, 16
type of a Banach space, 44
- of an operator, 44

uniform class of sets, 25
uniformly approximable function, 25

weak convergence, 7

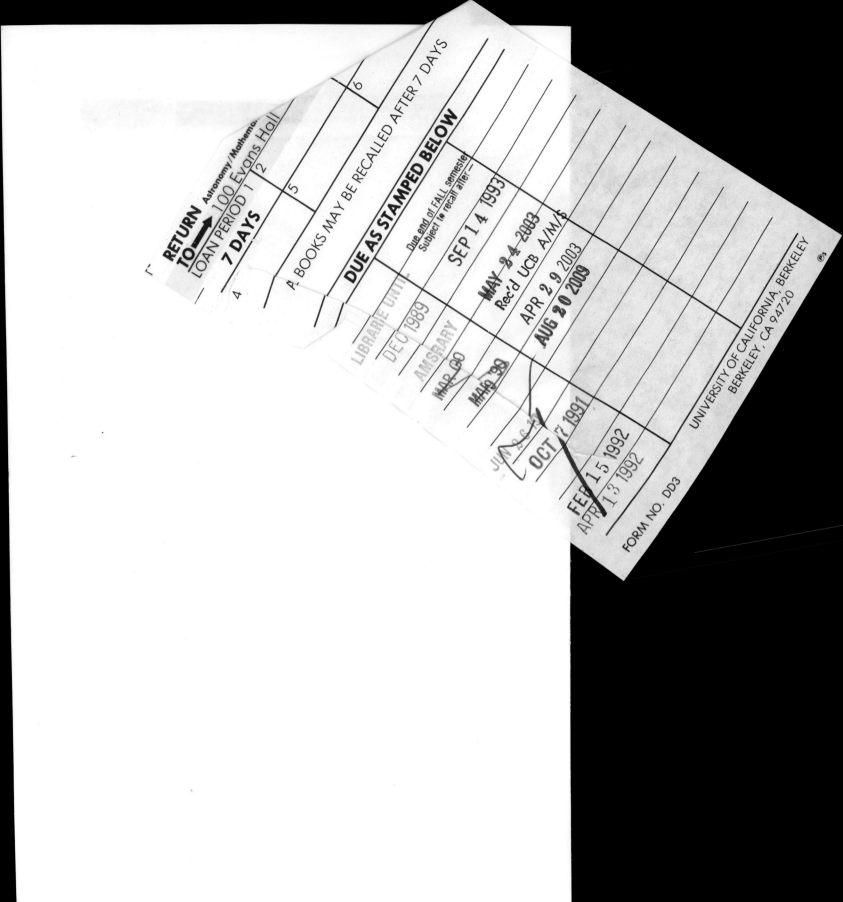